Introduction to
Network Technologies

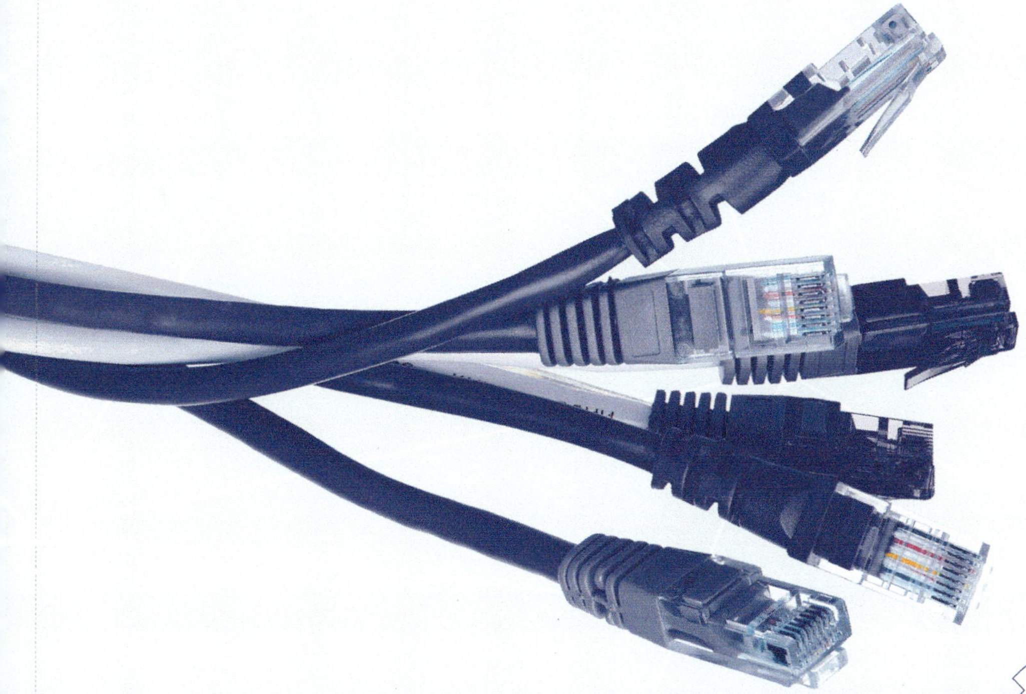

Introduction to Network Technologies is intended to be an educational resource for the user and contains procedures commonly practiced in industry and the trade. Specific procedures vary with each task and must be performed by a qualified person. For maximum safety, always refer to specific manufacturer recommendations, insurance regulations, specific job site and plant procedures, applicable federal, state, and local regulations, and any authority having jurisdiction. The *electrical training ALLIANCE* assumes no responsibility or liability in connection with this material or its use by any individual or organization.

© 2016 *electrical training ALLIANCE*

This material is for the exclusive use by the IBEW-NECA JATCs and programs approved by the *electrical training ALLIANCE*. Possession and/or use by others is strictly prohibited as this proprietary material is for exclusive use by the *electrical training ALLIANCE* and programs approved by the *electrical training ALLIANCE*.

All rights reserved. No part of this material shall be reproduced, stored in a retrieval system, or transmitted by any means whether electronic, mechanical, photocopying, recording, or otherwise without the express written permission of the *electrical training ALLIANCE*.

1 2 3 4 5 6 7 8 9 – 16 – 9 8 7 6 5 4 3 2

Printed in the United States of America

CREATIVE DIGITAL PRINTING

Contents

Chapter 1   Introduction to Networking xii
Networks Classified by Function 2
Network Architectures 3
Logical and Physical Networks 4
Network Types 5
Automation Networks 14
Standards Organizations 15

Chapter 2   Overview of Networking Components 20
Unicast, Broadcast, and Multicast Traffic 22
Network Hardware 25
Physical Media 32

Chapter 3   Understanding the OSI Model 50
The OSI Model 52
The TCP/IP Model 63
Devices Mapped on the OSI and TCP Models 64
Protocols Mapped on the OSI and TCP/IP Models 67

Chapter 4   Ethernet 70
Ethernet Standards 72
Ethernet Components 73

Chapter 5   Understanding Wireless Networking 88
Basic Wireless Components 90
Networking Standards and Characteristics 92
Network Security Methods 96
Wireless Networks 101
Point-to-Point Wireless 104

Contents

Chapter 6 IPv4 108
Components of an IPv4 Address 110
Binary IPv4 Address...117
IPv4 Address Subnetting 121
Manual Versus Automatic Assignment of IPv4 Addresses.................................128

Chapter 7 IPv6 134
IPv6 Addresses 136
Components of an IPv6 Address 139
The Dual IP Stack... 143
Manual Versus Automatic Assignment of IPv6.. 144

Chapter 8 Networking Protocols 148
Physical and Data-Link Layer Protocols.. 150
Network Layer Protocols... 151
Transport Layer Functions 155
Ports... 158
Application Layer Protocols 160

Chapter 9 Network Access with Switches............ 170
Multiple-Computer Connections 172
Physical Ports 172
Hubs and Switches 175
Unmanaged and Managed Switches.. 179
Switch Speeds 183
Security Options 185

Chapter 10 Routing Networks................................... 188
Multiple Network Connections... 190
Traffic Routing on a Network 193
Transmission Speeds 198
Router Configuration.. 199

Contents

Chapter 11 Resolving Names to IP Addresses 208
Types of Names Used in Networks 210
Types of Name Resolution 215
Steps in Name Resolution 223

Chapter 12 Network Security 228
Risks on the Internet 230
Intranets 231
Firewalls 235
Perimeter Networks 238
Extranets 242

Chapter 13 Wide Area Network Connectivity 246
Connectivity Methods Used in Homes and SOHOs 248
Connectivity Methods in Enterprises 252
Remote Access Services 255
RADIUS 259

Chapter 14 Troubleshooting 262
The Command Prompt 264
TCP/IP Configuration Check with ipconfig 265
Connectivity Troubleshooting with ping 268
Router Identification
    with tracert 272
Routed Path Verification with pathping 272
TCP/IP Statistics with netstat 275
Telnet 278

Chapter 15 Network Fault Tolerance 282
Malware 284
Availability 288
Disaster Preparation 301

Contents

Chapter 16   Management and Administration 304
SNMP Communications 306
Subnet Calculations 307
Network Monitoring Software 311
Network Process Flowcharts 312
Network Documentation 312

Glossary 321

Index 339

Introduction

Today most organizations rely on computer systems for performing daily work activities, communication, management, and even building and process control. To make sure that each employee in the organization has proper access to the resources in the computer systems and to accomplish their tasks, the systems are connected to a network.

Connecting the systems in a network also will help in reducing the expenses on the resources that are used by the organization. However, a minor technical problem or fault within a single computer system can lead to breakdown of the entire network. Therefore, a network technician helps in the establishment and maintenance of the network.

When computers or devices are unable to communicate on the network, down time and lost business processes typically result. Network technicians must be able to quickly identify any communication problems and resolve them. To do this efficiently and expediently, technicians need to understand how the network works under normal conditions.

About This Book

This book is for current or aspiring network technicians or Electrical Workers seeking knowledge in the fundamentals of networking. The contents will provide focused coverage of fundamental skills. Those individuals who wish to learn about networking or are already working in networking and want to fill in some gaps on fundamental networking can achieve substantial knowledge from the contents of this book. The knowledge gained can be used as a foundation for more advanced studies.

Chapters explain how a network techician performs complex professional work with computer hardware, software, and network systems. This entails installing and supporting network servers including operating systems and applications software. A technician also installs and supports personal computers and network operations; provides hardware and software planning and evaluation; provides problem-solving and training for end-users; and ensures a system's efficiency and integrity. Sometimes a technician provides research and support for new technologies. All of these tasks are covered in detail within the textbook.

Introduction to Network Technologies

Features

Facts offer additional information related to Network Technologies.

Figures, including photographs and artwork, clearly illustrate concepts from the text.

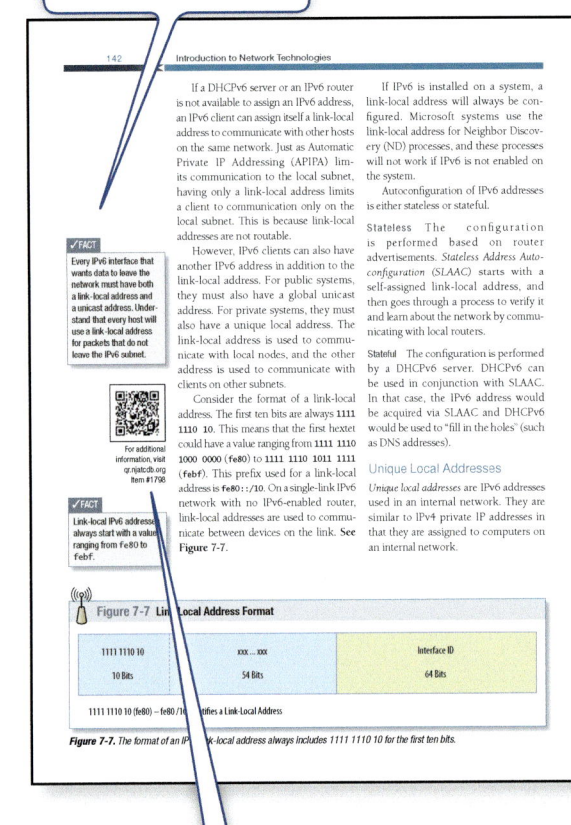

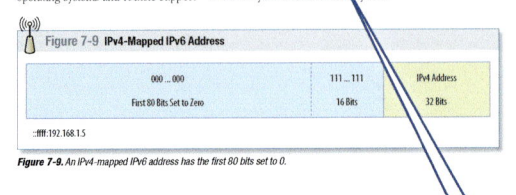

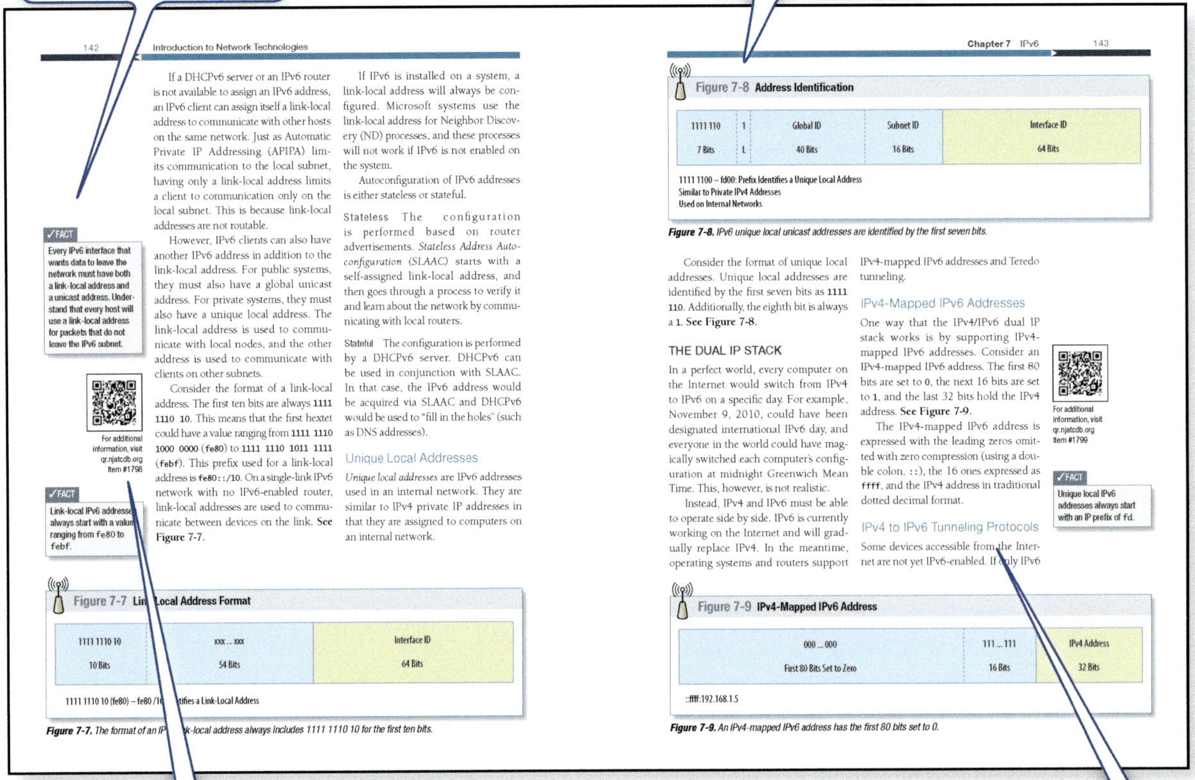

Blue **Headers** and **Subheaders** organize information within the text.

Quick Response Codes (QR Codes) create a link between the textbook and the Internet. They can be scanned using Smartphone applications to obtain additional information online. (To access the information without using a Smartphone, visit qr.njatc.org and enter the referenced Item #.)

For additional information related to QR Codes, visit qr.njatcdb.org Item #1079

Introduction to Network Technologies

Features

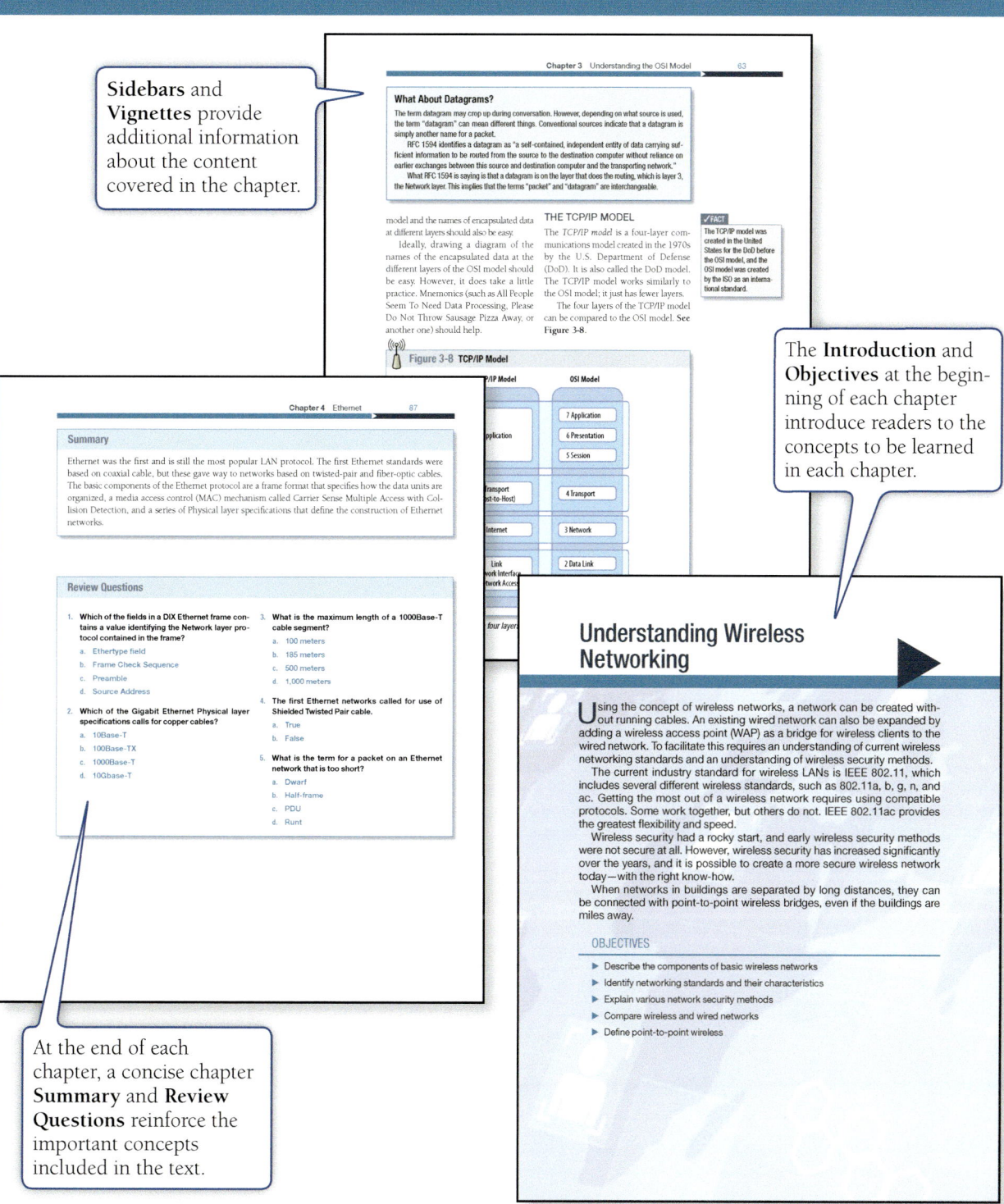

Sidebars and **Vignettes** provide additional information about the content covered in the chapter.

The **Introduction** and **Objectives** at the beginning of each chapter introduce readers to the concepts to be learned in each chapter.

At the end of each chapter, a concise chapter **Summary** and **Review Questions** reinforce the important concepts included in the text.

Acknowledgments

ACKNOWLEDGMENTS

Minnesota Statewide Limited Energy JATC

QR CODES

faqs.org

Federal Communications Commission

ICANN (Internet Corporation for Assigned Names and Numbers)

IEEE

IEEE Standards Association (IEEE-SA)

International Brotherhood of Electrical Workers (IBEW)

International Organization for Standardization (ISO)

Internet Assigned Numbers Authority (IANA)

Internet Engineering Task Force

National Electrical Contractors Association (NECA)

Telecommunications Industry Association (TIA)

The Internet Engineering Task Force (IETF®)

The Internet Society (Network Working Group)

Wi-Fi Alliance®

About the Author

Craig Zacker is the author or co-author of dozens of books, manuals, articles, and web sites on computer and networking topics. He has also been an English professor, an editor, a network administrator, a webmaster, a corporate trainer, a technical support engineer, a minicomputer operator, a literature and philosophy student, a library clerk, a photographic darkroom technician, a shipping clerk, and a newspaper boy. He lives in a little house with his beautiful wife and a neurotic cat.

About the Technical Reviewer

Brandon Nelson is the senior instructor for the Minnesota Statewide Limited Energy JATC and a member of IBEW Local 292. He has been involved in computer networking from a practical perspective for more than 20 years and holds numerous information technology (IT) industry certifications and state electrical licenses. He is a highly respected educator with the ability to view a computer network not only from the perspective of an engineer, but also that of the electrical contractor and systems integrator.

Introduction to Networking

Just about any computer in use today is on a network. Computer networks are so common that it is easy to take them for granted. However, there are many components and technologies working together behind the scenes to ensure that a networked computer can access resources on the network.

To begin learning about networking, it is important to know the names of many of the physical and logical components of a network. It is also important to identify the components included in very small networks, and how additional components are added as a network grows. Finally, it is essential to gain familiarity with the standards organizations that help ensure all of these computers can work together, no matter who manufactured them or where they are operating.

OBJECTIVES

- ▶ Classify networks by their function
- ▶ Identify various network architectures
- ▶ Describe logical and physical network elements
- ▶ List the components of a home computer network
- ▶ Identify the components of small office and home office networks
- ▶ List the components of a large office network
- ▶ Describe an enterprise network and its components
- ▶ Identify a standards organization and its role in the networking industry

CHAPTER 1

TABLE OF CONTENTS

Networks Classified by Function. 2
    Local Area Networks 2
    Wide Area Networks 2

Network Architectures. 3
    Peer-to-Peer Networks 4
    Client-Server Networks 4

Logical and Physical Networks. 4

Network Types 5
    Home Computer Networks 5
    Small Office and Home Office
    Networks 7
    Large Office Networks. 10
    Large Enterprise Networks 11

Automation Networks 14

Standards Organizations. 14
    Internet Engineering Task Force 15
    World Wide Web Consortium 16
    Institute of Electrical and
    Electronics Engineers 16
    International Telecommunication
    Union 16

Summary. 16

Review Questions 17

NETWORKS CLASSIFIED BY FUNCTION

A *network* is a group of computers and other devices connected together. This is the simplest possible definition of what a network is. However, a network that consists of computers in the same building is vastly different from one that connects computers in different cities or countries. This is the difference between a local area network (LAN) and a wide area network (WAN).

Local Area Networks

The home network and the Small Offices and Home Offices (SOHO) network are both considered local area networks. A *local area network (LAN)* is a group of connected computers in the same geographical location, sharing the same level of connectivity. LANs can be relatively small or very large. An organization can conceivably have hundreds or thousands of computers connected to a LAN.

The primary defining characteristic of a LAN is the speed at which the computers are capable of communicating. LANs have relatively fast network connectivity, when compared to WANs or the Internet. Common speeds of wired LANs today are 100 megabits per second (Mbps) or 1,000 Mbps (also called 1 gigabits per second, or Gbps), while speeds of 54 Mbps or 300 Mbps are typical for wireless LANs.

A LAN is an internal network, usually wholly owned by the organization that constructs it. Most LANs will have connectivity to the Internet through a router, but the LAN itself is internal. Traffic back and forth through to the Internet is filtered for security purposes, using a firewall. However, traffic within the LAN itself is usually not filtered. The internal network is considered a high trust area, so any traffic on the network is allowed.

Wide Area Networks

A *wide area network (WAN)* consists of two or more LANs in separate geographical locations that are connected. The connection between the LANs is almost always slower than the speed of the LANs themselves. For example, consider an organization headquarters that has a high-speed 1,000 Mbps (1 Gbps) LAN connecting all the computers and other devices. Similarly, the regional office also has a high-speed 100 Mbps LAN. **See Figure 1-1**.

The T1 WAN link connects the two LANs at a much slower connection speed of 1.544 Mbps. Although a speed of 1.544 Mbps is much quicker than a dial-up speed of 56 Kbps, it is significantly slower than the internal speeds of the LANs (100 Mbps and 1 Gbps).

Megabit and Gigabit

LAN speeds identify how much data they can transfer. Mbps is short for megabits per second, and a megabit represents a million bits. A LAN with a speed of 100 Mbps can transfer data at a rate of 100 million bits per second. A gigabit LAN (1 Gbps or 1,000 Mbps) transfers data at a rate of 1 billion bits per second.

Occasionally, data is measured in bytes instead of bits. A byte consists of 8 bits. When bytes are mentioned, a capital "B" is used. For example, a system may have 4 gigabytes (GB) of random access memory (RAM). This is commonly listed as 4 GB. It is not accurate to list this as 4 Gb (with a lowercase "b"). Similarly, it is not accurate to list a 100 Mbps LAN as 100 MBps (with a capital "B").

When a WAN connects two networks, users are able to access resources in the other LAN. For example, users in the Virginia Beach LAN can access files on servers in the Las Vegas LAN. Similarly, users in Las Vegas can access servers in Virginia Beach.

Unlike LANs, organizations typically do not own the WAN connections used to connect their sites. Many organizations lease the WAN links from a service provider. This is similar to people leasing phone lines for telephone access. It is not practical for users to run individual phone lines to each person who is called. Similarly, it is usually too expensive for most organizations to run cables to different locations.

There are a variety of WAN technologies available to connect remote sites. Leasing lines from a telecommunications company is one type of WAN, but it can be expensive. Leased lines usually carry more data than a typical phone connection. For example, a leased line connection called a T1 runs at 1.544 Mbps, which is enough bandwidth for 24 separate telephone connections. For WAN use, however, the entire bandwidth is often devoted to a single data connection, but it still runs far slower than even the most modest LAN. The cost of a T1 is not quite 24 times the price of a standard telephone line, but it often runs into the hundreds of dollars per month.

In addition to leased telephone lines, there are many other WAN technologies available, many of which offer variable amounts of bandwidth on demand, so that an organization can save money by paying only for the bandwidth it uses.

NETWORK ARCHITECTURES

Connecting computers together forms the broad definition of a network, but how those computers work together is

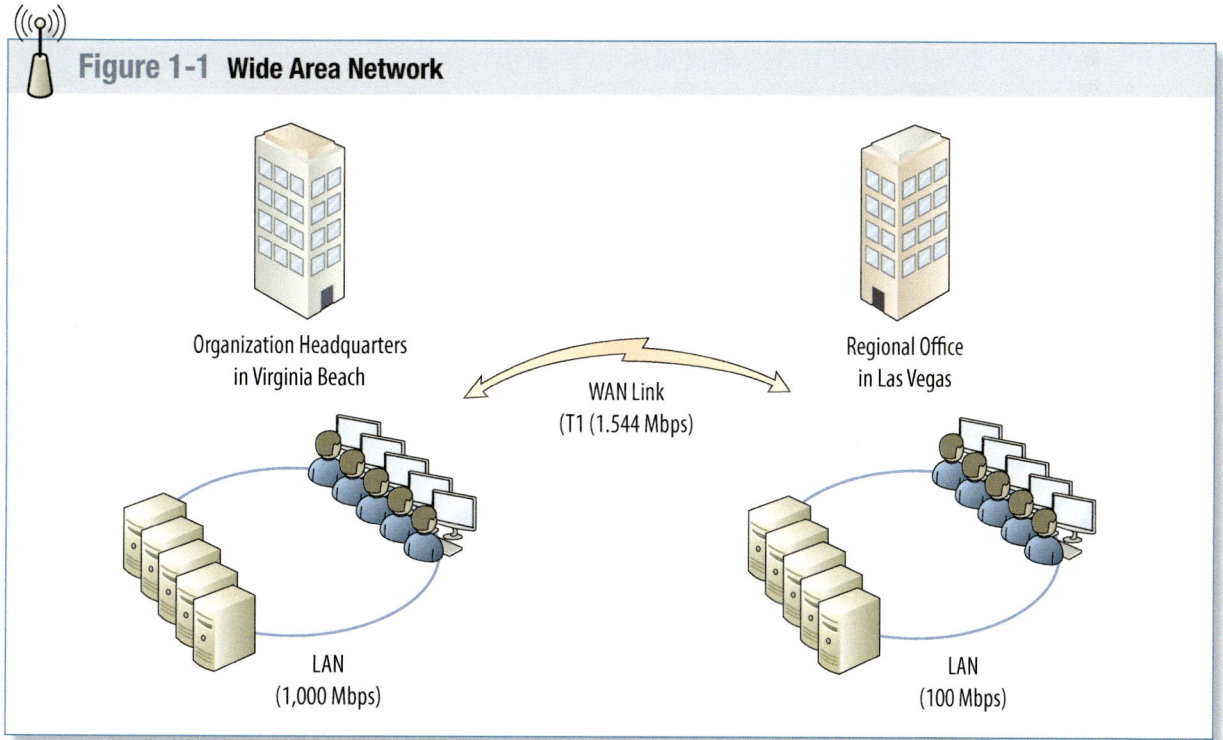

Figure 1-1. A WAN connects two or more LANs.

a critical part of understanding the networking paradigm.

Peer-to-Peer Networks

When the computers on a network perform the same functions for each other, they are said to be peers, and a network consisting of all peers is called (not surprisingly) a peer-to-peer network. A *peer-to-peer network* is a group of connected computers, each of which can act as a server to the others for transfer of the files stored on it. Computers on a network provide services to each other; this is their primary function as network citizens. When Ralph wants to use his computer to print a document, he can send it over the network to Alice's computer, which has a printer connected to it.

In this relationship, Alice's computer is functioning as a print server, while Ralph's computer is a client. This is a temporary relationship that exists only for the purpose of printing. At the same time, Alice might want to access a document stored on Ralph's hard drive. In this relationship, Ralph's computer functions as a file server, and Alice's computer is the file client. This is the essence of a peer-to-peer network; each computer can function both as a client and as a server, depending on the functions each user requires.

Client-Server Networks

On a client-server network, the relationships are basically the same, but the computers are dedicated to their functions. A *client-server network* is a group of connected computers in which one or more, called servers, perform the function of transferring files to the others, called clients. For example, instead of Ralph and Alice storing their document files on each other's computers, they both store their files on a dedicated file server.

The file server is not a workstation with a user sitting in front of it. Instead, it is a specialized computer, typically stored in a secured location such as a data center, which is equipped to handle the data storage needs of dozens or hundreds of users. Client-server networks can have a variety of servers providing different services to the clients on the network.

In a client-server environment, the servers are typically run by administrators, who are responsible for securing data on the servers and maintaining reliability of the servers. Servers typically have more hardware resources than workstations (such as processing power, memory, and storage), and servers might also have high-availability technologies (such as redundant power supplies and elaborate storage arrays).

LOGICAL AND PHYSICAL NETWORKS

Networks connect computers together. These connections can use cables, wireless technologies, or both. Networks can be discussed in both logical and physical terms.

> **Clients and Servers are Defined by Their Roles**
>
> It is important to note that the differences between a peer-to-peer and a client-server network lie mainly in the roles that the users and administrators designate for the various computers on the network. All of the personal computer operating systems used on networks today (including Windows, UNIX/Linux, and MacOS) are capable of functioning as both clients and servers.

The logical organization of a network identifies its overall design. It differentiates between LANs and WANs. The *logical design* of the network is a high-level description of how information moves through the entire network without reference to its physical components like switches, routers, and firewalls. By contrast, the physical network infrastructure includes the details of the physical components. The physical components are the devices and cabling.

When attempting to grasp logical network organization concepts, it is important to understand the different types of network designs found in home networks, small offices, larger offices or organizations, and enterprises.

It is also important to understand how devices in a logical structure work to fully understand how data moves through a network. With an understanding of how the data moves through the network, maintaining and troubleshooting the network is easier when problems occur.

NETWORK TYPES

A network can be as simple as two computers in a home that share an Internet connection, or as elaborate as a corporate internetwork with thousands of computers scattered around the world.

Home Computer Networks

Most home computers are part of a network today. At the very least, home computers have the capability of being connected to the Internet, which is a massive network of networks. **See Figure 1-2**.

The home computer has access to the Internet through a modem to an *Internet service provider (ISP)*, a company or organization that provides access to the Internet for households and businesses. This could be a cable modem used in a broadband connection or any one of several other technologies, including the traditional dial-up connection. Broadband connections are widely available through cable television systems, fiber-optic lines,

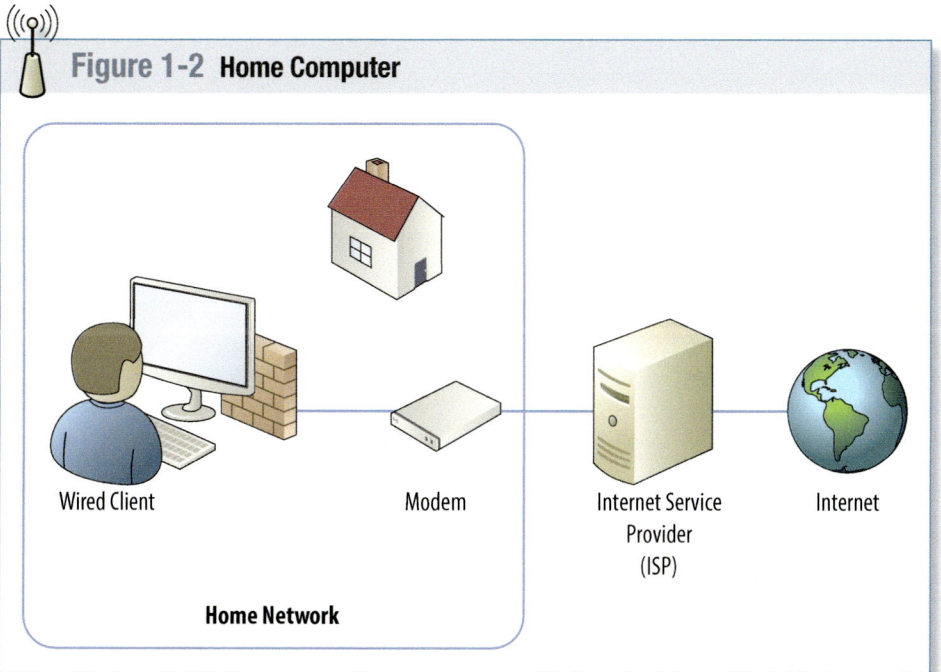

Figure 1-2. A home computer may be connected to the Internet.

and even phone connections such as DSL and 3G/4G data services.

Even if a broadband connection is not available, home users can still connect to the Internet through a phone line, also known as a dial-up system. Dial-up connections perform much slower, but are used in some areas where broadband connections are not available. Internet access via satellite is also available in more remote areas, providing better connections than dial-up, but it is still not comparable in speed to broadband connections.

When home users add computers into their home, the ultimate goal typically is to network these computers. Users on the network are then able to share resources. For example, a typical home network with computers connected to each other and to the Internet may utilize both wired and wireless connections. **See Figure 1-3**.

The wired user is connected to a wireless router directly with a cable, and another user is connected via a wireless connection. A wireless printer is added so that any users with access to the wired network can share printing

Enable the Local Firewall

When a computer connects directly to the Internet through an ISP (without going through an internal router or wireless access point), it is at significant risk. The computer has a public Internet Protocol (IP) address and is accessible from any other computer on the Internet, anywhere in the world. Attackers often prowl the Internet looking for unprotected computers. Enabling the software firewall on the computer provides a layer of protection.

capabilities. An ISP provides connectivity to the Internet, just as it would for a single user. A single cable modem connects to the ISP, and then the cable modem connects to a wireless router.

Without a network, each computer would need to connect to the Internet separately, incurring individual access charges. However, the single Internet connection can be shared by adding the wireless router. A great benefit of wireless technology is that it is not necessary to install cables to each computer.

Most wireless routers include additional capabilities. For example, it is common for the wireless routers used in most home networks to include the following:

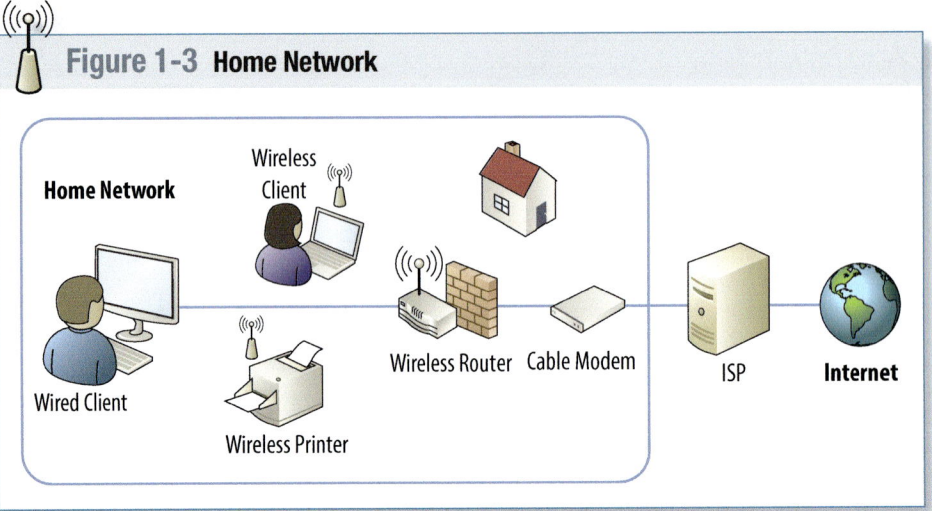

Figure 1-3. A typical home network can be created by connecting several computers.

Wireless Access Point (WAP) The core purpose of the wireless device is to support connectivity for wireless clients. The WAP provides this connectivity.

Routing Capabilities A built-in router will route data from the internal network to the Internet and from Internet data back to the internal network.

Network Address Translation (NAT) *Network Address Translation (NAT)* is the process where the public IP addresses used on the Internet are translated to private IP addresses on the internal network, and vice versa. If NAT translates the public IP addresses used on the Internet to private IP addresses on the internal network, and vice versa. If NAT were not used, each internal computer would need a public IP address. Additionally, each computer would be directly on the Internet and exposed to unnecessary risks. NAT hides the internal computers from Internet attackers.

Dynamic Host Configuration Protocol (DHCP) DHCP provides clients with IP addresses and other TCP/IP configuration information. The other Transmission Control Protocol/Internet Protocol (TCP/IP) information includes the address of the DNS server and the address of the router that provides a path to the Internet. The router address is also known as the default gateway.

Firewall A WAP provides basic firewall capabilities. A *firewall* is part of a computer network that blocks unwanted traffic from the Internet, providing a layer of protection for internal clients.

Small Office and Home Office Networks

A *small office and home office network (SOHO)* is a network providing services to a small office on commercial or residential property. A SOHO is very similar to the sophisticated home network. They are both considered LANs. SOHOs have access to the Internet and can have wireless clients, wired clients, or a combination of both. **See Figure 1-4**. SOHOs typically have up to 10 workers, but may have as many as 100.

The primary difference between a SOHO and a home network is that a SOHO will typically have a server to provide additional capabilities for the office. For example, the server can be used as a file server to store files used within the business.

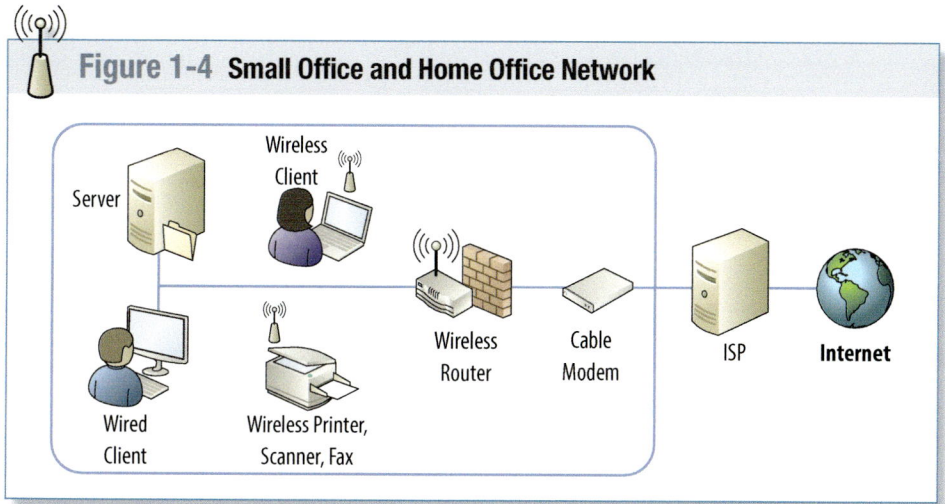

Figure 1-4 Small Office and Home Office Network

Figure 1-4. *A SOHO network can include wireless clients, wired clients, or both.*

Although some offices will have a server, it is not essential. Important files can be stored on a primary user's computer and shared with other users from there if needed. However, when storing important files on multiple computers, it becomes more difficult to back up those files.

Additionally, a business might have a wireless multifunction printer that can print, scan, and fax documents to meet the needs of the business. It is not essential to have a wireless printer, but these are becoming more popular in SOHOs because they are easier to share between network users.

The WAP used in a SOHO can be the same as the WAP used in the home network, and it can provide many of the same capabilities to the office as a WAP provides for a home network. This includes routing, NAT, DHCP, and a firewall.

Workgroups and Domains

A SOHO typically includes from one to ten workers and is usually configured in a peer-to-peer arrangement called a workgroup. A *workgroup* is a collection of networked computers that share a common workgroup name. The default name of a Microsoft workgroup is simply Workgroup. In a workgroup, user accounts are located on each computer.

Consider an office with four users. Each of the users has his or her own computer, and an additional server is available to them. For Sally to log on to her computer, she needs a user account on her computer. However, this account will not work on Bob's computer, Alice's

> **Secure Wireless Networks**
>
> It is very important to lock down wireless networks with the best security available. The primary method used to secure wireless networks is WPA2 (or 802.11i). If the network is not locked down, an attacker can compromise it using a simple laptop with a wireless network interface card (NIC) while driving by in a car. This "war driving" technique enables an attacker to tap into the network and access the network's resources if the network is not secured.

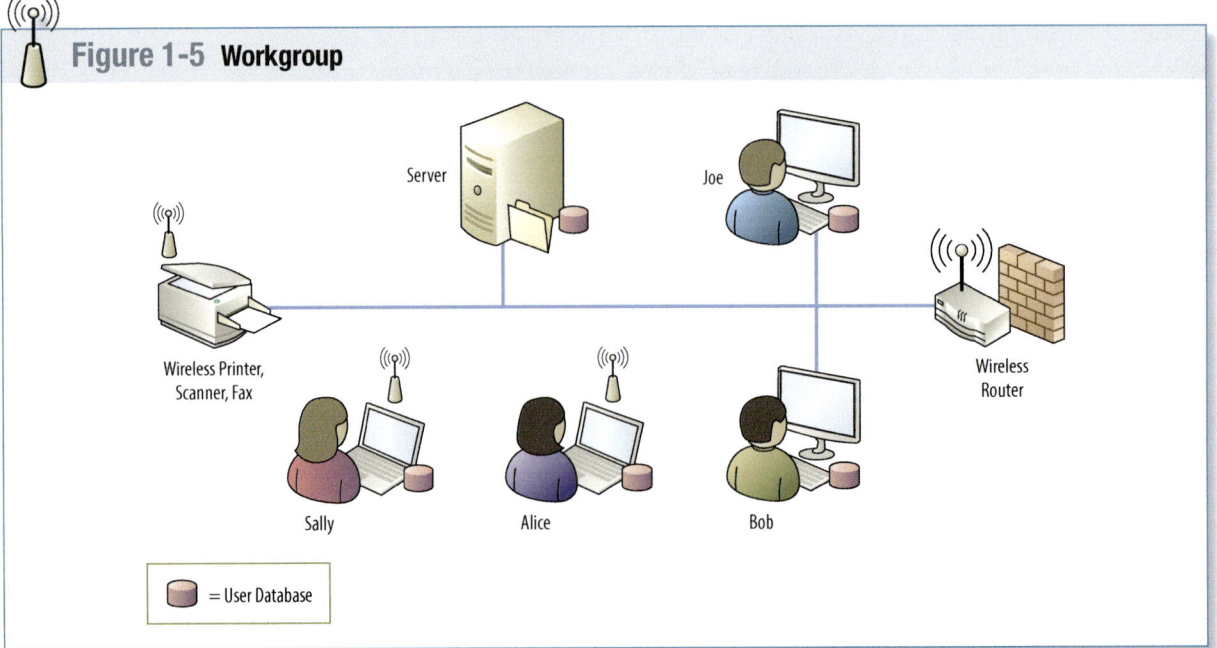

Figure 1-5. *A SOHO can operate as a workgroup.*

computer, or Joe's computer. If Sally needs to log on to any other computer in the workgroup, she must have a separate account on that computer. **See Figure 1-5**.

Even though Internet access is not shown here, a SOHO configured as a workgroup will typically have Internet access. The focus here is the internal LAN. In this scenario, there are five separate user account databases—one on the server and one on each of the four computers. Each user must have a separate user account and password to log on to each of the five computers.

However, most users in a SOHO will typically log on to only one computer on the network, and will need only one user account. To require users to remember five usernames and five passwords would probably break a cardinal rule of security. Users would probably start writing down the usernames and passwords.

When offices get larger than ten computers (or whenever offices need to have more centralized user and computer management), a domain configuration is often introduced. A *domain* is a set of network resources to which certain users are given access. A Windows server can be added and promoted to a domain controller, or an existing server can be promoted to a domain controller.

In Microsoft domains, the domain controller hosts Active Directory Domain Services (AD DS). AD DS uses objects to represent network resources, such as users and computers.

Consider a SOHO configured as a domain. It has eight users with nine computers connected to the LAN. The server has been promoted to a domain controller and is hosting AD DS. Instead of requiring users to memorize passwords for each computer, each user has a single account hosted on the domain controller. **See Figure 1-6.**

This arrangement supports single sign-on (SSO), in which a user needs to sign on to the network only once. AD DS grants all access to domain resources for the user with this single account.

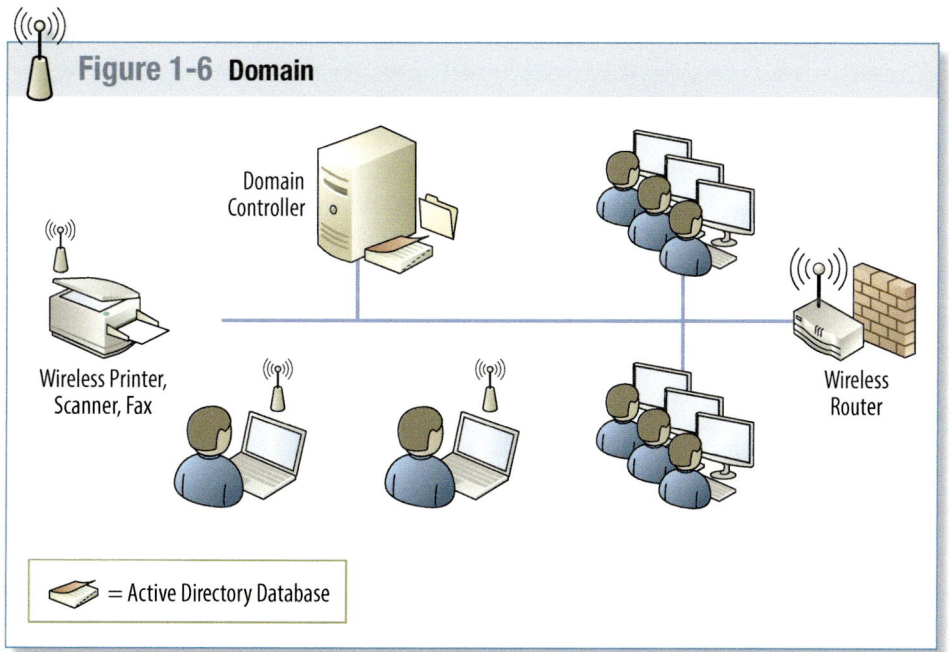

Figure 1-6. A SOHO can be configured as a domain.

Additionally, the user can log on to the domain from any workstation on the network.

By default, AD DS authorizes domain users to log on to any computer in the domain except for domain controllers. Administrators receive the right to log on to domain controllers as well. However, it is possible to restrict users from logging on to other computers within the domain if necessary.

Even though the server has been promoted to a domain controller, it can still perform other functions on the network. For example, a domain controller can still host files as a file server.

Benefits of Domains and Domain Controllers

Promoting a server to a domain controller provides several benefits beyond SSO. These include the following:

Simplified Management To manage accounts in a domain, administrators use a group of centralized tools. For example, the Active Directory Users and Computers feature is used to perform common administration tasks for all the users and computers in the domain.

Group Policy Administrators use Group Policy to configure, control, and manage users and computers. Group Policy can be used to configure password-protected screen savers for all computers in the domain. The setting can be configured one time in Group Policy, and AD DS replicates the setting to all the computers in the domain. It does not matter if the organization has 20 users or 20,000 users; the setting is configured once, and Group Policy does the rest. Thousands of settings can be configured using Group Policy.

Built-In Redundancy and Fault Tolerance With at least two domain controllers in a domain, they automatically replicate the domain data to each other. When an account is added on one domain

---

When to Switch from a Workgroup to a Domain

There is no maximum number of computers that determines when networks must change from a workgroup to a domain. The decision is based on preference and usability. However, most offices switch over when the number of users reaches between 10 and 20. Multiple reasons encourage the switch.

The primary reason to switch is when users must remember multiple user accounts to perform assigned jobs. The domain provides SSO capabilities that require users only to remember a single user account to log on.

A secondary reason is to help administrators reduce workloads. A domain provides centralized administration through AD DS. It also includes advanced administration tools such as Group Policy. Group Policy enables an administrator to configure a setting once and apply it to some or all of the network's computers and users.

Another reason is to allow more concurrent connections from other devices on the LAN. In older operating systems such as Windows XP, each computer was restricted to ten concurrent connections. For example, if a computer shared a printer, only ten other users could send print jobs to it at a time. The eleventh connection was refused. This worked the same way if a computer hosted a shared application. Ten users could connect, but the eleventh connection was refused. This became a logical reason to switch to a domain when the office had more than ten computers. However, Windows 7 Professional and Ultimate editions support 20 concurrent connections, and Windows 8 and 8.1 have no such limitation.

controller, it is copied to the other. If a user changes a password, AD DS makes the change everywhere. This ensures that a redundant copy of the AD DS database provides fault tolerance. If one domain controller develops a fault or fails, the domain continues to operate because the other domain controller will carry the load.

Microsoft domains require a Domain Name System (DNS) server. DNS is used primarily to resolve computer names to IP addresses, but it is also used to locate domain controllers within a domain. Without DNS, or if DNS fails, Active Directory fails.

Large Office Networks

Large offices include more people, more end-user computers, and more users. Although it is possible to network thousands of people in a single LAN, additional steps must be taken to improve the performance of the LAN. The primary difference is that groups of computers are subdivided into different subnets. A *subnet* is a group of computers separated from other computers by one or more routers.

Consider a larger office that includes multiple subnets, and each pair of subnets is joined by a router. The computers are separated on the different subnets so that each subnet has less traffic. In this example, subnet A has only servers, while other subnets have users. Placing the servers on separate subnets is common in larger networks. **See Figure 1-7**.

Traffic on a network is similar to traffic on roads and highways. When there are fewer cars, traffic runs more smoothly. When there are more cars, traffic becomes congested, and the potential for collisions increases. Traffic flow can be improved by adding more roads and highways, providing multiple paths to common destinations, and widening commonly used roads.

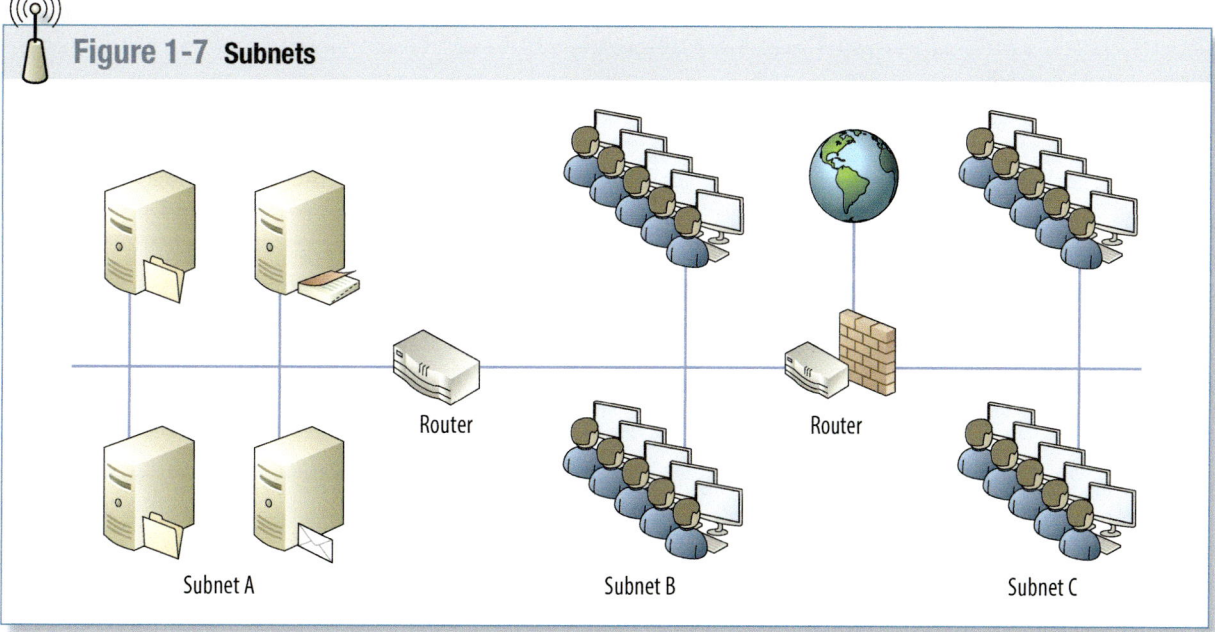

Figure 1-7. A LAN in a large office may be separated into subnets.

Similarly, more computers on a network results in more network traffic and more congestion. Performance can be improved by adding subnets to control and limit traffic in different areas.

Just as cars can have collisions on a road, data packets sent on a shared network medium can collide, resulting in collisions. When two computers on the same subnet send data at the same time, the data collides and is unreadable. Both computers must then send the data again. They both wait a random amount of time and send the data again. If the network is very busy, the data can collide again when it is resent.

Of course, every time computers must resend data, the network becomes that much busier, since there is more traffic. More traffic results in more collisions, and more collisions result in even more traffic. If the network is not optimized, the network performance can slow to a crawl, just as traffic slows in rush-hour traffic on a busy city street.

Large Enterprise Networks

There is no formal definition of an enterprise, but it generally implies an organization with multiple locations.

> **Routers and Switches**
>
> Switches or hubs connect computers to each other within a subnet. Routers connect the subnets together.
>
> For example, a switch can connect all of the servers in subnet A. Similarly, a second switch can connect all the computers in subnet B, and a third can connect the computers in subnet C.
>
> Switches connect computers within a subnet. Routers connect subnets.

Occasionally, some sources define an enterprise as an organization with more than 250 users, to differentiate it from a large office, while other sources define it as more than 5,000 users.

From an information technology (IT) professional's perspective, the biggest difference between a large office and an enterprise is the number of IT professionals supporting the network. Some offices with as many as 50 users are supported by only one or two administrators. These administrators do a little of everything.

In contrast, an enterprise may have dozens of IT professionals, with many of them having specialized knowledge. Some may be experts on e-mail systems such as Microsoft Exchange. Others may be experts on database systems such as Microsoft SQL Server. End-user helpdesk professionals are experts on Windows 8.1 and other desktop operating systems and provide direct support to the users.

Another significant difference with enterprise networks is the method used to connect the different locations. Instead of just a single LAN in a single location, the organization is connected using WAN technologies. WANs can connect large offices to large offices, or they can connect smaller branch offices to the larger main offices.

Finally, many workers in an enterprise are mobile. For example, salespeople often travel to meet customers. These mobile workers still need to access resources on the main network. Remote access technologies enable mobile workers to connect to the main network from remote locations. While remote access technologies are more common in enterprises, they can be used anywhere, including SOHOs.

Branch Offices

Large organizations often have branch offices. This enables the organization to have a broader reach and enables their employees to be closer to their customers.

Branch offices are often much smaller than the main headquarters of the organization. They have fewer people and limited local computing resources. Individuals will have workstation computers, but the branch office might not have any servers on site. However, employees still need to access organizational resources such as servers at the headquarters location. It is common for a branch office to be connected to either a headquarters or a regional office using a WAN link.

Consider a branch office connected to the main headquarters of the organization, and another branch office connected to the regional office. An organization can have as many branch offices as desired. However, each WAN link costs additional money. **See Figure 1-8**.

Because a branch office has fewer people, it often has limited support. In other words, the headquarters location might have many IT professionals, but a branch office might not have any. Instead, the branch office might rely on the headquarters IT staff to provide remote support via the phone or by using remote technologies.

Several remote assistance technologies are available to provide assistance to users at distant locations. For example, Windows includes the Remote Assistance feature, which enables administrators in one location to take control of a user's desktop (with the user's permission) to resolve a problem or to show the user how to accomplish a task.

Network Remote Access

Many organizations also set up remote access capabilities. *Remote access* enables individuals working outside the company to access resources internal to the

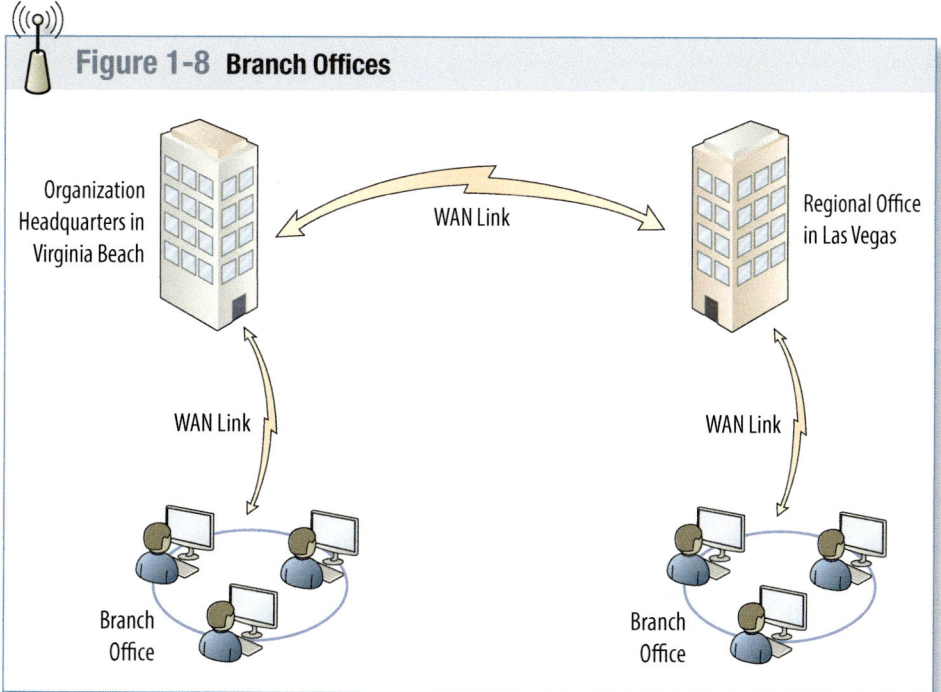

Figure 1-8. *Branch offices can be connected in an organization.*

company. Following are the two primary methods of remote access:

Dial-Up A client uses a modem and phone line to connect to a remote access server that also has a modem and a phone line. After authenticating with the server, the server provides connectivity to the internal network. A dial-up remote access server is accessible to any client that has access to phone lines. However, long-distance charges can be a problem. **See Figure 1-9**.

Virtual Private Network (VPN) A *Virtual Private Network (VPN)* is a group of connected computers that provides access to an internal network using a public network such as the Internet. The client accesses the Internet using any available means, and then connects to the company's VPN server, which is also connected to the Internet. After authenticating with the server, the VPN server provides connectivity to the internal network. The VPN server is accessible to any client that has access to the Internet, which can often be less expensive than a long-distance telephone connection. **See Figure 1-10**.

AUTOMATION NETWORKS

A *building automation system (BAS)* can consist of many different products, but virtually all of them utilize some form of network signaling to provide monitoring and control from a central location. At its most basic, a BAS controls the heating, ventilation, and air conditioning system for a building, but modern smartbuildings can have systems that also control lighting, security, environmental, and elevator systems.

Early BAS products used proprietary networking systems—technologies developed from pneumatic controls to analog devices to today's all-digital controls. Many older buildings still retain these older technologies, because retrofitting new equipment can be complicated and costly.

Proprietary systems eventually gave way to early standards, such as BACnet. Today, there are a variety of modern standards, many of which are compatible with the data networking

> **Remote Access or VPN Server**
>
> The terms "remote access server" and "VPN server" have subtle differences. In short, a VPN server is a remote access server. However, not all remote access servers are VPN servers. Some remote access servers can use dial-up technologies only. If a remote access server uses dial-up only, it is not correct to call it a VPN server.

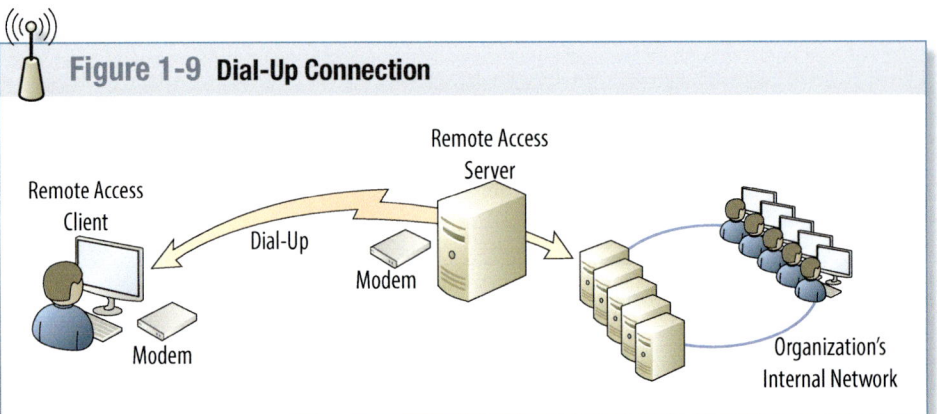

Figure 1-9. Remote access can be achieved via a dial-up connection.

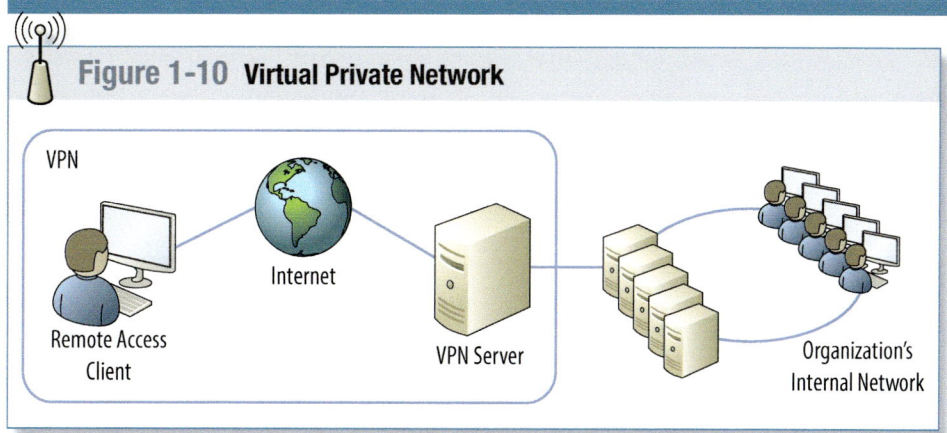

Figure 1-10. Remote access can also be achieved via a VPN connection.

technologies. Wireless BAS products are becoming increasingly popular, because they make it possible to add new systems without extensive construction.

When constructing a new building, it is relatively easy to find products that are compatible with standard IP networks, eliminating the need for additional wiring and proprietary installations. In older buildings, however, systems of various types might coexist or specialized products might bridge the older technologies to the new.

STANDARDS ORGANIZATIONS

Most networking technologies are based on published standards, so that the networking products produced by one manufacturer can interact with those of another. There are several standards organizations that are important in networking, because these organizations develop different types of standards to meet specific needs. For example, the Internet Engineering Task Force (IETF) has created standards for Internet communications. Without a central authority creating standards used by all developers, there is no way the Internet would be the valuable global resource it is today.

These organizations include the following:

- Internet Engineering Task Force (IETF)
- World Wide Web Consortium (W3C)
- Institute of Electrical and Electronics Engineers (IEEE)
- International Telecommunication Union (ITU)

Internet Engineering Task Force

The IETF defines Internet communications standards. Its goal is to make the Internet work better. It does so by creating high-quality, relevant technical documents used by designers, managers, and users of the Internet.

TCP/IP is the protocol suite used on the Internet. It is also the primary protocol suite used within Microsoft networks. The IETF has produced a wide range of documents that define how the different protocols are used.

Most of the documents published by the IETF are known as *Requests For Comments (RFCs)*. RFCs are written and then released to the world for comments. Many RFCs are assigned to the Standards Track category and go through a lengthy ratification process. Following are the four primary stages for an RFC in the Standards Track category:

Proposed Standard (PS) An RFC starts at the PS stage, the first official stage where the standard is introduced. Many standards never progress beyond this level.

Draft Standard (DS) The second official stage is DS. At this stage, the standard has been tested and verified to work as expected. It is on the track to become an actual standard, but is not there yet.

Standard (STD) The final stage of an RFC is STD. RFCs at this stage are widely used.

Best Current Practice (BCP) BCP is a single-stage alternative to the previous stages. A BCP provides operational specifications.

When the IETF releases an RFC, it assigns it a number. This number stays with the RFC, and the document with that number never changes. If a change is needed, the IETF publishes a new RFC with a new number. Some RFCs are Informational or Experimental, and do not follow the standard track stages.

As an example, RFC 791, "Internet Protocol," defines the 32-bit IPv4 protocol. Even though this document was published in 1981, it is still in use today. The IETF recognized that the Internet was running out of IPv4 addresses, so it tasked a working group with creating a solution. The working group first came up with RFC 1819 (commonly called the 64-bit IPv5). However, comments on RFC 1819 made it apparent that if 64-bits were used for IPv5, then the Internet would probably run out of IP addresses again in about ten years.

RFC 1819 was scrapped. The IETF ultimately released RFC 2460, "Internet Protocol, Version 6," defining the 128-bit IPv6 protocol. RFC 2460 is on the Standards Track and currently has a status as a DS.

World Wide Web Consortium

The W3C defines standards for the World Wide Web (WWW). As an example, the *Hypertext Transfer Protocol (HTTP)* was defined by the W3C. HTTP is the primary protocol used to transfer WWW information over the Internet. Membership of the W3C consists of organizations rather than individuals.

Most web pages are created in a *Hypertext Markup Language (HTML)* format. A markup language like HTML uses tags within a text file to set the format and layout of a web page. The web started as an Internet-based hypermedia initiative for global information and grew into what it is today. Tim Berners-Lee invented HTML and the World Wide Web. (His proper title is Sir Timothy Berners-Lee, since he was knighted by Queen Elizabeth in 2004.) Sir Tim Berners-Lee is currently the director of the W3C.

Note that although the WWW runs on the Internet, it is not the Internet itself. The Internet is a huge network of millions of networks and includes all of the infrastructure hardware belonging to those networks. The WWW is one of many methods used to access information over the Internet. The Internet also supports transferring files using the File Transfer Protocol (FTP) and sharing information through newsgroups, such as Usenet newsgroups.

For additional information, visit qr.njatcdb.org Item #1786

Institute of Electrical and Electronics Engineers

The IEEE is a professional association dedicated to advancing technical innovation. It has more than 375,000 members in 160 different countries. A primary function of the IEEE is defining lower-level network standards.

IEEE standards are identified as IEEE (pronounced "I triple E") with a number. The IEEE has defined many different standards. For example, IEEE 802.3 defines various standards for wired Ethernet networks. IEEE 802.11 defines various standards for wireless networks.

International Telecommunication Union

The ITU is a United Nations agency that includes members from 192 countries. It is focused on information and communication technology issues. It has contributed to shared global use of the radio spectrum, as well as international cooperation in assigning satellite orbits. It has also helped improve telecommunication infrastructure throughout the world.

Many of the telephony standards used by computers today have been defined by the ITU. This includes standards used for modem communications and video conferencing.

Summary

The logical network organization identifies the overall layout of LANs and WANs. A local area network (LAN) is a group of computers and computing devices in a single high-speed layout. It can include one or more subnets. A wide area network (WAN) is a group of two or more LANs connected with a slower WAN link. WAN links can also connect branch offices.

Review Questions

1. What should be enabled on a computer that has direct connection to the Internet?
   a. Firewall
   b. Router
   c. Switch
   d. VPN

2. A WAP often provides access to the Internet.
   a. True
   b. False

3. What is the term for a group of computers connected in a single location?
   a. LAN
   b. VLAN
   c. VPN
   d. WAN

4. A network is connected using high-speed components rated at 1 Gbps. Gbps stands for gigabits per second.
   a. True
   b. False

5. Users in the network must remember an average of five usernames and passwords to access different computers. How can the number of passwords be reduced?
   a. Change the network to domain
   b. Change the network to a workgroup
   c. Create a VPN
   d. Create a WAN

6. A LAN is two or more networks connected together over a large geographical distance.
   a. True
   b. False

7. A WAN is a group of computers connected together in a network.
   a. True
   b. False

8. An employee is able to connect to the employer's private network over the Internet. What is the employee using?
   a. Domain controller
   b. LAN
   c. VPN
   d. WAP

9. Which of the following is a type of remote access servers?
   a. Domain controller
   b. ISDN
   c. VPN
   d. WAP

10. All RFCs are known as standards.
    a. True
    b. False

Overview of Networking Components

Every computer network includes various components that connect the computers on a network. Although it is important to understand how each of these components works individually, it is easier to grasp the details by first understanding how they work together.

A big-picture view of all the components and how they work is key to understanding network technologies. A good starting point is to overview the basic transmission methods used in networks (such as unicast, broadcast, and multicast) and then progress to an understanding of basic hardware components.

A network uses protocols as the rules of communication. Basic network connection topologies include the star, the bus, and the ring.

OBJECTIVES

- ▶ Compare unicast, broadcast, and multicast traffic
- ▶ Identify various network hardware
- ▶ Describe the elements of various physical media

CHAPTER 2

TABLE OF CONTENTS

Unicast, Broadcast, and Multicast Traffic 22
    Unicast Traffic 22
    Broadcast Traffic 23
    Multicast Traffic 25

Network Hardware 25
    Hubs 26
    Switches 27
    Bridges 28
    Routers 30

Physical Media 32
    Twisted-Pair Cable 32
    Coaxial Cable 36
    Fiber-Optic Cable 36
    Basic Topologies 38
    Cabling Standards 44

Summary 48

Review Questions 49

UNICAST, BROADCAST, AND MULTICAST TRAFFIC

Before digging too deep into the different physical devices used on networks, it is important to understand the different types of data transmission. Data is transmitted to and from hosts on IPv4 networks using one of the following three transmission types:

- Unicast
- Broadcast
- Multicast

An *Internet Protocol (IP) address* is a unique identifier for each computer that directs data to its destination. It is similar to how a postal service uses home and business addresses to route letters and other correspondence to their destinations. As long as the address is correct, the correspondence arrives. Similarly, computers and other network devices use IP addresses to get data where it needs to go. A typical IPv4 address used in an internal network might resemble 192.168.1.10.

The primary IEEE standard used in local area networks (LANs) is Ethernet, defined by Institute of Electrical and Electronics Engineers (IEEE) standard 802.3 (and the associated subsections of 802.3). Ethernet is a group of technologies used to connect computers using wired media such as twisted-pair and fiber-optic connections. Wireless connections are defined by other standards, such as 802.11.

Unicast Traffic

Unicast traffic is traffic sent from one computer to one other computer. On a typical organization's network, most other computers will not receive the unicast traffic that is not addressed to them. For example, when a group of computers is connected to a switch, the data sent from one computer goes to the switch, and the switch then sends it only to the destination computer.

If a hub is used to connect the computers, the traffic will go to all the computers connected to that hub. This is because the hub is not as sophisticated as a switch. However, the network interface adapters in the computers recognize when unicast traffic is not addressed to them, and they do not process the data.

A *protocol data unit (PDU)* is a unit of data packaged for transport on a network. Depending on the communications protocol being discussed, the traffic a PDU carries might be called a frame, a datagram, a segment, or a packet.

Consider a network consisting of four computers. If Bob's computer sends a unicast packet to Sally's computer, the packet will not reach Joe's or Maria's computer or will not be processed by these other computers. Bob's computer transmits the unicast packet. Only Sally's computer will receive and process the unicast transmission. **See Figure 2-1.**

When the packet reaches a computer, the network interface adapter examines the packet to determine whether it is addressed to it. Even if the traffic reaches one of the other computers (such as Joe's or Maria's), the network interface card (NIC) will determine that the traffic is not addressed to the computer, and the packet will not be processed.

Different devices on the network (such as routers and switches) also examine the packet to ensure it reaches its ultimate destination.

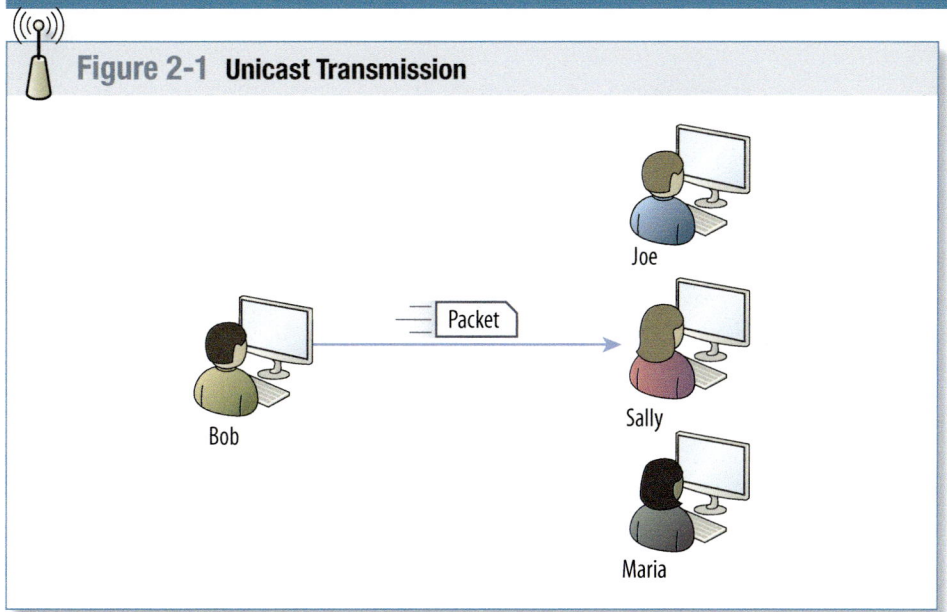

Figure 2-1. *A unicast transmission is one to one.*

Broadcast Traffic

Broadcast traffic is traffic transmitted by one computer that goes to all of the computers on a subnet. Notice the clarification, though. A broadcast packet does not go to all computers in the world; it goes to all the computers on a subnet.

Recall that a subnet is a group of computers separated from other computers by one or more routers. Another way of saying this is that broadcast traffic goes to all computers on the same side of a router.

For example, consider a scenario in which Bob's computer is broadcasting a

---

Packets, Frames, Datagrams, and PDUs

Data is packaged together in a specific format before it is transmitted. Often, the term packets refers to this packaged data. Although the term packets is common, it is not always technically accurate.

The Open Systems Interconnection (OSI) model has seven layers: Application, Presentation, Session, Transport, Network, Data Link, and Physical. Technically, a packet only refers to data at the Network layer of the OSI model. A packet is only one of many protocol data units (PDUs), and PDUs are defined by the information they contain. For example, at the Data Link layer, the PDU is a frame. Data at the Transport layer is referred to as a segment. Data at the upper three layers (Application, Presentation, and Session) is usually simply referred to as data for the purposes of networking.

The ability to differentiate between all of the different PDUs will become more important while progressing through the study of networks. For now, the terms packet and packets refer to any data transferred on the network.

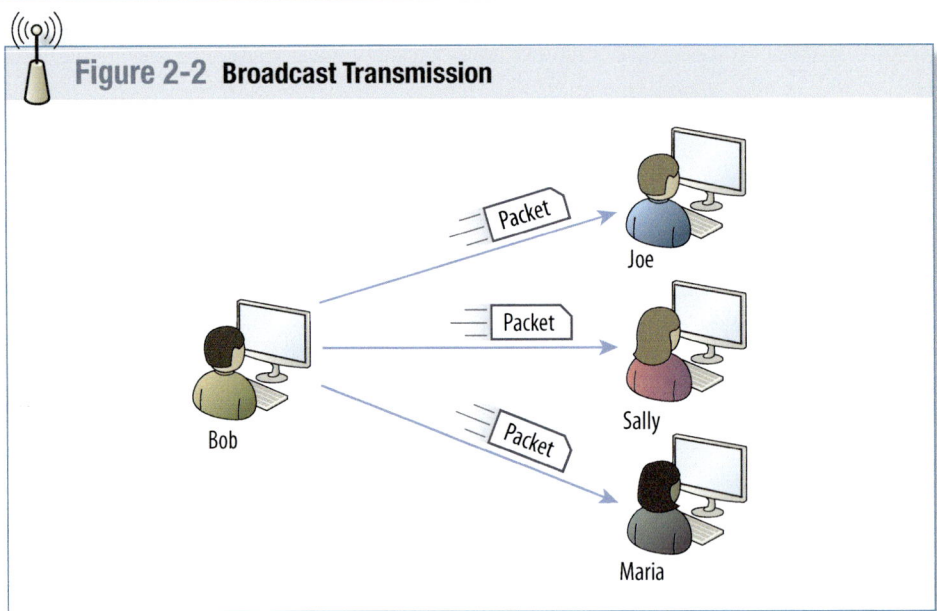

Figure 2-2. *A broadcast transmission is one to all.*

packet on the network, and each of the computers on the subnet will receive and process the packet. **See Figure 2-2.**

Notice that there is a slight difference in the language that defines a unicast packet and a broadcast packet. This difference is important.

- Unicast traffic is one-to-one traffic between two computers.
- Broadcast traffic is traffic sent from one computer to all other computers on a subnet (not the entire network).

All of the computers on a subnet will receive a broadcast packet. However, if the network has more than one subnet, all of the computers on the network will not receive the packet. In other words, broadcast traffic does not cross subnets barriers. Routers separate subnets, so another way of saying this is that routers do not forward broadcasts.

Computers are connected within a subnet using hubs or switches. Both hubs and switches *do* pass broadcast traffic.

Routers and Broadcast Exceptions

What is consistent with almost all rules is that there are exceptions. This is true with routers and broadcast transmissions.

First, remember the rules about routers. Routers do not pass broadcasts. Routers separate subnets, and broadcasts in one subnet will not reach computers in another subnet.

Except…it is possible to program a router to pass certain broadcasts.

For example, consider the Dynamic Host Configuration Protocol (DHCP). A DHCP server provides IP addresses and other information to DHCP clients. Both the DHCP clients and DHCP server use a special type of broadcast known as a BootP broadcast. Routers can be programmed to pass these broadcasts on User Datagram Protocol (UDP) ports 67 and 68. This allows a single DHCP server to serve multiple DHCP clients even if they are on separate subnets.

Think of a broadcast similar to how one person in a room can yell something and everyone in the room can hear it. Compare this to a unicast message, where one person whispers something so that only one other person hears it.

Multicast Traffic

Multicast traffic is traffic transmitted from one computer to many other computers. When a computer joins a multicast group, the network adapter is internally configured to process traffic using the multicast group's IP address. Now, when traffic is multicast to the multicast group, any computers that have joined the multicast group will receive and process the packet. Multicast traffic will pass to different subnets.

Consider a scenario in which multicast traffic sent by Bob's computer will reach multiple computers. In this scenario, Joe and Maria's computers have joined the multicast group, and they will receive the traffic. Sally's computer has not joined the multicast group and will not receive the traffic. **See Figure 2-3.**

In most network configurations, the multicast traffic will not even reach Sally's computer. However, even if it does, the network adapter will determine that the traffic is not destined for Sally's computer, and the packet will not be processed.

Internet Group Multicast Protocol (IGMP) is the primary protocol used to transmit and process multicast traffic.

NETWORK HARDWARE

Computers are connected within a network using several networking components. Computers have network interface adapters. Cables connect the wired network interface adapters in the computers to network devices such as switches. Routers connect the different subnets on a network.

Network technicians must understand basic terms that are often used in the field.

Collision Domain A *collision domain* is a group of devices on the same network segment that are subject to collisions. Collisions occur when two devices on

> ✓ FACT
> Anycasts send traffic to one of many computers on a list and are more efficient than broadcasts.

> ✓ FACT
> Networks can be either wired or wireless. Wired networks have cables, but wireless networks connect using radio frequency transmissions.

Figure 2-3 Multicast Transmission

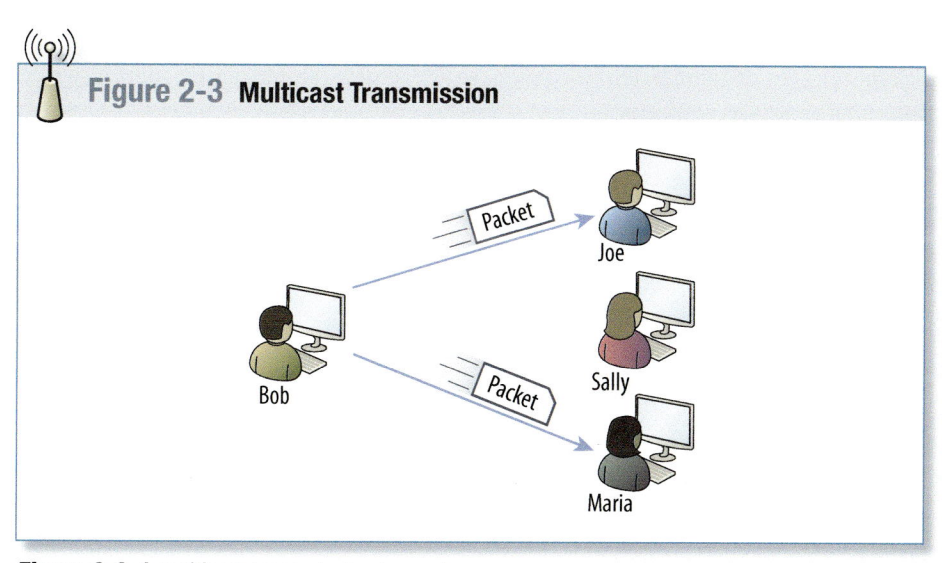

Figure 2-3. A multicast transmission is one to many.

the same segment send traffic at the exact same time. In other words, only one device can send data at any given time. If a collision occurs, both devices must then resend the data. Collisions were normal in Ethernet networks, but they have been largely eliminated with the use of switches.

Broadcast Domain A *broadcast domain* is a group of devices on a network that can receive broadcast traffic from each other. In other words, if one device sends a broadcast packet, all of the other devices in the broadcast domain will receive it. Broadcasts are sometimes necessary, but they generate a lot of traffic, so it is useful to limit the number of computers in a broadcast domain.

Network administrators use different devices to create separate collision domains and separate broadcast domains. Although the following sections cover many devices, it is important to understand how switches and routers are related to collision and broadcast domains.

Computer networks utilize various devices and components. Each device or component performs specific functions on the network. Network devices and components include:

- Hubs
- Switches
- Bridges
- Routers
- Physical media (such as cables)

Hubs

A *hub* is a device that provides basic connectivity for other devices in a network. Although these were once common devices on Ethernet networks, switches have replaced them in most networks today. A hub does not have any intelligence. Any data sent in to the device though one hub port is forwarded out through all of the other ports.

A *port* in this context is a physical connection between devices. To connect a computer to a wired network, one end of a cable is plugged into a hub port, and the other end is plugged into the network interface adapter on the computer.

Consider a scenario in which a four-port hub has different computers connected to the ports. If Bob's computer sends a unicast packet to Sally's computer, the same packet will also reach Joe's and Maria's computers. The network adapters on Joe's and Maria's computers will recognize the packet is not destined for them and discard it, so the traffic will not be processed by their computers. However, the traffic can cause collisions if either Joe or Maria is trying to send data at the same time. **See Figure 2-4.**

This is a result of all the computers that are connected to the hub being in the same collision domain. Hubs forward broadcasts, so all of these computers are also in the same broadcast domain.

In the case of a four-port hub, broadcast traffic and collisions are a problem. However, when hubs were popular in production environments, they would often have 24 or more ports. Additionally, it was common to daisy-chain multiple hubs together on the same network, which would create larger broadcast and collision domains, and more traffic problems.

> ✓ FACT
>
> When referring to a hub or switch, the term *physical port* refers to a physical component—that is, a cable jack. However, ports can also be logical, as in the case of the Transmission Control Protocol (TCP) and User Datagram Protocol (UDP) protocols, in which a port is a number used to identify a protocol or service.

Figure 2-4 Hub

Collision Domain and a Broadcast Domain

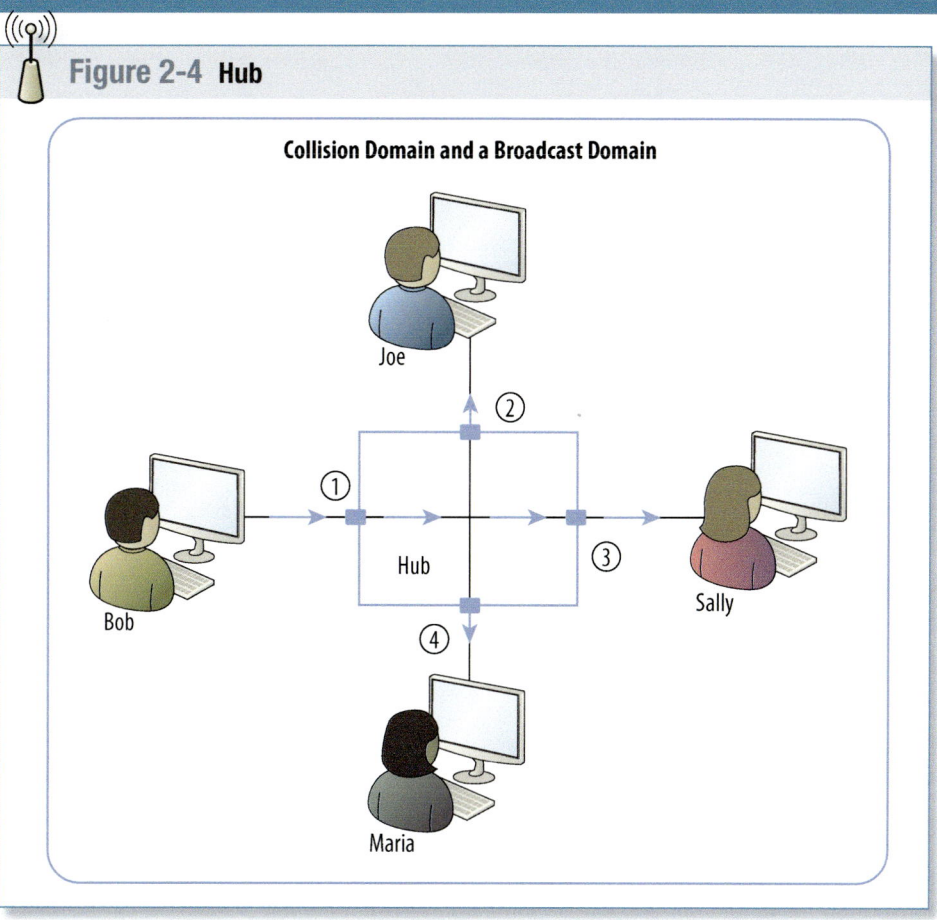

Figure 2-4. A four-port hub has different computers connected to it.

Switches

A *switch* is a device that connects the computers in a network segment together. Switches create separate collision domains. A switch passes broadcast traffic to all connections, so it does not create separate broadcast domains.

Physically, switches connect computers to a network in the same way as hubs. However, switches have intelligence that improves the performance of a network by isolating the computers into separate collision domains.

✓ **FACT**
Switches enable the creation of smaller collision domains, the reduction of collisions, and the improvement of network performance.

✓ **FACT**
Switches have multiple capabilities that can provide significant performance enhancements for any network.

Consider a scenario in which a switch replaces the hub. Now, when Bob sends a unicast packet to Sally, the traffic reaches only Sally's computer.

The packet will not reach either Joe's or Maria's computer. **See Figure 2-5.**

The switch dynamically determines the destination port as it receives traffic. In the first example, it determines that the traffic should flow from port 1 to port 3. However, if Maria's computer sends unicast traffic to Bob's computer, the switch makes a different determination and instead sends traffic from port 4 to port 1.

Since the switch can dynamically determine through which port to send traffic, it has effectively separated the computers into four separate collision domains, each consisting of two devices: a computer and a switch port. Since a switch forwards broadcasts to all connected devices, the computers connected to the switch are said to be in a single broadcast domain.

By processing incoming packets, switches learn which computer is connected to which port. In short, the switch tracks the locations of the computers. As each computer sends packets on the network through the switch, the switch identifies the computer and the port that it is using. It maintains a table identifying the computers and their ports.

Bridges

A *bridge* is a network device that connects two or more network segments. Any of the segments can have one or more computers on it. For example, one segment could have 10 computers connected together with a hub, and another segment could have another five computers connected together with a different hub.

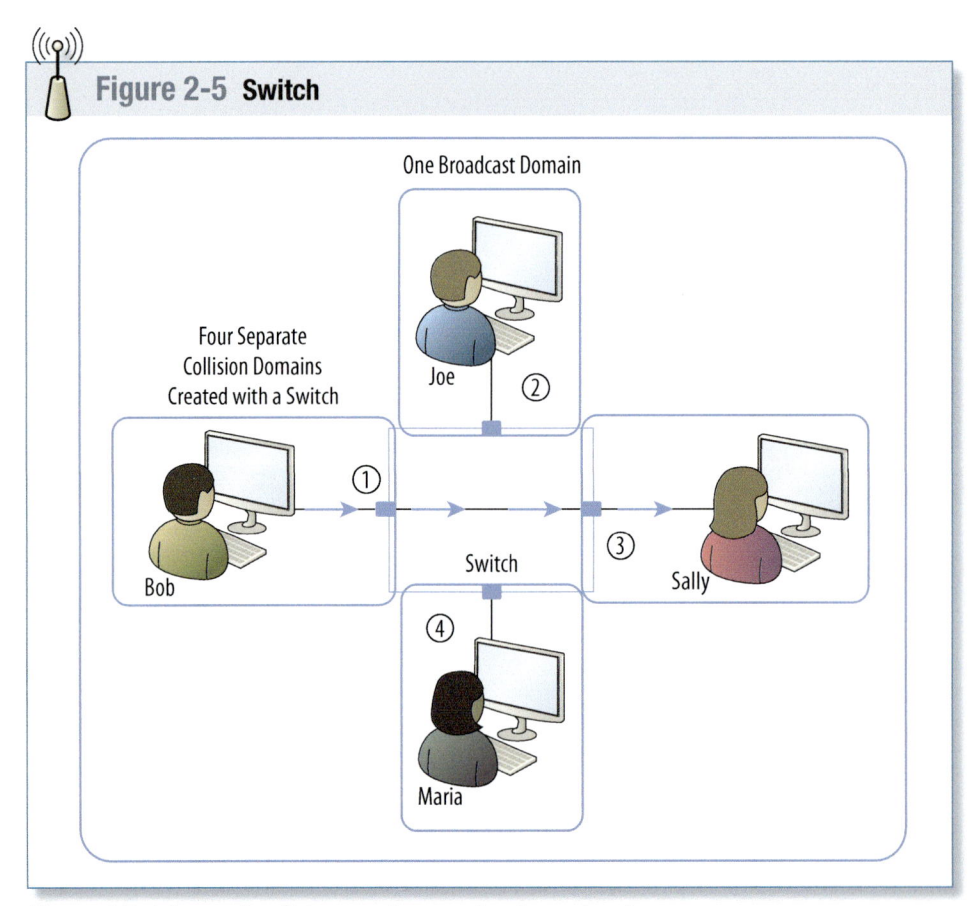

Figure 2-5. *A four-port switch connects computers to a network.*

Bridges are not common on networks anymore, but they still might be in use. A bridge is similar to a switch in that it learns which port a computer is connected to and internally switches traffic to the right port. The separate ports create separate collision domains. In fact, a switch may be nothing more than a bridge with many ports.

However, multiple computers are connected to each port on a bridge. Separate hubs connect these computers together, and then the hubs connect to the bridge. The alternative is to connect each of the hubs together to create a single collision domain.

For example, four 24-port hubs are used to connect 96 computers in a single network. This results in a collision domain of 96 computers, on which 96 computers are all competing to send their data on the network.

Instead, a bridge can be used to connect these four hubs. The bridge creates four separate collision domains of 24 computers each. Computers on each of these separate collision domains are only competing with 23 other computers to send their data, instead of 95 other computers. **See Figure 2-6.**

Another benefit of bridges is that they can connect dissimilar physical topologies. For example, one port can connect computers using twisted-pair cables, and another port can connect computers using fiber-optic connections. Both wired and wireless bridges also exist. Wireless bridges are commonly used to connect a wireless access point (WAP) to another type of device on a wired network.

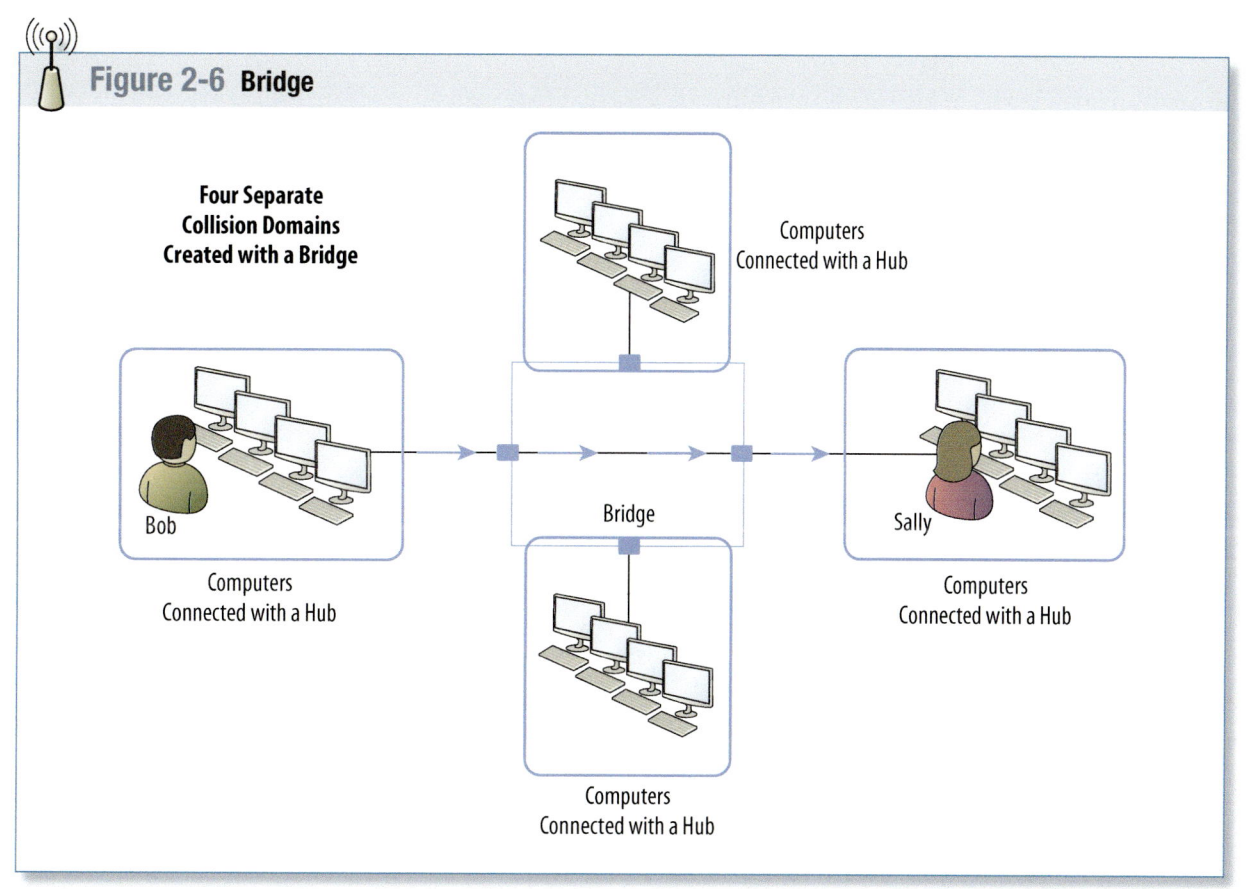

Figure 2-6 Bridge

Figure 2-6. A four-port bridge creates four collision domains.

Routers

A *router* is a device that connects networks together. Routers do not pass broadcast traffic. Routers create both separate collision domains and separate broadcast domains.

Routers move packets between networks or subnets. Switches (or hubs) connect the devices within the subnet, and routers connect the subnets. Remember from the discussion earlier in this chapter that switches track the computers connected to them. However, routers do not track individual computers; instead, they track networks or subnets within a network.

Consider a scenario in which subnet A with two computers is connected to a hub, and subnet B has four computers connected to a switch. These two subnets can be connected with a router. **See Figure 2-7.**

All of the computers connected via the hub are in one collision domain. Each of the users connected to the switch makes up another collision domain, for a total of four. Last, the connection between the router and the switch creates a sixth collision domain. There are two subnets (subnet A and subnet B), and each subnet is a separate broadcast domain.

Routers direct, or route, traffic between networks. When a router receives a packet, it identifies the best path for the packet to take in order for it to arrive at the final destination. In Figure 2-7, there are only two subnets, so the router does not have to make many decisions.

However, consider a scenario in which a multiple-subnet network is connected via several routers. The routers learn the locations of all the subnets and determine the best path to take to get traffic to its destination. **See Figure 2-8.**

> ✓ FACT
> Routers track subnets within a network. In comparison, switches track computers on a subnet.

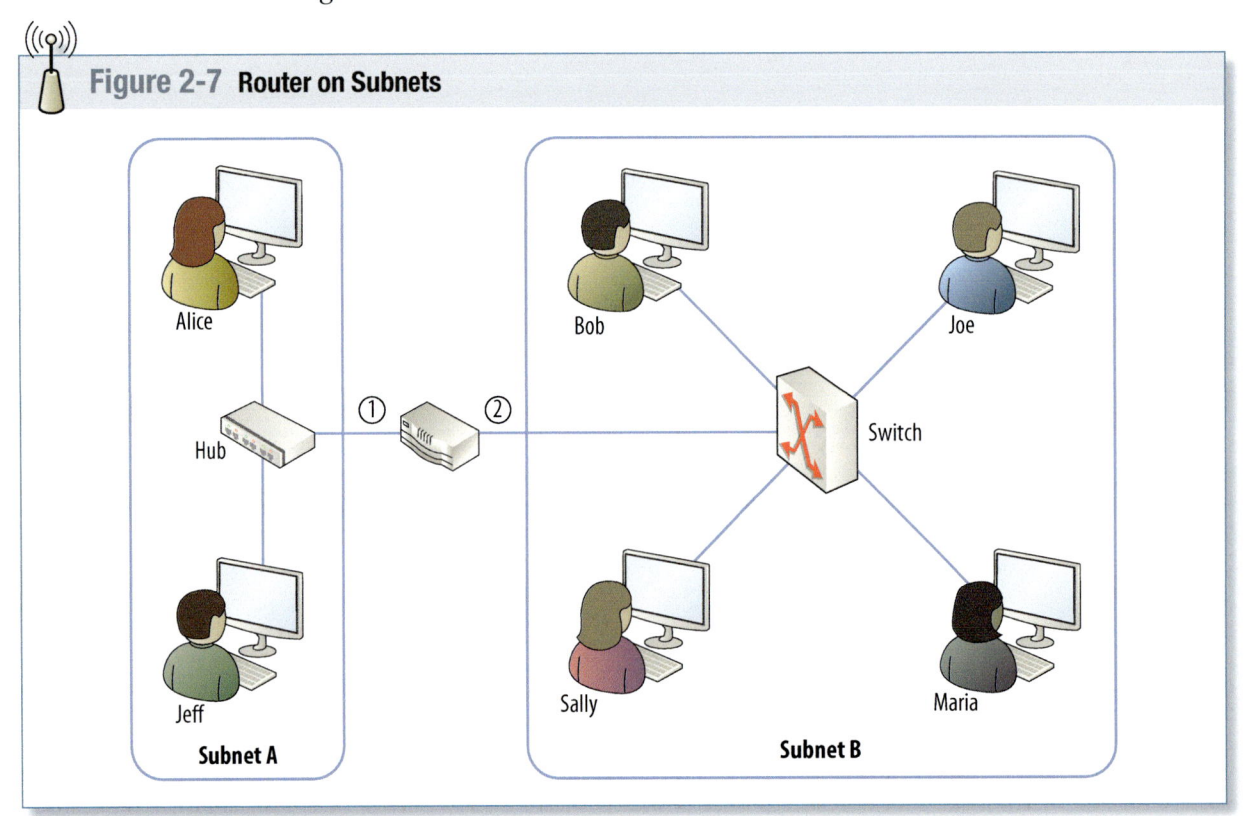

Figure 2-7 Router on Subnets

Figure 2-7. *Two subnets can be connected by a router.*

Routers and Default Gateways

Each computer on a subnet is configured with the IP address of a router. However, in this context, it is not called a router, but is instead called the default gateway.

In the early days of TCP/IP, routers were referred to as gateways. The term default gateway is a vestige of this earlier usage. Today, the term gateway typically refers to a network system that functions as an interface to another network that uses different protocols.

Computers on subnet A use the IP address of port 1 of the router as their default gateway. Computers on subnet B use the IP address of port 2 of the router as their default gateway. **See Figure 2-7.**

Multiple routers with a subnet in the center (labeled as a server subnet). The server subnet connects with four separate subnets. However, computers can be configured with only one default gateway. **See Figure 2-8.**

The default gateway provides the path to the rest of the network that the computer should use. Usually, this path provides access to the Internet. Therefore, computers in the server subnet will most likely be configured with the IP address of port 1 of router 1.

Figure 2-8 Multiple Subnets

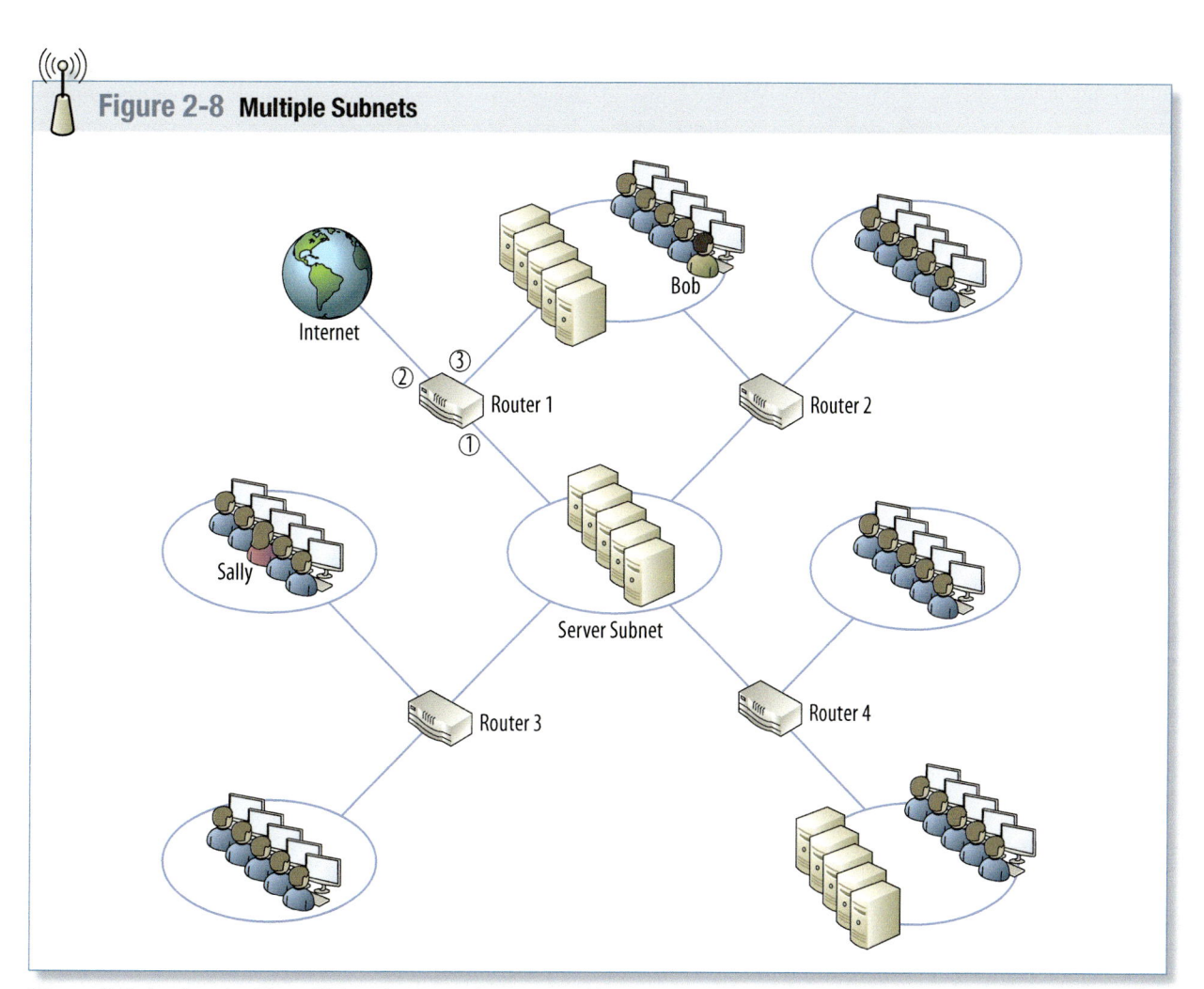

Figure 2-8. A network with multiple subnets can be separated by routers.

For example, what if Sally wants to access the Internet? How many routers will she have to go through to get to the Internet? How about Bob? Notice that Sally has to go through two routers (router 3 and router 1), while Bob has to go through only one router. Although that is easy to see in the diagram, the routers do a lot of work to learn these paths.

One way that routers determine the best path to take is by talking to each other. Routers can use different types of routing protocols that help them learn the network and the paths to different subnets. Some routing protocols are used only in internal networks, while other routing protocols are used only on the Internet.

PHYSICAL MEDIA

A data network must have a network medium that connects the computers and other devices together. Until relatively recently, that network medium was a cable of some sort. Most of the cables used throughout the history of data networks have been copper-based. One or more strands of electrically-conductive copper are encased within a non-conductive sheath. The computers transmit signals by applying electric voltages to the copper conductors in a pattern that the receiving systems interpret as binary data.

For the most part, the history of copper-based data networks mirrors the history of the Ethernet protocol. As Ethernet has evolved, it has increased in speed and changed the types of copper cables defined in its Physical layer specifications.

Although copper cables are relatively efficient transmitters of signals, they are prone to a number of shortcomings, including attenuation, (that is, the gradual weakening of the signal over distance), and disruption by noise and electromagnetic interference (EMI). *Electromagnetic interference (EMI)* is disruption to the functioning of an electrical device caused by other electrical equipment. When installing copper cables, contractors must be careful not to exceed recommended cable lengths and to avoid sources of EMI.

To avoid these problems altogether, it is also possible to build a network using another type of cable, called fiber-optic. *Fiber-optic cable* is cable that uses transparent fibers to transmit light instead of electrical signals. A fiber-optic cable consists of a glass core surrounded by a slightly different kind of glass called the cladding. The core and cladding are then covered by other things that provide mechanical protection to this hair-thin strand of glass. The signals on a fiber-optic network are beams of light generated by a light-emitting diode or laser. Fiber-optic networks are much less affected by attenuation than copper networks, so they can span much greater distances. They are also completely immune to EMI, making them a viable medium in environments where copper networks cannot function.

Twisted-Pair Cable

A *twisted-pair cable* is a cluster of thin copper wires, with each wire having its own insulating sheath. Individual pairs of insulated wires are twisted together, gathered in groups of four pairs (or more), and encased in another insulating sheath. **See Figure 2-9**.

There are many different types of twisted-pair cables designed to accommodate various environmental, legal, and performance requirements.

Unshielded twisted-pair (UTP) is a cable that contains four wire pairs within a jacket with no additional shielding. UTP cable is the standard medium for

Figure 2-9 Twisted-Pair Cable

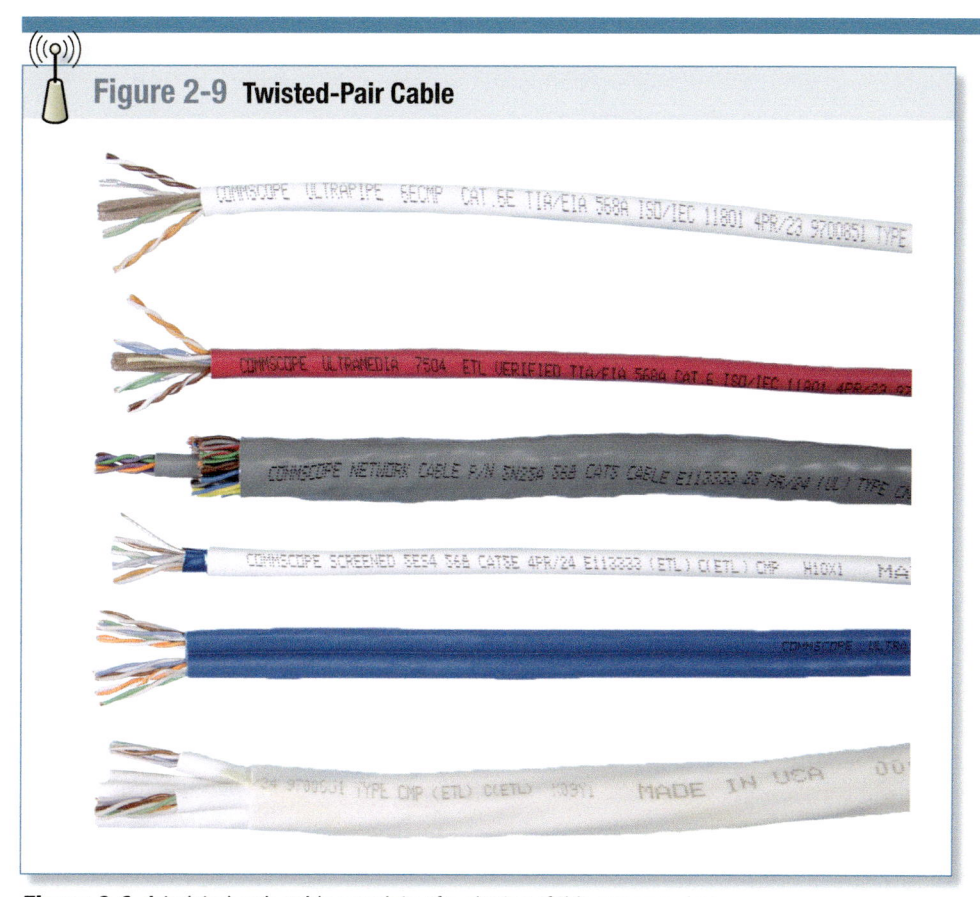

Figure 2-9. A twisted-pair cable consists of a cluster of thin copper wires.

copper-based Ethernet LANs. UTP grew to replace coaxial cable for Ethernet networks for two main reasons. First, the eight separate wires in a UTP cable make it much more flexible and easy to install in walls and ceilings. Second, voice telephone systems also tend to use UTP, which means that the same contractors installing telephone cables can often install the data networking cables as well.

Shielded twisted-pair (STP) cable is twisted-pair cable that has a foil or mesh shield surrounding all four pairs. It is useful for environments with greater amounts of EMI. Some installations may even call for screened twisted-pair (ScTP), which not only has a shield around all four pairs, but also includes a shield around each pair.

Some now-obsolete Data Link layer protocols (such as Token Ring) had Physical layer specifications that called for STP cables. Today, however, Ethernets networks only use them in special situations that require additional protection from EMI.

UTP Categories

UTP cable comes in a variety of different grades, called categories in the cabling standards published by the Telecommunications Industry Association/Electronic Industries Alliance (TIA/EIA). The categories define the signal frequencies that the various cable types support, along with other characteristics, such as resistance to certain types of interference. The higher the category number, the higher the cable quality, and the higher the price.

Cable categories that network administrators are likely to encounter are:

- *Category 3 (CAT3)*—Long the standard for telephone communications, the first UTP-based Ethernet networks (called 10Base-T) used CAT3 cables. CAT3 is no longer supported for new network installations.

- *Category 5 (CAT5)*—Designed for 100Base-TX Fast Ethernet networks, CAT5 cabling was dropped from the latest version of the TIA/EIA cabling standards.

- *Category 5e (CAT5e)*—Still rated for frequencies up to 100 megahertz, CAT5e cable is designed to support full duplex transmissions over all four wire pairs, as on 1000Base-T Gigabit Ethernet networks.

- *Category 6 (CAT6)*—Designed to support higher frequencies than CAT5e, CAT6 cables easily handle 1000Base-T Gigabit Ethernet traffic and, with special installation considerations, 10Gbase-T.

- *Augmented Category 6 (CAT6A)*—Created for 10Gbase-T installations with cable segments up to 100 meters long. CAT6A cables use larger conductors and leave more space between the wire pairs, meaning that the outside diameter of the sheath is larger than a CAT6 cable. CAT6A is capable of supporting 10 gigabit Ethernet at 500 MHz. CAT6a was added to the TIA wiring standards in 2008.

- *Category 7 (CAT7)*—Not officially ratified by the TIA, the CAT7 standard calls for a fully shielded and screened cable design. CAT7 is capable of supporting 10 gigabit Ethernet at 600 MHz.

- *Category 7a (CAT7a)*—CAT7A, also not ratified by the TIA, is a fully shielded and screened cable that extends the frequency range and provides full support for 10Gbase-T. CAT7A is capable of supporting 10 gigabit Ethernet at 600 MHZ.

- *Category 8 (CAT8)*—CAT8 is a fully shielded and screened cable that extends the frequency range and provides full support for 40Gbase-T. CAT8 was added in 2016 with an addendum to the current standard.

Figure 2-10 Twisted-Pair Cable Connector

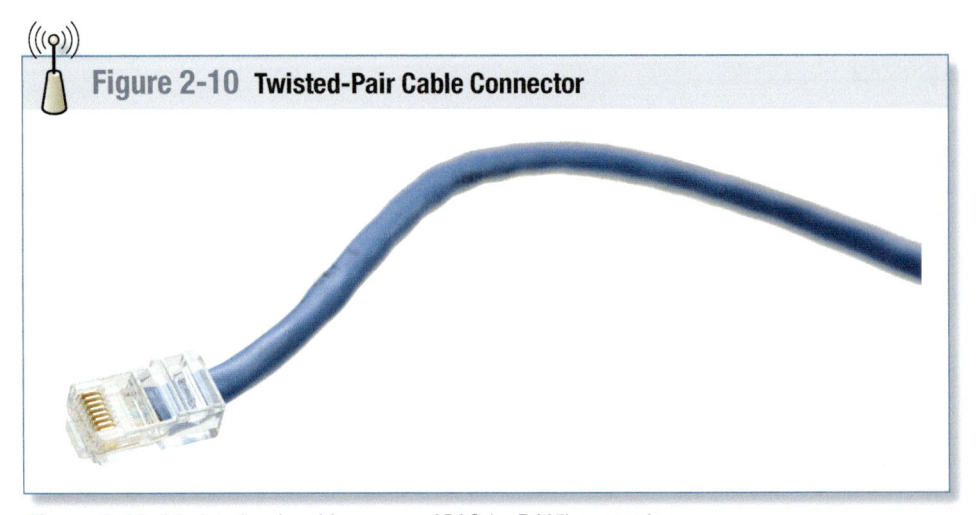

Figure 2-10. *A twisted-pair cable uses an 8P8C (or RJ45) connector.*

Although CAT3 and CAT5 cables have been dropped from the official standards, this does not mean that administrators no longer encounter them. There are many CAT5 installations still in operation, and CAT5 cable products are still available to maintain them.

CAT3 is decidedly less prevalent in the installed base, as most networks have long since been upgraded to at least Fast Ethernet (at 100 megabits per second), which requires at least CAT5.

Twisted-Pair Connectors

Twisted-pair cables use modular connectors that are most commonly referred to by the designation RJ45, but which should properly be called 8P8C under the latest standards. Network interface adapters, wall plates, patch panels, and other networking components (such as switches and hubs) all have female connectors. The patch cables used to connect everything together have male connectors. **See Figure 2-10.**

Twisted-Pair Wiring

Twisted-pair cables can be wired as either a straight-through cable or as a crossover cable.

Straight-Through Cable Wires are connected to the same pins on both connectors of a straight-through cable. A *straight-through cable* connects computers to networking devices, as in the case of a connection from a computer to a hub, or a computer to a switch.

Consider the wiring diagram of a straight-through cable. Just as the name implies, the connections are straight through end to end, and each wire is connected on the same pins on both ends. The colors of the cable are based on the T568B standard. **See Figure 2-11.**

Figure 2-11 Straight-Through Cable

T568A Straight-Through Ethernet Cable

T568B Straight-Through Ethernet Cable

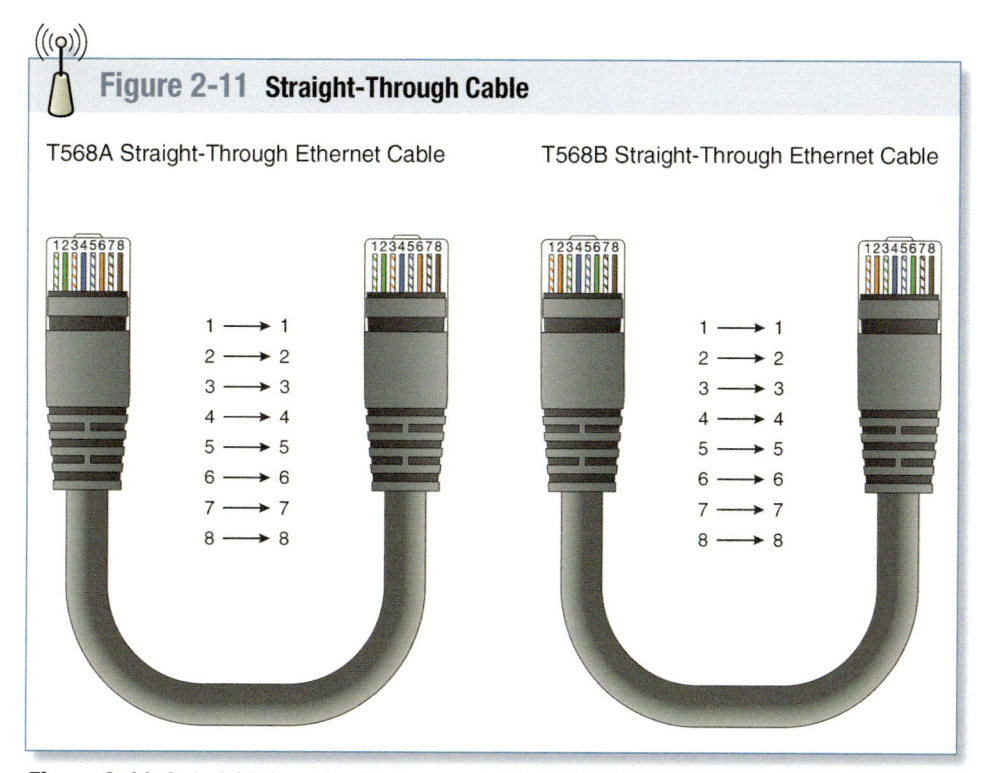

Figure 2-11. A straight-through cable has connections straight through end to end, with the cable color-coded based on either the T568A or T568B standard.

Crossover Cable Specific wires are crossed on opposite connectors of the crossover cable. A *crossover cable* connects similar devices to each other. For example, a crossover cable might connect any two networking devices together such as the following:

- A switch and a switch
- A switch and a hub
- A computer and a computer
- A computer and a router

Consider the wiring diagram for a crossover twisted-pair cable. The straight-through cable has the pairs connected from the same pins on one side to the same pins on the other side. However, the crossover cable crosses over some key wires so that transmitted signals on one side go to receive on the other side. **See Figure 2-12.**

> ✓ FACT
>
> Many modern routers and switches autosense the connection. In other words, if the connection needs a crossover cable, the connection is changed internally.

Figure 2-12 Crossover Cable

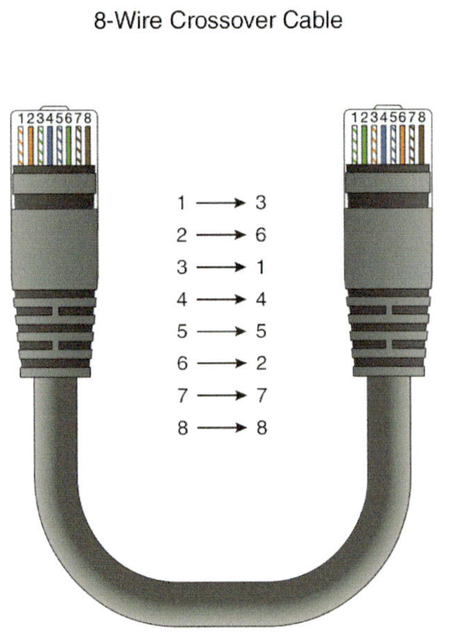

8-Wire Crossover Cable

1 → 3
2 → 6
3 → 1
4 → 4
5 → 5
6 → 2
7 → 7
8 → 8

Figure 2-12. A typical crossover cable reverses the transmit and receive pairs to allow for two way communication between like devices (such as a switch).

It is easy to identify a crossover cable by placing both connectors of the same cable side by side. If the orange and green pairs are swapped, it is a crossover cable.

Coaxial Cable

Prior to the introduction of UTP cables, Ethernet networks called for coaxial cable of various types. A *coaxial cable* consists of a central copper conductor (which carries the signals) surrounded by a layer of insulation. Surrounding the insulation is a second conductor made of copper mesh (which functions as the ground), and the whole assembly is encased in a sheath. The comparatively large amount of insulation in coaxial cables makes them quite resistant to EMI.

All of the coaxial cable types have been removed from the TIA/EIA network cabling standards, and it is rare to encounter coaxial Ethernet networks in the field. But other kinds of coaxial cable are still used in cable television installation, and they are an important part of the history of data networking.

Fiber-Optic Cable

Recall fiber optic is cable that uses transparent fibers to transmit light instead of electrical signals. It is a form of network medium that is completely different from copper and that avoids nearly all of copper's shortcomings. However, it does have a few shortcomings of its own.

Fiber-optic cables transmit pulses of light through a strand of glass that actually contains two slightly different types of glass called the core and the cladding. This glass is surrounded by a protective coating, and then that is surrounded by other things designed to provide physical protection. There are a wide variety of fiber-optic constructions that are each

Figure 2-13 Fiber-Optic Cable

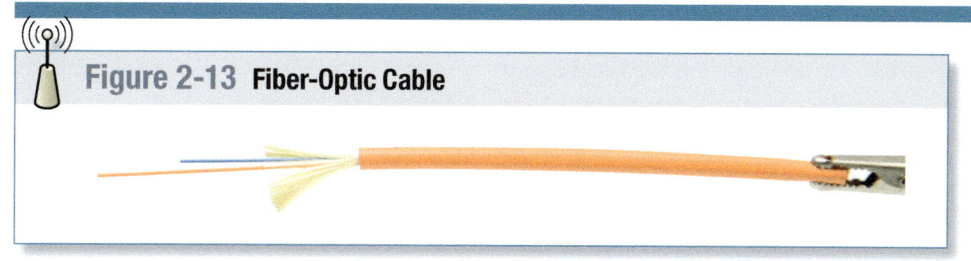

Figure 2-13. A fiber-optic cable transmits pulses of light through a strand of glass.

designed for use in specific environments. **See Figure 2-13.**

Unlike electrical voltages, light pulses are completely unaffected by EMI, which means that fiber-optic cables can be installed near light fixtures, heavy machinery, or other environments in which copper would be problematic. Fiber-optic cable is also much less prone to attenuation than any copper cable.

Most of the twisted-pair media used for networking today are limited to cable segments no longer than 100 meters. Fiber-optic cables can span distances as long as 120 kilometers and are immune to outdoor conditions, which makes them ideal for installations that span long distances or connect campus buildings together.

Fiber-optic cables are also inherently more secure than copper, because it is extremely difficult to tap into the cable and intercept the signals it carries without disturbing them.

Fiber-optic cables have been available as an alternative to copper for decades. Even the early 10 Mbps Ethernet standards included a fiber-optic specification. However, for local area networking, fiber-optic has never been more than a marginal solution, used only in situations where copper cables were untenable. The primary reasons for this are that fiber-optic cables are much more expensive than copper, and their installation is much more difficult, requiring a completely different skillset (and toolset).

Fiber-Optic Cable Types

There are two types of fiber-optic cable: singlemode and multimode. They differ in core size and in the means used to generate the light pulses they carry. Fiber-optic cable sizes are measured in microns (millionths of a meter, represented by the symbol μm). Sizes consist of two numbers that refer to the diameter of the core, followed by the diameter of the cladding surrounding it. **See Figure 2-14.**

Multimode fiber-optic is more common in LANs than singlemode, largely

Figure 2-14 Fiber-Optic Cable Characteristics

CABLE TYPE	CORE/CLADDING DIAMETER	LIGHT SOURCE
Singlemode	8.3/125 μm	Laser
Multimode	62.5/125 μm or 50/125	Light emitting diode (LED) or VCSEL

Figure 2-14. Fiber-optic cables are identified with two numbers. The first number identifies the size of the core, and the second number identifies the diameter of the cladding surrounding the core.

because the cable is less expensive and has a smaller bend radius, which means that it can be bent more sharply around corners. Multimode fiber typically uses an LED light source with an 850 or 1,300 nanometer (nm) wavelength.

Singlemode cables use a laser light source at a wavelength of 1,310 or 1,550 nm. Because of its expense, administrators rarely choose singlemode fiber for LANs, preferring it for long runs instead.

Fiber-Optic Connectors

Unlike twisted-pair and coaxial cables, there are a large number of different connectors available for fiber-optic cables. Choosing which type of connector to use is a matter of compatibility with existing equipment, as well as personal preference. All fiber-optic connectors perform basically the same function, precisely joining two ends of connectors together, face-to-face, so that light pulses can pass from one cable segment to another.

Basic Topologies

Selecting the correct cable type is an important factor in the network deployment process. But once the cables are in hand, it is not recommended to simply start connecting computers together with no thought given to a pattern. The physical topology of a network is the pattern used to connect the computers and other devices together.

When deploying a LAN, the topology used is directly related to the type of cable selected. Each Physical layer specification for a Data Link layer LAN protocol has installation requirements that must be observed if the network is to function properly, and the topology is one of those requirements.

A WAN solution for a large network with offices at multiple distant sites can use any one of several topologies to connect the offices together, depending on the organization's communication requirements and its budget.

Star Topology

The *star topology* is a configuration in which each computer or other device is connected by a separate cable run to a central cabling nexus (that is, a switch or a hub). It is the most common LAN topology in use today. Compared to the other LAN topologies, the star topology affords a good deal of fault tolerance. The failure of a cable affects only the connection of one node to the network. The switch or hub does provide a central point of failure, however, which can bring down the entire network, but failure of these devices is comparatively rare. **See Figure 2-15.**

All of the UTP-based Ethernet networks in use today call for the star topology, as do many fiber-optic-based LAN protocols.

The hub or switch at the center of a star network provides a shared medium,

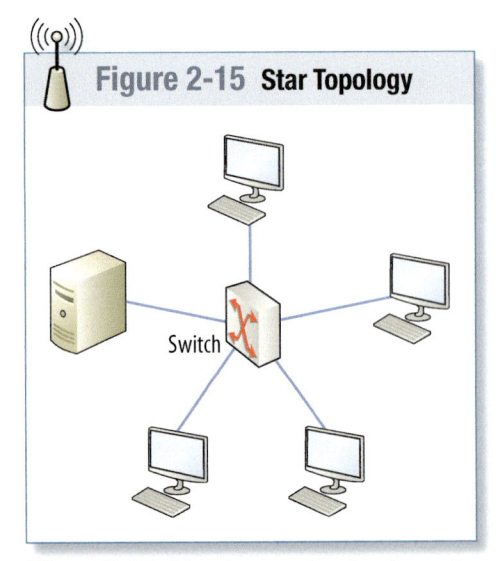

Figure 2-15. *A local area network using a star topology connects computers through a central cabling nexus.*

just like a bus or ring topology. When a computer transmits a broadcast message, the hub or switch receiving it over one port propagates the message out through all of its other ports to all of the nodes on the network.

A basic star network can be as large as the number of ports available in the switch or hub. When a switch or hub is fully populated, it is possible to expand the network further by adding another switch or hub and connecting it to the first. This creates what is known as a hierarchical star topology (or sometimes a branching tree topology or extended star). The network can be expanded in this way, within certain limits specified in the Ethernet standards. **See Figure 2-16.**

Bus Topology

A *bus topology* is a configuration in which each computer is connected to the next one in a straight line. The first two Ethernet standards called for coaxial cable in a bus topology. The first, Thick Ethernet, used a single length of cable up to 500 meters long, with individual transceiver cables connecting the computers to the main trunk. Thin Ethernet used the smaller RG-58 cable in lengths attaching the T connectors on

Figure 2-16 Hierarchical Star Topology

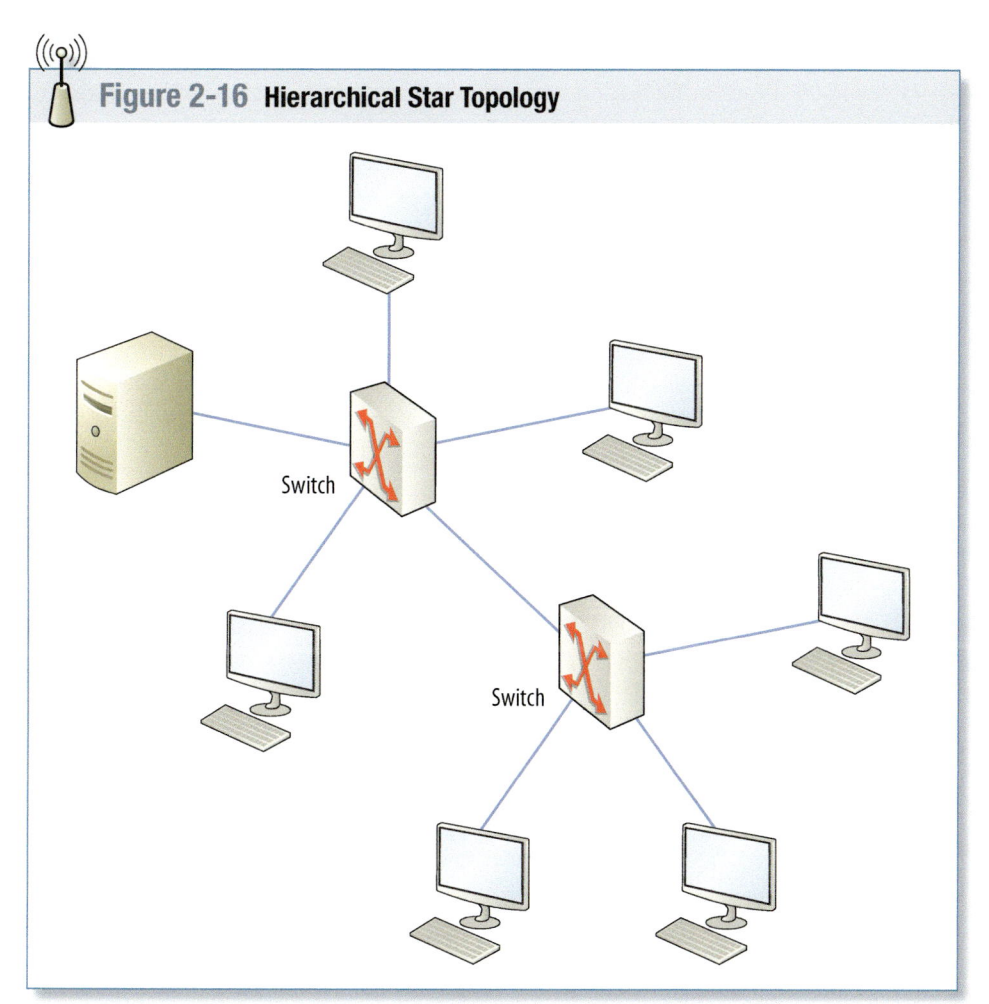

Figure 2-16. *A LAN can incorporate a hierarchical star topology.*

Figure 2-17 Bus Topology

Thick Ethernet

Thin Ethernet

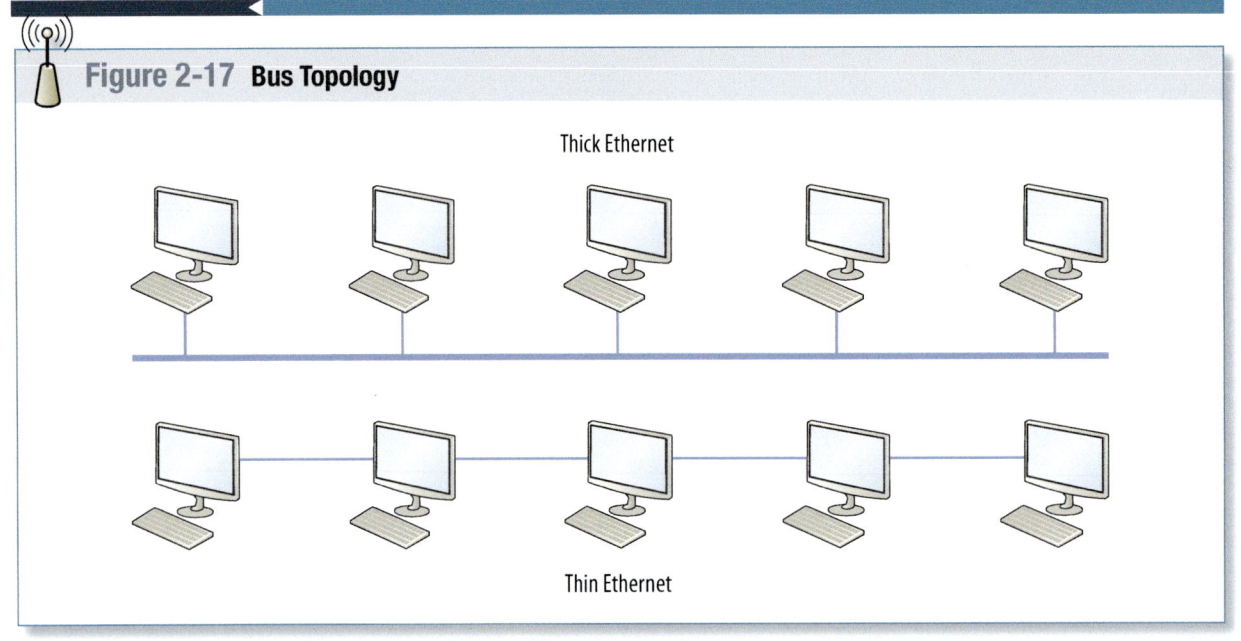

Figure 2-17. Thick and Thin Ethernet networks may utilize a bus topology.

the computers' network interface adapters, forming a bus up to 185 meters long. **See Figure 2-17.**

On a bus network, when any computer transmits data, the signal travels down the cable in both directions. Both ends of the bus must be terminated with resistor packs that negate the signals arriving there. On a bus that is not properly terminated, signals reaching the end of the cable tend to echo back in the other direction and interfere with any newer signals that the computers transmit.

The inherent weakness of the bus topology is that a single cable failure can disrupt communications between entire sections of the network. A broken cable splits the bus into two halves, preventing the nodes on one side from communicating with those on the other. In addition, both halves of the network are left with one end in an unterminated state, which prevents computers on the same side of the break from communicating effectively.

The only Ethernet networks that used the bus topology were those wired with coaxial cable. The decline of Ethernet on coaxial cable also meant the decline of the bus topology, which is no longer used today.

Ring Topology

A *ring topology* is a configuration in which a signal transmitted by a computer in one direction circulates around ring, eventually ending up back at its source, like a bus with the two ends joined together. The Data Link layer protocol traditionally associated with the ring topology is the now-obsolete Token Ring protocol. **See Figure 2-18.**

Early Token Ring networks were cabled together in an actual ring, with each computer connected to the next. However, Token Ring was also the origin

Figure 2-18 Ring Topology

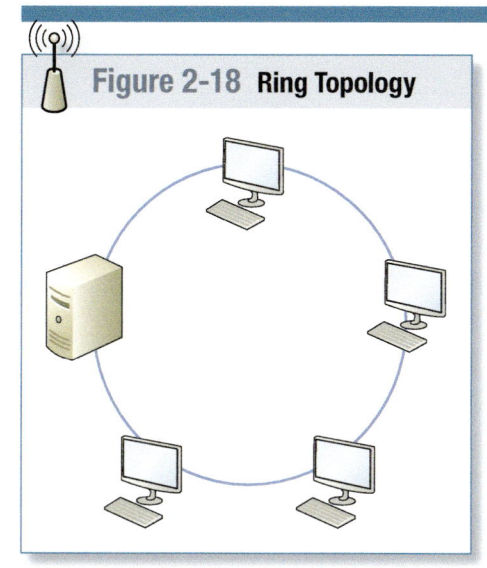

Figure 2-18. A LAN using a ring topology sends transmitted signals around a ring.

of the hybrid topology, in which a single network contained the attributes of two different topologies.

In the majority of Token Ring networks (including nearly all of the few left in operation), the ring is a logical topology implemented in the wiring of the network. Physically, most Token Ring networks take the form of a star topology, with a cable running from each computer to a central cabling nexus called a multistation access unit (MAU). The MAU implements the logical ring by transmitting signals to each node in turn, and waiting for the node to send them back before it transmits to the next node. Thus, while the cables are physically connected in a star, the wires inside the cables take the form of a ring. This is sometimes referred to as a star ring topology. **See Figure 2-19.**

One protocol (also obsolete) that does allow for a physical ring topology is Fiber Distributed Data Interface (FDDI). FDDI is a 100 Mbps fiber-optic solution that was fairly popular on backbone networks back in the days when Ethernet only ran at 10 Mbps. Because a ring is not fault tolerant—a cable break anywhere on the network prevents packets from circumnavigating the ring—FDDI calls for a double ring topology. Traffic on the double ring flows in opposite directions, and computers are connected

Figure 2-19 Logical Ring Topology

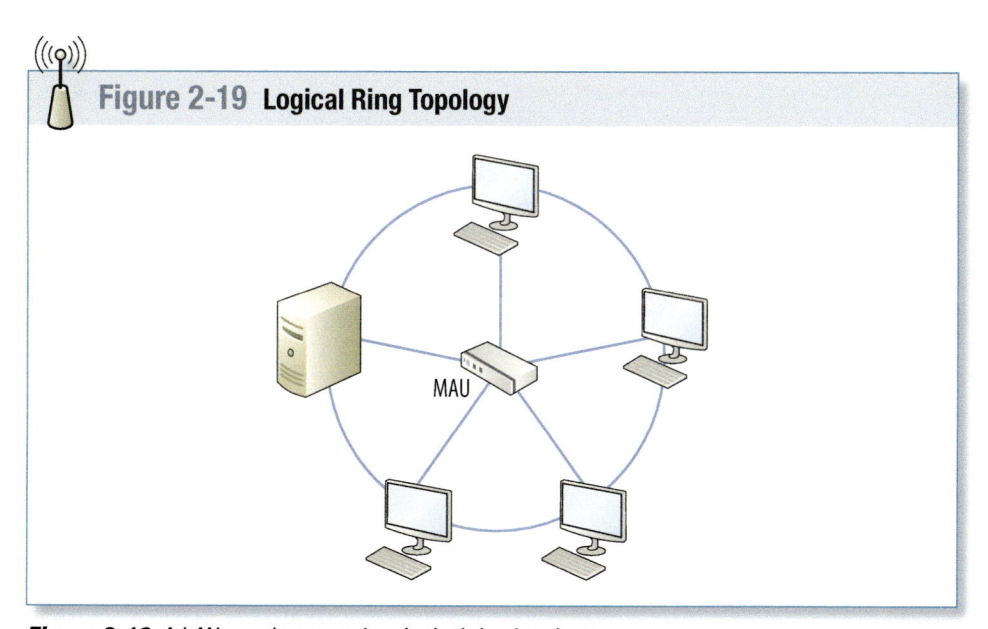

Figure 2-19. A LAN may incorporate a logical ring topology.

Figure 2-20 Double Ring Topology

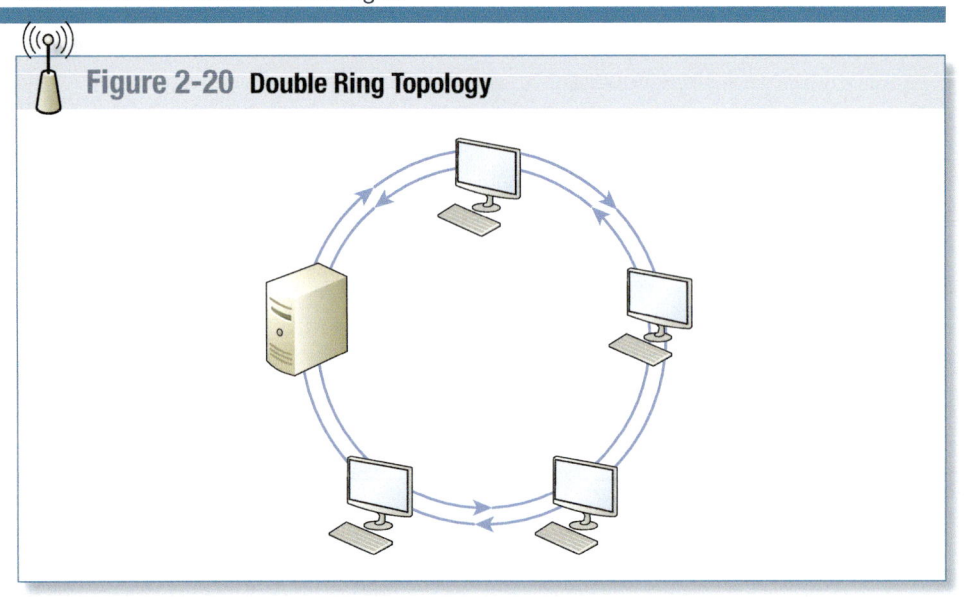

Figure 2-20. A local area network using a double ring topology provides fault tolerance.

to both rings. This way, the network can tolerate a cable break by diverting traffic to the other ring. **See Figure 2-20.**

Mesh Topology

A *mesh topology* is a network configuration in which each computer is connected to two or more other computers. As mentioned earlier, Data Link layer LAN protocols do not provide much flexibility when it comes to choosing a network topology. The characteristics of specific cable types impose exacting limitations on how to install them. However, designing an internetwork topology—such as when installing multiple LANs at one location or connecting remote networks with WAN links—presents a good deal more freedom.

Use routers to connect networks together (forming an internetwork), and then the topology created when linking networks together is not subject to the same restrictions as the creation of a LAN with a specific cable type.

For example, if an organization has several branch offices located around the country, WAN links can be installed to connect them in any desirable way. By connecting each branch office to the company headquarters, point-to-point connections are used to build a star WAN topology. In the same way, a bus or a ring topology could be created, (although these models are rare in WAN implementations because of their lack of fault tolerance). **See Figure 2-21.**

Under these circumstances, another option is to create some form of mesh topology, in which each office is connected to two or more other offices. This can be an expensive solution, but its redundancy enables the network to continue functioning, despite the failure of one or more links. There are two types of mesh topology: a partial mesh (in which each site has a point-to-point link to at least two other sites) and a full mesh (in which each site is connected to every other site). **See Figure 2-22.**

In both types of mesh topologies, there are at least two routers on each network. In the event that a router is

Figure 2-21 Point-to-Point Connections

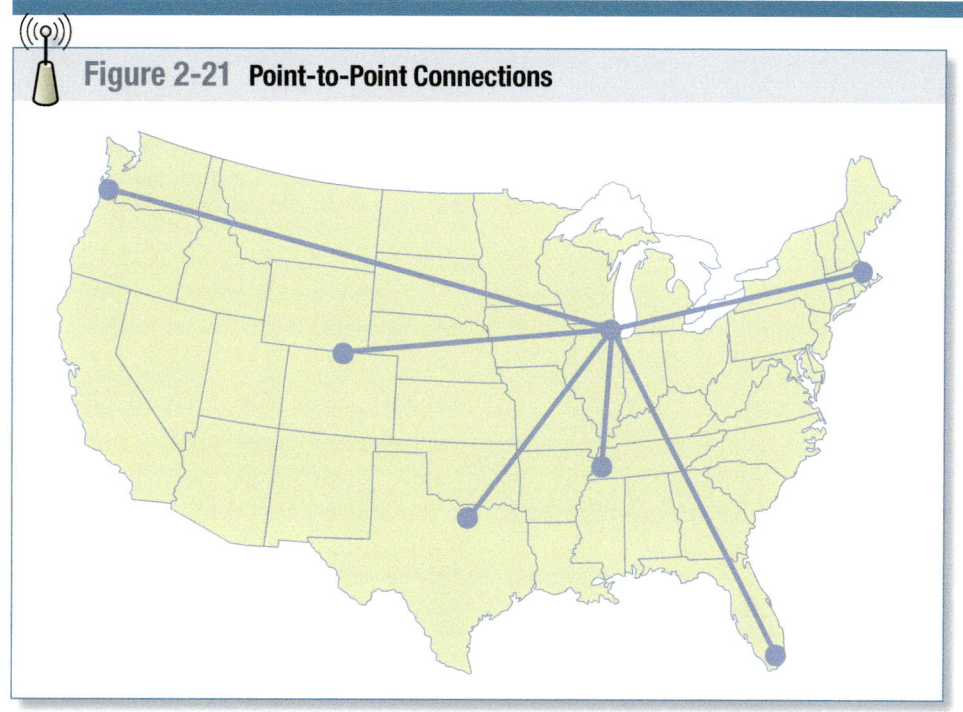

Figure 2-21. A star topology can be built using WAN point-to-point connections.

Figure 2-22 Mesh Topology

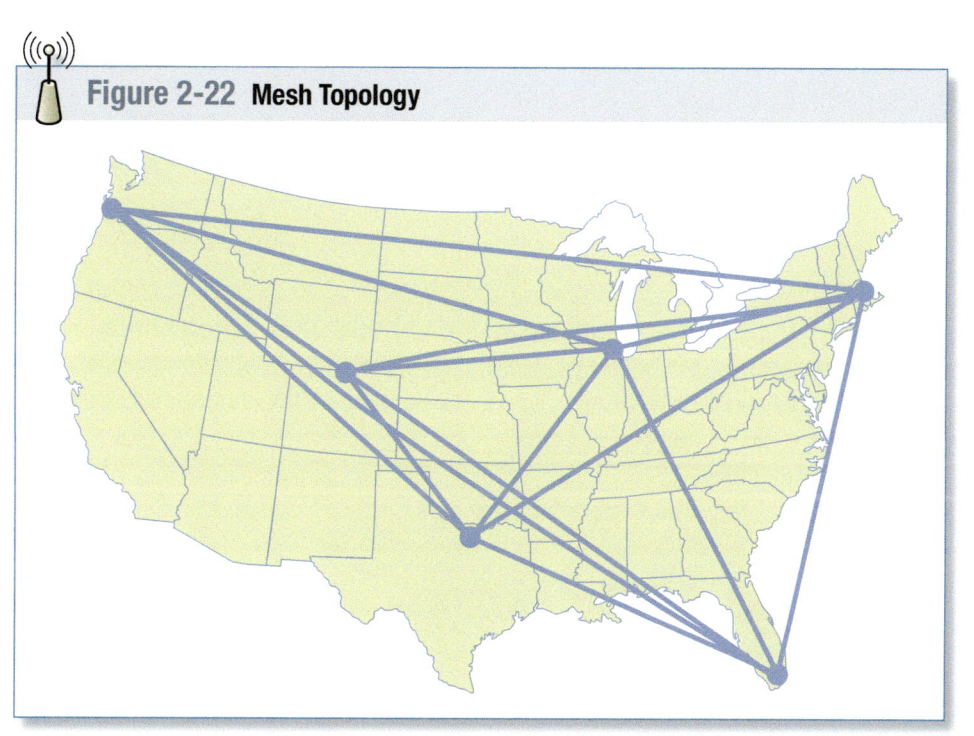

Figure 2-22. The full mesh topology is built using WAN point-to-point connections, so each site is connected to every other site.

unable to send data to a specific destination, it can forward the data to another router for transmission.

Point-to-Point Topology

A *point-to-point topology* is a simple configuration in which one computer is directly connected to another. Obviously, this topology limits the network to two computers. **See Figure 2-23.**

It is possible to connect two computers together using a single Ethernet cable, creating the simplest possible LAN, but this can require a crossover cable. In most cases, even two-node LANs connect using a switch or a hub instead.

Point-to-point connections are also possible using wireless networks. In wireless LAN parlance, two computers connected directly together using wireless network interface adapters form what is called an ad hoc network.

The point-to-point topology is more commonly found in wide area networking, which consists of all point-to-point links. For example, when connecting a home computer to the Internet, a point-to-point WAN connection is established between the home computer and the Internet service provider's (ISP's) network. Corporate networks also use point-to-point WAN links to connect LANs in remote offices.

Cabling Standards

Until 1991, there were no standards defining the nature of the cabling used for LANs, other than the Physical layer specifications in the Data Link layer protocol standards. This often resulted in hardware incompatibilities and confusion for cable installers. It was eventually recognized that the networking industry needed a standard defining a cabling system that could support a variety of networking technologies.

To address this need, the American National Standards Institute (ANSI) and the Telecommunications Industry Association (TIA), a division of the Electronic Industries Alliance (EIA), along with a consortium of telecommunications companies, developed a set of documents that define the best practices for designing, installing, and maintaining networks using twisted-pair and fiber-optic cabling.

Wired LAN Standards

Originally called the "ANSI/TIA-568 Commercial Building Telecommunications Cabling Standard," this document like most ANSI documents are revised periodically and will continue to be revised as technology changes. The latest revision to the standard occurred in 2016 and incorporates changes to three of the five documents included in the

Figure 2-23 Point-to-Point Topology

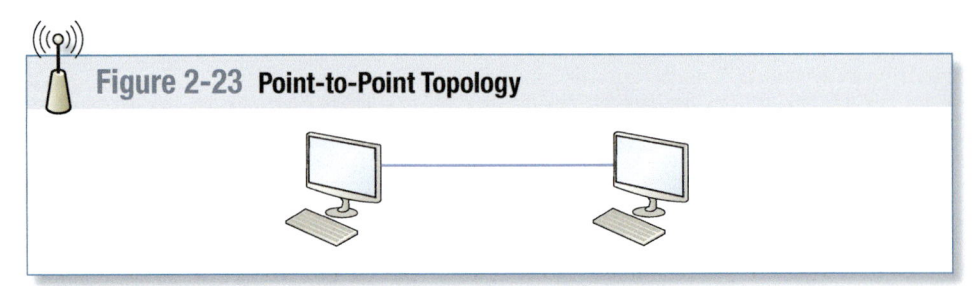

Figure 2-23. A point-to-point topology connects two computers.

standard. Prior to the update the documents were labeled ANSI/TIA-568-C.0 through ANSI/TIA-568-C.4. The latest update reflects a naming convention change to three of the documents and will eventually change for all five documents. The latest naming convention is now ANSI/TIA-568-D. The latest revision of the standard consists of the following five documents:

- *ANSI/TIA-568-D.0*—"Generic Telecommunications Cabling for Customer Premises"
- *ANSI/TIA-568-D.1*—"Commercial Building Telecommunications Infrastructure Standard"
- *ANSI/TIA-568-C.2*—"Balanced Twisted-Pair Telecommunication Cabling and Components Standard"
- *ANSI/TIA-568-D.3*—"Optical Fiber Cabling Components Standard"
- *ANSI/TIA-568-C.4*—" Broadband Coaxial Cabling and Components Standard"

The 568-D standard defines a structured cabling system for voice and data communications in office environments that has a usable life span of at least 10 years, supporting the products of multiple technology vendors and using twisted-pair or fiber-optic cable.

In addition to the ANSI/TIA-568-D standard, there are other TIA wiring standards that provide guidelines for specific types of cabling within and between the subsystems listed here. Any contractor hired to perform an office cable installation should be familiar with these standards and should be willing to certify that his or her work conforms to these standards.

> **Choosing the Correct Wiring Standard**
>
> One of the most widely known elements of the TIA/EIA-568 cabling standard is a diagram that specifies which color wires in a twisted-pair cable should be connected to which pins in the connectors. Twisted-pair cables are supposed to be wired "straight through," meaning that each pin at one end of a cable must be connected to the same pin at the other end. If the pins are not wired this way, the cable does not function properly.
>
> There are two pinout standards in common use in today's networks, called T568A and T568B. The primary difference between the two is that the colored pairs go to different pins. Specifically, the orange and green wire pairs are swapped. In practice, it does not matter which standard is used, as long as the same standard is used at both ends of the cable.

Wireless LAN Standards

The wireless LAN equipment on the market today is based on the 802.11 standards published by the IEEE, the same LAN/MAN Standards Committee—IEEE 802—that publishes the 802.3 Ethernet standards. Because the same standards body produces them both, the 802.11 wireless technology fits neatly into the same layered structure as the Ethernet specifications.

The 802.3 standards divide the Data Link layer of the OSI model in two, with the Logical Link Control (LLC) layer on top and the media access control (MAC) layer on the bottom. The LLC layer is defined in a separate standard: IEEE 802.2. A wireless LAN uses the same LLC layer as an 802.3 Ethernet network, with the 802.11 documents defining the

For additional information, visit qr.njatcdb.org Item #1787

physical layer and MAC layer specifications. **See Figure 2-24.**

As with the 802.3 Ethernet standard, the IEEE has updated and expanded on the 802.11 specification several times over the years, increasing the maximum transmission speed of the network and altering the frequencies and modulation techniques. **See Figure 2-25.**

Figure 2-24 IEEE Standards Mapping

IEEE 802.2 – Logical Link Control		Data-Link
IEEE 802.3 – Media Access Control	IEEE 802.11 – Media Access Control	
IEEE 802.3 Physical Layer Specifications	IEEE 802.11 Physical Layer Specifications	Physical

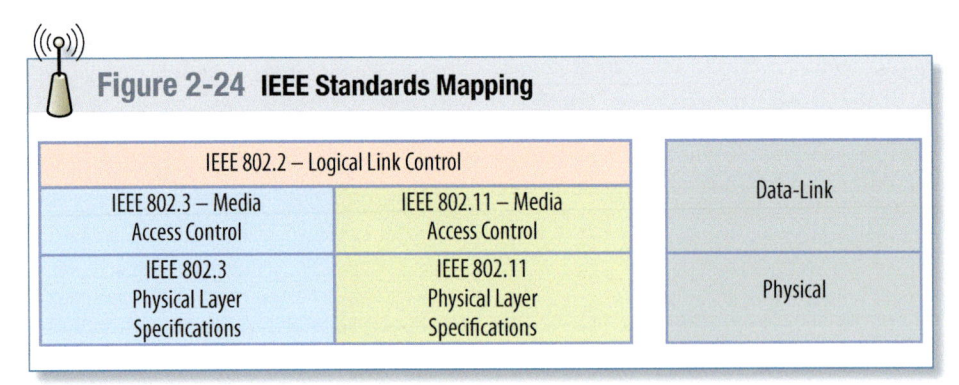

Figure 2-24. IEEE standards can be mapped to two levels in the OSI model.

Figure 2-25 IEEE 802.11 Standards

STANDARD	FREQUENCY (GHZ)	TRANSMISSION RATE (MBPS)	MODULATION TYPE	RANGE (INDOOR/OUTDOOR) (METERS)
802.11-1997	2.4	1, 2	DSSS, FHSS	20/100
802.11a-1999	5	6 to 54	OFDM	35/120
802.11b-1999	2.4	5.5 to 11	DSSS	38/140
802.11g-2003	2.4	6 to 54	OFDM, DSSS	38/140
802.11-2007	Republication of the base standard with eight amendments			
802.11n-2009	2.4 and 5	7.2 to 288 (@20 MHz)		
15 to 600 (@40 MHz)	OFDM	70/250		
802.11y-2008	3.7	6 to 54	OFDM	5,000+
802.11ac (Draft)	5	433 to 867 (@80 MHz)		
867 to 6.93 Gbps (@160 MHz)	QAM			

Figure 2-25. IEEE updates the 802.11 standard as newer technologies become available.

Wi-Fi

The term Wi-Fi™ has entered the daily lexicon of mobile computer users, as increasing numbers of businesses and public places provide wireless LANs for the use of their customers. However, Wi-Fi is a privately-owned trademark; it is not a name sanctioned by the IEEE for 802.11 networks or equipment.

The name Wi-Fi is a trademark owned by an organization of hardware and software manufacturers called the Wi-Fi Alliance. The group operates an interoperability certification program for wireless LAN equipment and allows certified products to carry a special logo indicating their participation. Not all manufacturers submit their products for testing, however, which does not necessarily mean that they are not compatible.

Summary

All networks require the use and knowledge of many basic networking components. A network technician must understand the basics of unicast, broadcast, and multicast transmissions. In addition, a network technician must understand the basics on how hubs and switches connect computers, and how routers connect networks. Networks are connected using different types of media such as twisted-pair cable, fiber-optic cable, and wireless. Networks are wired using different topologies. The properties of network media and their installation are governed by industry standards published by various organizations.

Review Questions

1. **What type of traffic always goes to all devices in a subnet?**
   a. Allcast
   b. Broadcast
   c. Multicast
   d. Unicast

2. **A switch blocks broadcasts.**
   a. True
   b. False

3. **What is the difference between a switch and a router?**
   a. A switch connects devices together, and a router connects subnets together.
   b. A switch connects subnets together, and a router connects devices together.
   c. Nothing. They are the same.
   d. Switches do not pass broadcasts, but routers do.

4. **Bridges can connect dissimilar physical topologies.**
   a. True
   b. False

5. **A crossover cable is used to connect a computer to a switch.**
   a. True
   b. False

6. **Which of the following standards define how twisted-pair cables should be wired?**
   a. Extranet wiring practices
   b. IEEE 802.3
   c. RFC 791
   d. T568B

Understanding the OSI Model

The Open Systems Interconnection (OSI) model is one of the most referenced models in networking. It includes seven layers with specific activities, protocols, and devices working on each. Many network exams test on the knowledge of the different elements of the OSI model, and some hiring managers quiz potential network employees on their knowledge. The TCP/IP model is similar, but includes only four layers instead of seven.

OBJECTIVES

- ▶ Describe the OSI model
- ▶ Explain the TCP/IP model
- ▶ Identify devices mapped on the OSI and TCP/IP models
- ▶ Locate protocols mapped on the OSI and TCP/IP models

CHAPTER 3

TABLE OF CONTENTS

The OSI Model 52
    Application Layer 53
    Presentation Layer 54
    Session Layer. 54
    Transport Layer 55
    Network Layer 57
    Data Link Layer 58
    Physical Layer 60
    The Complete Picture 60
    Packets and Frames 62

The TCP/IP Model 63

Devices Mapped on the OSI and TCP Models. 64
    Physical Layer 66
    Data Link Layer 66
    Network Layer 66
    Application Layer 67

Protocols Mapped on the OSI and TCP/IP Models. 67

Summary. 68

Review Questions 69

THE OSI MODEL

The *Open Systems Interconnection (OSI) model* is a general framework or set of guidelines for data handling and network communication. It also identifies the framework of the TCP/IP protocols and the hardware used on networks. There is no single standard or compliance test for the OSI model itself. Instead, many standards have been created based on the different elements of the model.

One of the primary goals of the OSI model is operating system independence. In other words, the OSI model enables computers running any operating system to communicate with other computers using the same or different operating systems. The OSI model includes seven specific layers. **See Figure 3-1**.

The OSI model was created by the International Organization for Standardization (ISO). Primary advantages to the OSI model include:

Layers Interact Only with Adjoining Layers The Transport layer only interacts with the Session layer above it and the Network layer below it. It does not matter to the Session layer what applications are used at the Application layer, or what type of cable media is used to transmit the data on the Physical layer.

It Has Encouraged Creation of Industry Standards Functions at each layer are standardized. Development of network components by different vendors is simplified, and different operating systems are able to communicate with other.

Network Communication Processes Are Segmented Instead of a single protocol that does everything, multiple protocols are used by networking systems to isolate functions. Troubleshooting is easier with an understanding of the OSI model.

It is important to know the names and layer numbers of each OSI model layer. Mnemonics are commonly used by network technicians to help memorize the OSI model. **See Figure 3-2**.

One method (All People Seem To Need Data Processing) starts at layer 7 and moves down to layer 1. The other method (Please Do Not Throw Sausage Pizza Away) starts on layer 1 and moves up to layer 7. There are many other mnemonics used by technicians to memorize the OSI layers. The technique employed is not as important as using some method to memorize it.

Just knowing the names and numbers of the layers is not enough. A basic understanding of what happens at each layer is also important.

✓ **FACT**
The seven-layer OSI model was created by the International Organization for Standardization (ISO). ISO may look like a typo. However, it is not an acronym for the International Organization for Standardization. Instead, ISO is derived from *isos*, which is Greek for equal.

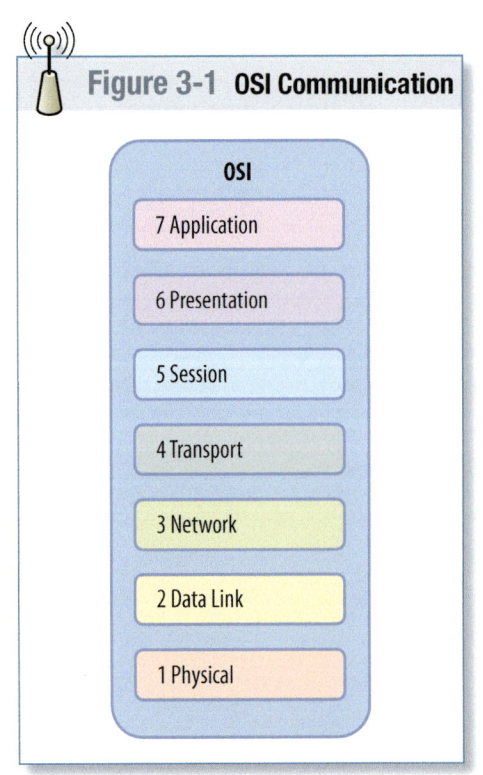

Figure 3-1 OSI Communication

OSI
- 7 Application
- 6 Presentation
- 5 Session
- 4 Transport
- 3 Network
- 2 Data Link
- 1 Physical

Figure 3-1. *The OSI model enables disparate computers to communicate with each other.*

Figure 3-2 Mnemonics

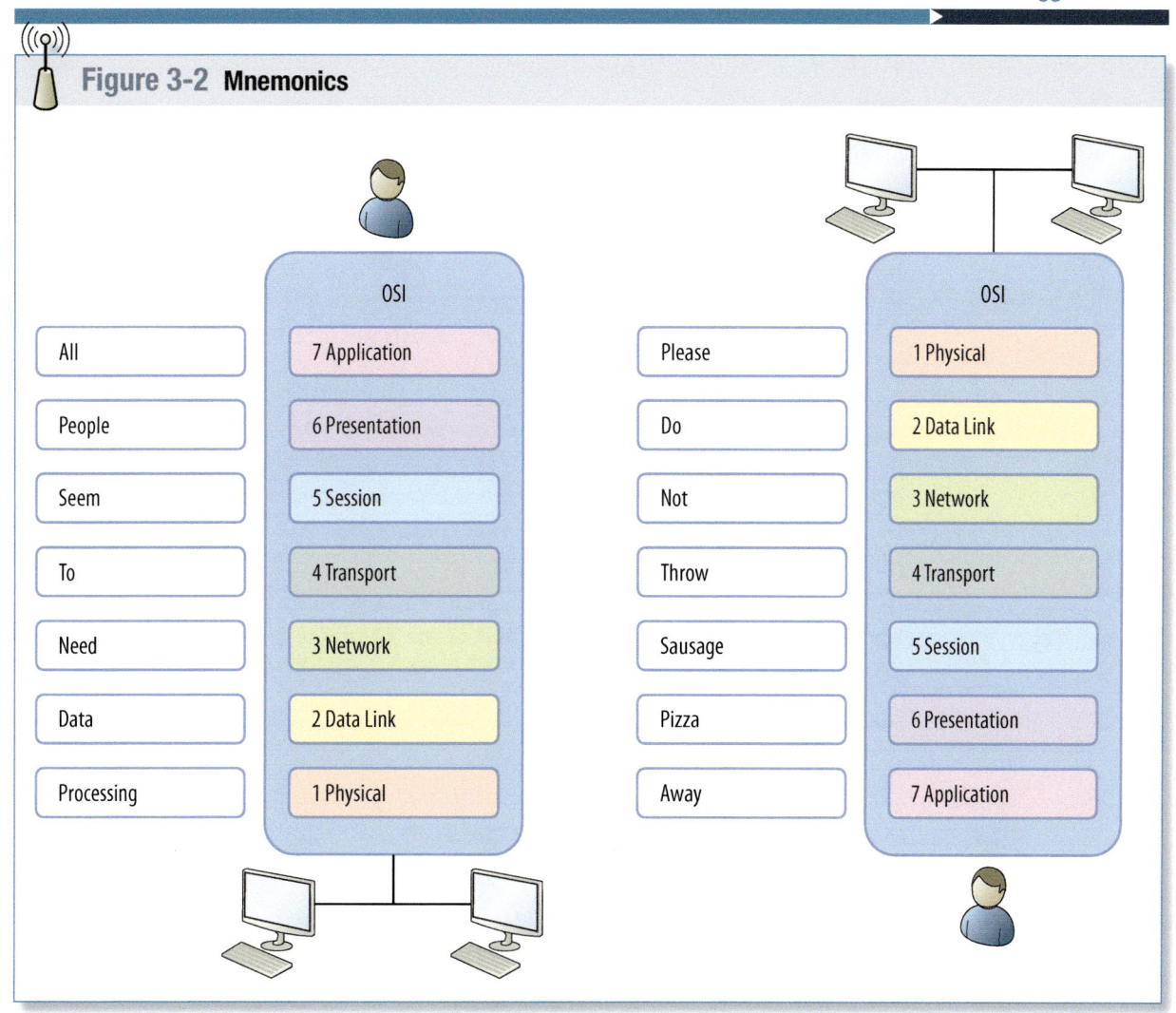

Mnemonic	OSI Layer	Mnemonic	OSI Layer
All	7 Application	Please	1 Physical
People	6 Presentation	Do	2 Data Link
Seem	5 Session	Not	3 Network
To	4 Transport	Throw	4 Transport
Need	3 Network	Sausage	5 Session
Data	2 Data Link	Pizza	6 Presentation
Processing	1 Physical	Away	7 Application

Figure 3-2. Mnemonics aid in familiarization with the OSI model.

Application Layer

The *Application layer* is layer 7 of the OSI Model, which interacts with the Presentation layer below it and the application running on the computer. Protocols operating at the Application layer include:

Domain Name System *Domain Name System (DNS)* is the primary name resolution service that the Internet and Microsoft networks use. DNS resolves hostnames to IP addresses. Passing the name of a server to DNS enables DNS to return the server's IP address. Microsoft networks also use DNS to locate servers running specific services.

Hypertext Transfer Protocol *Hypertext Transfer Protocol (HTTP)* is the primary protocol that clients use to transfer data to and from web servers on the Internet. Similarly, HTTPS is a secure version of HTTP that transmits data in an encrypted format.

File Transfer Protocol *File Transfer Protocol (FTP)* is a protocol that transfers files to and from an FTP server. Although not as popular as they once were, FTP servers are common on the Internet.

Trivial FTP *Trivial FTP (TFTP)* is a lightweight FTP protocol that transfers smaller files with less data overhead.

✓ FACT
DNS is a required service in Microsoft domains and is heavily used on the Internet.

> **✓ FACT**
> DHCP saves a lot of labor and is found in most networks.

TFTP often transfers files to network devices such as routers.

Dynamic Host Configuration Protocol *Dynamic Host Configuration Protocol (DHCP)* is a method of dynamically assigning TCP/IP configuration information to clients. A DHCP server assigns IP addresses to systems. It can also assign the subnet mask, the address of the default gateway (a router), the address of a DNS server, the domain name, and more.

Lightweight Directory Access Protocol *Lightweight Directory Access Protocol (LDAP)* is a protocol that transmits queries and replies to and from a directory service, such as Microsoft's Active Directory Domain Services (AD DS).

Post Office Protocol *Post Office Protocol (POP3)* is an e-mail protocol that clients use to retrieve e-mail from POP3 servers. POP3 servers are commonly hosted by Internet service providers (ISPs).

Simple Mail Transfer Protocol *Simple Mail Transfer Protocol (SMTP)* is the primary protocol that transmits e-mail messages to and between mail servers. Clients use SMTP to send e-mail to an e-mail server, and then the server uses it to relay the messages to other mail servers. A typical Internet e-mail client application uses a POP3 e-mail server to receive e-mail and an SMTP server to send it.

Internet Message Access Protocol *Internet Message Access Protocol (IMAP)* is another system of rules that clients use to receive e-mail messages. An IMAP server enables clients to store and manage e-mail on the server. Users can download the e-mail onto a computer or organize the e-mail in different folders on the server.

Simple Network Management Protocol *Simple Network Management Protocol (SNMP)* is a protocol that systems use to manage network devices such as routers and switches. SNMP can detect and report problems before they become significant.

Server Message Block *Server Message Block (SMB)* is a file transfer protocol that Microsoft networks use. Its primary functions are file and printer sharing.

The application layer does not refer to end-user applications. For example, applications such as Web browsers are not part of this layer and are not actually part of the OSI model at all.

Presentation Layer

The *Presentation layer* is layer 6 of the OSI Model, which interacts with the Session and Application layers by acting as a translator and determining how to format and present the data.

A common method of formatting data is by using the American Standard Code for Information Interchange (ASCII) table. The ASCII table includes 128 codes to display characters such as numbers, letters, and symbols. **See Figure 3-3**.

Many other codes beyond ASCII are defined at the Presentation layer. The Extended Binary Coded Decimal Interchange Code (EBCDIC) extended the ASCII table from 128 to 256 characters. File types such as MP3, .JPG and .GIF have their own codes defined on this layer.

The Presentation layer is also responsible for data compression and decompression, as well as data encryption and decryption. Multimedia transferred over the Internet is often compressed to conserve bandwidth.

Session Layer

The *Session layer* is layer 5 of the OSI Model and is responsible for establishing,

Figure 3-3 Partial ASCII Table

CHARACTER	DECIMAL	HEXADECIMAL	OCTAL	HTML
A	65	41	101	&#65
B	66	42	102	&#66
C	67	43	103	&#67
a	97	61	141	&#97
b	98	62	142	&#98
c	99	63	143	&#99
1	49	31	061	&#49
2	50	32	062	&#50
3	51	33	063	&#51

Figure 3-3. ASCII codes are commonly used to represent text in computers.

maintaining, and terminating sessions. A session is simply a lasting connection between two networking devices. For example, a chat program on one computer establishes a session with another computer to exchange the data.

The Session layer manages the connections. It starts the session, manages the traffic during the session, and terminates the session when appropriate.

The Session layer also ensures that data from different functions at the Application layer are kept separate. This becomes critical when a computer is running multiple applications, or when applications require more than one resource.

For instance, a user might be having a chat session in one window, downloading music in another, and reading e-mail in a third. The system establishes and maintains three sessions for three different applications. The Session layer ensures that resources are available for each session and kept separate from each other.

The Session layer also tracks the mode of transmission used by the computers. Computers utilize three modes to transmit data including:

Simplex Systems can send data only one way. This mode is not commonly used today in networking applications.

Half-Duplex Systems can send data both ways, but only one way at a time. This is similar to a walkie-talkie, where one user can press a button to talk but cannot receive any transmissions while the button is pressed.

Full-Duplex Systems can send data and receive it at the same time. Systems use separate methods to send and receive data.

The Session layer coordinates the communication and determines which mode to use. Two network protocols that operate on this layer are the Network Basic Input/Output System (NetBIOS) and Remote Procedure Calls (RPC).

Transport Layer

The *Transport layer* is layer 4 of the OSI Model and is responsible for handling transmission services such as flow control, reliability, and error checking. Transport layer protocols can divide data into smaller chunks, or segments, and then re-assemble the received data at the destination.

> ✓ FACT
> Protocol data units traveling on the Transport layer are referred to as segments.

For example, imagine wanting to mail all of the volumes of *Harry Potter* to a friend in another state, but only being able to use envelopes. This would require tearing the pages from the books and mailing them all separately. The friend would then have to re-assemble the books from the contents of all the envelopes.

This is similar to how systems manage data at the Transport layer. Huge, megabyte-sized files cannot travel over the network. Instead, the Transport layer segments (or divides) these large files into smaller-sized pieces. The Transport layer protocol system transmits these smaller segments over the network and then re-assembles them when the destination system receives them. The Transport layer also manages the ordering of the segments so that upon arrival, packets can be re-assembled in the same order.

There are two primary protocols operating at the Transport layer:

Transmission Control Protocol (TCP) *Transmission Control Protocol (TCP)* is a set of rules that provides guaranteed delivery of data. It starts by establishing a session and will not transmit data until a session is established. TCP is commonly referred to as connection-oriented. This means that it establishes a session before transmitting data.

User Datagram Protocol (UDP) *User Datagram Protocol (UDP)* is a set of rules that provides a best-effort method of delivering data. It does not provide guaranteed delivery of data like TCP. UDP is referred to as connectionless. A *connectionless protocol* is one in which systems send data without first verifying a connection with the other system. Systems typically use UDP for media streaming and diagnostic messages. Instead of requiring additional overhead data in the form of messages used to establish the connection or session, UDP accepts that there may be some data loss and simply transmits the data.

The term port means different things depending on the context. Ports can be logical (numbers) or physical (connections on devices).

Logical ports such as TCP and UDP ports are simply numbers used to indicate how systems handle data when it reaches its destination. Many ports represent specific protocols such as port 80, which represents the well-known port of HTTP.

Physical ports are components of switches and routers into which cables are plugged. A switch learns what computer is connected to each port, and a router learns what network is connected to each port.

Port numbers identify the applications generating the data that the Transport layer protocols transmit. There are 65,536 possible TCP ports and another 65,536 possible UDP ports. Some protocols use both TCP and UDP ports, while others use only one or the other. In this context, a port is simply a number from 0 to 65,535 that the protocol uses for connection purposes. It does not represent a physical port.

The HTTP protocol uses the well-known port 80 by default. Visiting a website may entail the use of the HTTP address by itself, as in http://www.bing.com/. However, the port number could also be included, as in `http://www.bing.com:80`. `bing.com` is the website, and once the request reaches the server that is hosting it, the TCP port of 80 identifies the data as HTTP traffic. The web server passes the data to the service handling the application layer HTTP protocol.

In short, while the IP address gets traffic from one computer to another,

> ✓ **FACT**
> Although port 0 is a valid port, it is reserved for both TCP and UDP.

the port number identifies what application, service, or protocol should process the data.

Network Layer

The *Network layer* is layer 3 of the OSI Model and is responsible for determining the best route for data to travel to its destination. It usually uses routing protocols to build routing tables and uses a protocol such as Internet as the routed protocol. This protocol uses IP addresses at this layer to ensure that data can get to its destination.

Protocols operating at this layer include:

Internet Protocol v4 (IPv4) IPv4 is an addressing protocol that uses 32-bit addresses for the devices on the network. The TCP/IP suite uses IP addressing to get traffic from one computer to another. IPv4 addresses are commonly expressed in dotted decimal format, such as 192.168.1.1.

Internet Protocol v6 (IPv6) IPv6 is an addressing protocol that uses 128-bit addresses. IPv6 is intended to replace IPv4 and is now being used concurrently with IPv4 on networks throughout the world. IPv6 addresses are commonly expressed in hexadecimal format, such as `2001:0000:4137:9E76:3C2B:05AD:3F57:FE98`.

Address Resolution Protocol (ARP) *Address Resolution Protocol (ARP)* is a protocol that resolves IP addresses into the physical address or the Media Access Control (MAC) address. Systems use the IP address to route packets to the network interface adapter of their next stop on the way to their destinations. Switches use MAC addresses to track computers connected to different physical ports. While ARP resolves the IP address to a MAC address, the Reverse Address Resolution Protocol (RARP) does the opposite. RARP resolves MAC addresses to IP addresses.

Internet Group Multicast Protocol (IGMP) IGMP is responsible for managing the groups that receive multicast traffic. Multicast traffic goes from one computer to multiple computers.

Internet Control Message Protocol (ICMP) *Internet Control Message Protocol (ICMP)* is a system of rules that carries error messages and diagnostic reporting messages between systems. Several diagnostic tools such as Ping, PathPing, TraceRt, and others use ICMP.

Internet Protocol Security (IPsec) IPsec is a collection of security protocols that secure IP traffic by digitally signing and/or encrypting it before transmission. It also includes authentication mechanisms, which enables computers to ensure they communicate only with known entities.

Routing Information Protocol (RIP) RIP is a basic routing protocol that routers use on internal networks. Routers use RIP to communicate with each other and share information about the network. The current version of IPv4 is RIPv2, though OSPF has replaced it on most networks.

Open Shortest Path First (OSPF) *Open Shortest Path First (OSPF)* is a routing protocol that routers use to communicate with each other on internal networks. OSPF is more advanced than RIP, and more networks use it.

The Network layer includes two key physical devices. The primary device working at this layer is a router. Routers are the devices that perform IP-based routing functions. The router looks at the IP address and determines the best

> ✓ **FACT**
> Protocol data units at the Network layer are referred to as packets.

> ✓ **FACT**
> Valid hexadecimal characters are 0 through 9 and A through F.

> **✓ FACT**
> Layer 3 switches also operate on the Network layer (layer 3). Layer 3 switches combine the capabilities of layer 2 switches and routers.

> **✓ FACT**
> Protocol data units at the Data Link layer are referred to as frames.

path to the destination network. It then transmits data packets to the destination using this path.

RIP and OSPF are common routing protocols used by routers on internal networks. These two protocols determine the best route to a destination based on a metric (cost). The route with the best metric will have a lower cost and will be the selected route for IP.

This is similar to using a map on a highway. When traveling from point A to point B, a map helps in determining the best route. The map provides the routing information and helps to identify the best path to get to a destination. Since maps are static and show only paths, they do not show construction, accidents, or other events that can slow down traffic. Other online tools may be used to identify traffic congestion or areas of construction.

Routing protocols such as RIP and OSPF are dynamic protocols that can adjust to changing conditions on a network. Routers use these protocols to communicate with each other regularly. If network events occur that impact known routes or if new routes get added, the routing protocols ensure that all the routers quickly learn about the impact.

Data Link Layer

The *Data Link layer* is layer 2 of the OSI Model and is concerned with data delivery on a local area network (LAN). This is where LAN technologies such as Ethernet are defined. The Data Link layer consists of two sublayers:

Logical Link Control (LLC) IEEE 802.2 LLC interacts directly with the network layer. It is defined by the IEEE 802.2 standard. LLC provides flow control and error control and enables multiple protocols to work simultaneously.

Media Access Control (MAC) IEEE 802.3 MAC defines how systems place packets onto the physical media at the Physical layer. IEEE 802.3 defines Carrier Sense Multiple Access/Collision Detection (CSMA/CD), which is used to handle data collisions. Other Data Link layer protocols use different MAC mechanisms. For example, the IEEE 802.11 wireless LAN standard calls for a MAC mechanism called Carrier Sense Multiple Access with Collision Avoidance (CSMA/CA).

MAC addresses are also defined at the Data Link layer. The MAC address is commonly called a physical address, hardware address, burned-in address, or Ethernet address. This used to be a permanent address that was written into (or burned into) the read-only memory (ROM) chip on a network interface card (NIC), but today it is usually stored on the firmware of a network interface adapter.

Entering the command `ipconfig/all` at the command prompt on a Windows computer or server provides a lot of information, including the physical address (or MAC address) of the network

---

Other Devices Have MAC Addresses

Although entering the command ipconfig/all at the command prompt of a windows computer or server will show the Mac address, other devices also have MAC addresses. For example, each interface on a router has a separate MAC address. These MAC addresses at the Data Link layer are then mapped to an IP address assigned at the Network layer. The MAC address is 6 bytes long and consists of 12 hexadecimal characters (or six pairs of hexadecimal characters). Four bits represent each hexadecimal character. Four bits times 12 characters shows that the MAC address is 48 bits (6 bytes) long.

Figure 3-4 ipconfig/all Command

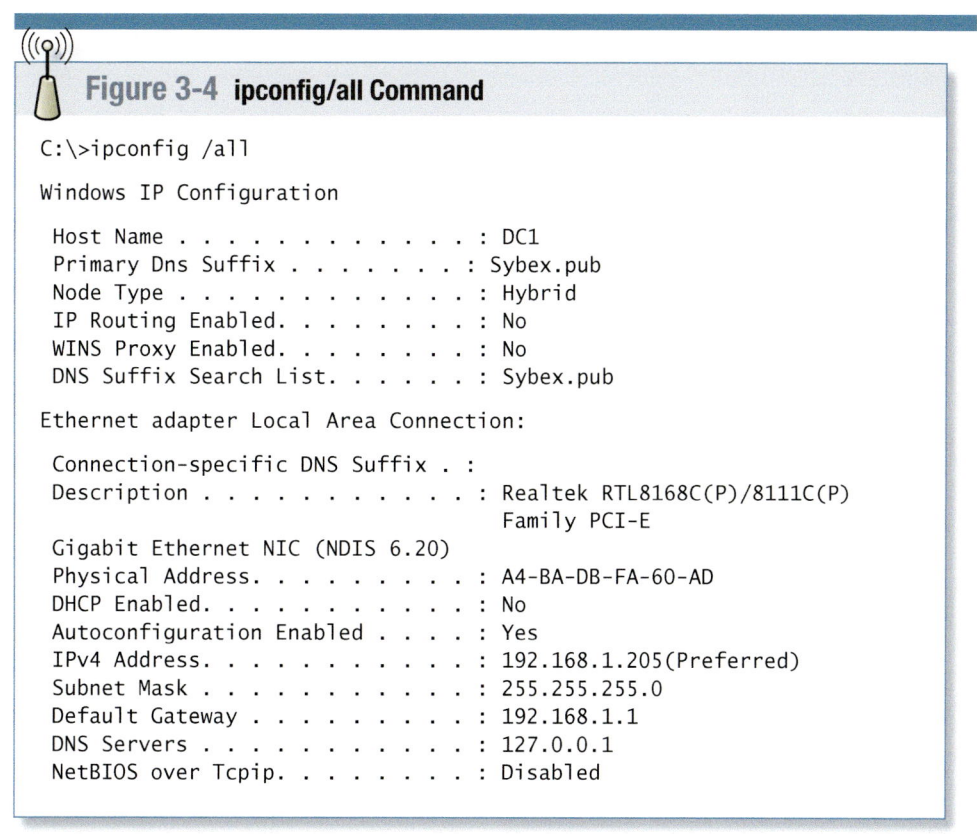

```
C:\>ipconfig /all

Windows IP Configuration

   Host Name . . . . . . . . . . . . : DC1
   Primary Dns Suffix  . . . . . . . : Sybex.pub
   Node Type . . . . . . . . . . . . : Hybrid
   IP Routing Enabled. . . . . . . . : No
   WINS Proxy Enabled. . . . . . . . : No
   DNS Suffix Search List. . . . . . : Sybex.pub

Ethernet adapter Local Area Connection:

   Connection-specific DNS Suffix  . :
   Description . . . . . . . . . . . : Realtek RTL8168C(P)/8111C(P)
                                       Family PCI-E
   Gigabit Ethernet NIC (NDIS 6.20)
   Physical Address. . . . . . . . . : A4-BA-DB-FA-60-AD
   DHCP Enabled. . . . . . . . . . . : No
   Autoconfiguration Enabled . . . . : Yes
   IPv4 Address. . . . . . . . . . . : 192.168.1.205(Preferred)
   Subnet Mask . . . . . . . . . . . : 255.255.255.0
   Default Gateway . . . . . . . . . : 192.168.1.1
   DNS Servers . . . . . . . . . . . : 127.0.0.1
   NetBIOS over Tcpip. . . . . . . . : Disabled
```

Figure 3-4. *The ipconfig/all command provides valuable information to the user including the IP Address and MAC Address.*

adapter. **See Figure 3-4.** Notice that it has an address of **A4-BA-DB-FA-60-AD**.

Every device on a network has a different MAC address. If MAC addresses on the network are not unique, the computers with the same MAC address cannot communicate on the network.

Organizations that manufacture network adapters are assigned an organizationally unique identifier (OUI) that is used for the first three bytes of the MAC. Three bytes worth of serial numbers are then added to this OUI to create the MAC. **See Figure 3-5.**

Protocols that operate at the Data Link layer include:

Ethernet Ethernet, defined in the Institute of Electrical and Electronics Engineers (IEEE) 802.3 standard, is the most commonly used LAN protocol at the Data Link layer. Virtually all personal computers sold today have an Ethernet adapter built into the motherboard.

IEEE 802.11 IEEE 802.11 defines the wireless LAN technology that is the most common alternative to Ethernet used today. Many portable computers and handheld devices include integrated IEEE 802.11 network interface adapters.

Figure 3-5 MAC Address

ORGANIZATIONALLY UNIQUE IDENTIFIER	MANUFACTURER SERIAL NUMBER
AA-BA-DB	FA-60-AD
Six hexadecimal characters (24 bits)	Six hexadecimal characters (24 bits)

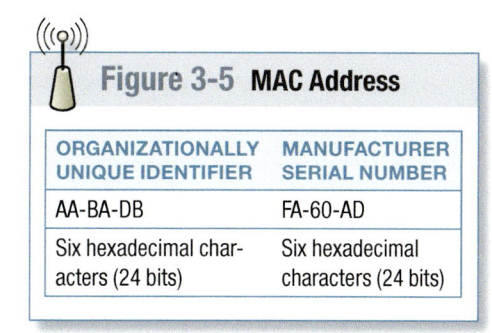

Figure 3-5. *A MAC address is divided into two parts, an organizational unique identifier and the manufacturer serial number.*

Token Ring IEEE 802.5 defines a token ring technology. The computers pass a logical token between themselves. A computer can communicate on the network only when it has the token. Using token passing for communication prevents collisions.

Point-to-Point Tunneling Protocol v4 (PPTP) *Point-to-Point Tunneling Protocol v4 (PPTP)* is a system of rules commonly used with virtual private networks (VPNs). VPNs provide remote users with access to a private network over a public connection such as the Internet.

Layer 2 Tunneling Protocol (L2TP) *Layer 2 Tunneling Protocol (L2TP)* is a protocol used with VPNs that often uses IPsec (as L2TP/IPsec) to encrypt the traffic.

Asynchronous Transfer Mode (ATM) ATM is a cell-based method of transferring data. Systems convert data into small fixed-sized cells and transfer them over the network. ATM is used in wide area networks (WANs).

Frame Relay Frame relay is another WAN technology. Systems convert data into variable-sized frames and transfer them over permanent virtual circuits.

Physical devices operating on the Data Link layer include bridges, switches, and network interface adapters.

Physical Layer

The *Physical layer* is layer 1 of the OSI Model and defines the physical specifications of the network, including physical media such as cables and connectors and basic devices such as repeaters and hubs. The Physical layer converts the data stream into zeros and ones (bits) and places them onto the physical media in the form of electrical pulses for copper cable, light pulses for fiber-optic cable, or radio signals for wireless technologies.

The Physical layer has some simple, yet unique, functions. It defines the physical characteristics of cables and connectors. It is also responsible for encoding signaling types, such as converting digital signals to analog signals.

On LANs, the Physical layer implementation is typically joined to the Data Link layer, as in the case of Ethernet. The IEEE 802.3 standard defines both the Data Link layer components and the Physical layer specifications.

The Complete Picture

The OSI model works when two different computers are interacting. *Encapsulation* is the overall process in which the data from the higher layer protocols is encapsulated, or packaged and incorporated by those at the lower layers. **See Figure 3-6.**

Imagine a user launching Chrome to access a web search engine such as Bing. The Application layer protocol accepts the data from Chrome, and the Session, Presentation, and Application layers work together to prepare the request to the Transport layer.

The Transport later then adds a TCP header. This header includes port information for the source and destination computers. Websites serve data using HTTP, and HTTP uses port 80, so the destination port is set as port 80. TCP assigns a port such as 49152 to Chrome as the source port. The source port is to ensure that the return traffic is returned to Chrome, which will display the page provided by the website.

At this point, two TCP ports have been assigned and a part of the TCP header has been added:

- Destination port: 80
- Source port: 49152

> ✓ **FACT**
> Network adapters also operate at the Physical layer.

> ✓ **FACT**
> Data traveling on the Physical layer is converted to bits, or ones and zeros (such as 110011010101).

Figure 3-6 Data and the OSI Model

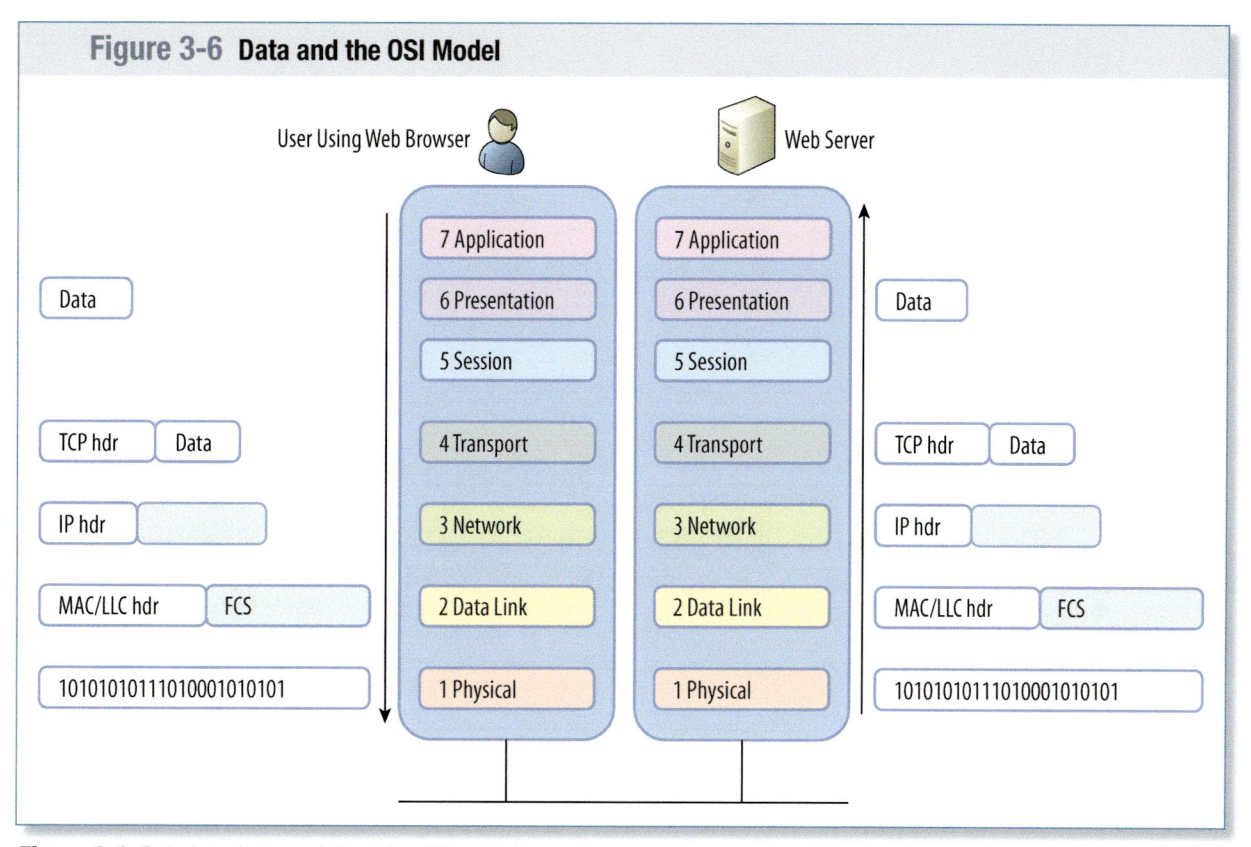

Figure 3-6. Data travels up and down the OSI model.

Next, the Network layer protocol adds IP addresses at the Network layer. The IP address of the computer running the web server is added as the destination IP address, and the IP address of the client's computer is added as the source IP address. This information becomes part of the IP header, which the system adds to the TCP header and the data. At this point, the following is present:

▶ Destination IP address and destination port: 80

▶ Source IP address and source port: 49152

Routers on the network use the destination IP address to route the packet to the destination subnet. When it arrives, the receiving computer uses the destination port to send the data to the service, application, or protocol associated with the port.

When the packet reaches the network where the destination computer is located, the Data Link layer discovers the MAC address of the destination computer. The system then adds the MAC address to the Data Link layer frame.

Finally, the Physical layer converts the data into ones and zeroes and places it on the wire. When the data reaches the destination computer, the process is reversed. The header information at each layers is stripped off, and the data inside is passed to the next layer.

The data will be passed to the service handling HTTP at the Application layer. The user's request is processed, and a web page is built and sent back. The entire encapsulation process is repeated on the server and then sent back to the computer that originally requested the data.

Packets and Frames

Although some professionals use the terms "packets" and "frames" interchangeably, this is not entirely accurate. The actual name depends on the layer of the OSI model.

Encapsulated data has different names at the different layers of the OSI model. **See Figure 3-7.**

Although there are multiple protocols throughout the OSI model, the primary focus is on the TCP and UDP protocols at the Transport layer and the IP protocol at the Network layer. The correct terms for each layer are:

Protocol Data Unit (PDU) Data units at layers 5, 6, and 7 are called protocol data units.

Segment A *segment* is a TCP unit of data at layer 4, the Transport layer. Remember that TCP can divide data into smaller segments at this layer.

Packet A *packet* is a unit of data at layer 3, the Network layer. This layer uses IP addresses to get the packets from the source to the destination.

Frame A *frame* is a unit of data at layer 2, the Data Link layer. MAC addresses are defined here.

Bits *Bits* are the data at the Physical layer, or ones and zeros.

Some sources identify a datagram as a unit of data at the Transport layer (layer 4) using the User Datagram Protocol. Since "datagram" is in the UDP name, this makes a lot of sense. It implies that the terms segment and datagram are interchangeable. However, official reference sources do not support this usage.

At this point, naming each of the seven layers of the OSI model and the corresponding layer numbers should be easy. Identifying the location of various protocols (such as TCP, UDP, and IP) on the OSI

> ✓ **FACT**
> Segment has two meanings. On a physical network, it is a common connection between multiple computers. On the OSI model, it is the data at layer 4.

Figure 3-7 Encapsulated Data

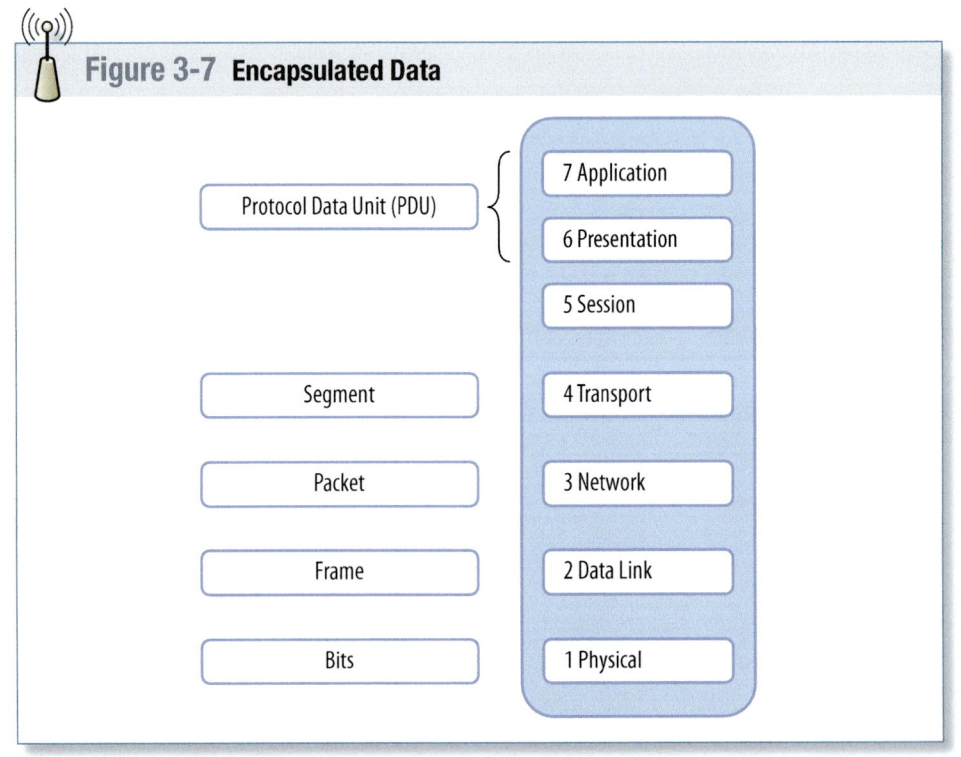

Protocol Data Unit (PDU)	7 Application
	6 Presentation
	5 Session
Segment	4 Transport
Packet	3 Network
Frame	2 Data Link
Bits	1 Physical

Figure 3-7. Encapsulated data has different names at different layers.

> ### What About Datagrams?
> The term datagram may crop up during conversation. However, depending on what source is used, the term "datagram" can mean different things. Conventional sources indicate that a datagram is simply another name for a packet.
>
> RFC 1594 identifies a datagram as "a self-contained, independent entity of data carrying sufficient information to be routed from the source to the destination computer without reliance on earlier exchanges between this source and destination computer and the transporting network."
>
> What RFC 1594 is saying is that a datagram is on the layer that does the routing, which is layer 3, the Network layer. This implies that the terms "packet" and "datagram" are interchangeable.

model and the names of encapsulated data at different layers should also be easy.

Ideally, drawing a diagram of the names of the encapsulated data at the different layers of the OSI model should be easy. However, it does take a little practice. Mnemonics (such as All People Seem To Need Data Processing, Please Do Not Throw Sausage Pizza Away, or another one) should help.

THE TCP/IP MODEL

The *TCP/IP model* is a four-layer communications model created in the 1970s by the U.S. Department of Defense (DoD). It is also called the DoD model. The TCP/IP model works similarly to the OSI model; it just has fewer layers.

The four layers of the TCP/IP model can be compared to the OSI model. **See Figure 3-8.**

> ✓ **FACT**
> The TCP/IP model was created in the United States for the DoD before the OSI model, and the OSI model was created by the ISO as an international standard.

Figure 3-8 TCP/IP Model

TCP/IP Model	OSI Model
Application	7 Application
	6 Presentation
	5 Session
Transport (Host-to-Host)	4 Transport
Internet	3 Network
Link (Network Interface or Network Access)	2 Data Link
	1 Physical

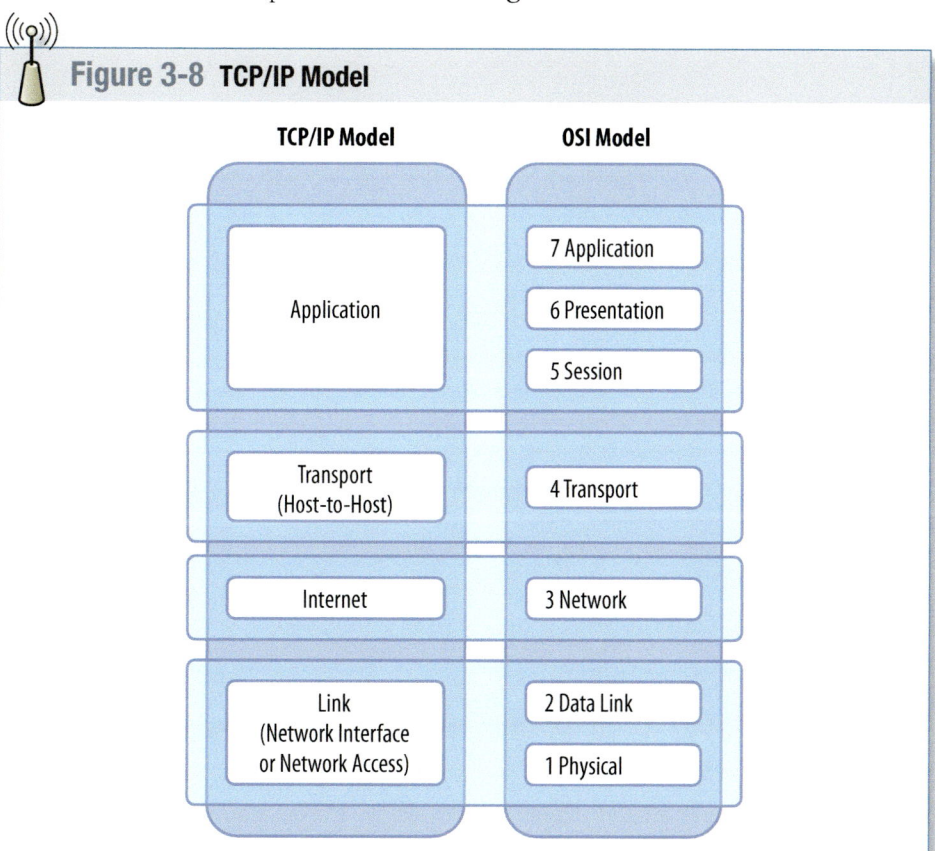

Figure 3-8. *The TCP model has four layers that map to the seven layers of the OSI model.*

✓ FACT
The Link layer is also known as the Network Interface or Network Access layer.

✓ FACT
The Transport layer is also known as the host-to-host layer.

For additional information, visit qr.njatcdb.org Item # 1789

For additional information, visit qr.njatcdb.org Item # 1790

Notice that the layers on the TCP/IP model correlate to layers of the OSI model. The TCP/IP Application layer maps to layers 5, 6, and 7 of the OSI model. The TCP/IP Transport layer maps to layer 4 of the OSI model. The TCP/IP Internet layer maps to layer 3 of the OSI model. The TCP/IP Link layer maps to layers 1 and 2 of the OSI model.

Application Layer Applications at this layer use protocols to access network resources. Protocols include DNS, SMB, HTTP, FTP, SMTP, POP3, IMAP4, and SNMP.

Transport Layer Protocols on this layer control data transfer on the network by managing sessions between devices. The two primary protocols are TCP and UDP.

Internet Layer Protocols on the Internet layer control the movement and routing of packets between networks. Protocols on this layer include IPv4, IPv6, IGMP, ICMP, and ARP.

Link Layer This layer defines how data is transmitted onto the media. It includes multiple protocols such as Ethernet, token ring, frame relay, and ATM.

DEVICES MAPPED ON THE OSI AND TCP MODELS

The OSI and TCP/IP models are reference points for the devices used on a network. These devices may include network adapters, hubs, switches, routers, and firewalls. **See Figure 3-9.**

Devices at the lower layers (such as layer 1, the Physical or Link layer) have very little intelligence. When moving up the layers, though, the devices become more and more sophisticated. For example, an advanced firewall on the Application layer (layer 7) can analyze traffic within a session and make decisions to block or allow the traffic.

A hub (on layer 1) is unable to make any decisions and simply transfers all

---

TCP/IP Model Layers

While studying different models, note that there may be different names given to the TCP/IP model layers.

Microsoft documentation typically labels these layers as Application, Transport, Internet, and Link. For example, Microsoft online resources identified as preparation materials for the Microsoft Technology Associates (MTA) Networking Fundamentals exam (98-366) use these labels. When preparing for this exam, know these labels with these names.

Some networking textbooks label these layers as follows:

▶ Application
▶ Host-to-host
▶ Internet, Internetwork, or Internet Protocol
▶ Network Access or Network Interface

Many consider the Internet Engineering Task Force (IETF) as the official source for these models, and reference RFC 1122 and RFC 1123. These documents identify the layers as follows:

▶ Application
▶ Transport
▶ Internet Protocol
▶ Link

Figure 3-9 Device Reference Points

TCP/IP Model	OSI Model	Devices
Application	7 Application	Proxy Servers, Advanced Firewalls
Application	6 Presentation	
Application	5 Session	
Transport (Host-to-Host)	4 Transport	
Internet	3 Network	Router, Layer 3 Switch
Link (Network Interface or Network Access)	2 Data Link	Switch, Bridge, Network Adapter
Link (Network Interface or Network Access)	1 Physical	Hub, Repeater (Amplifier), Modem, Cables, Network Adapter

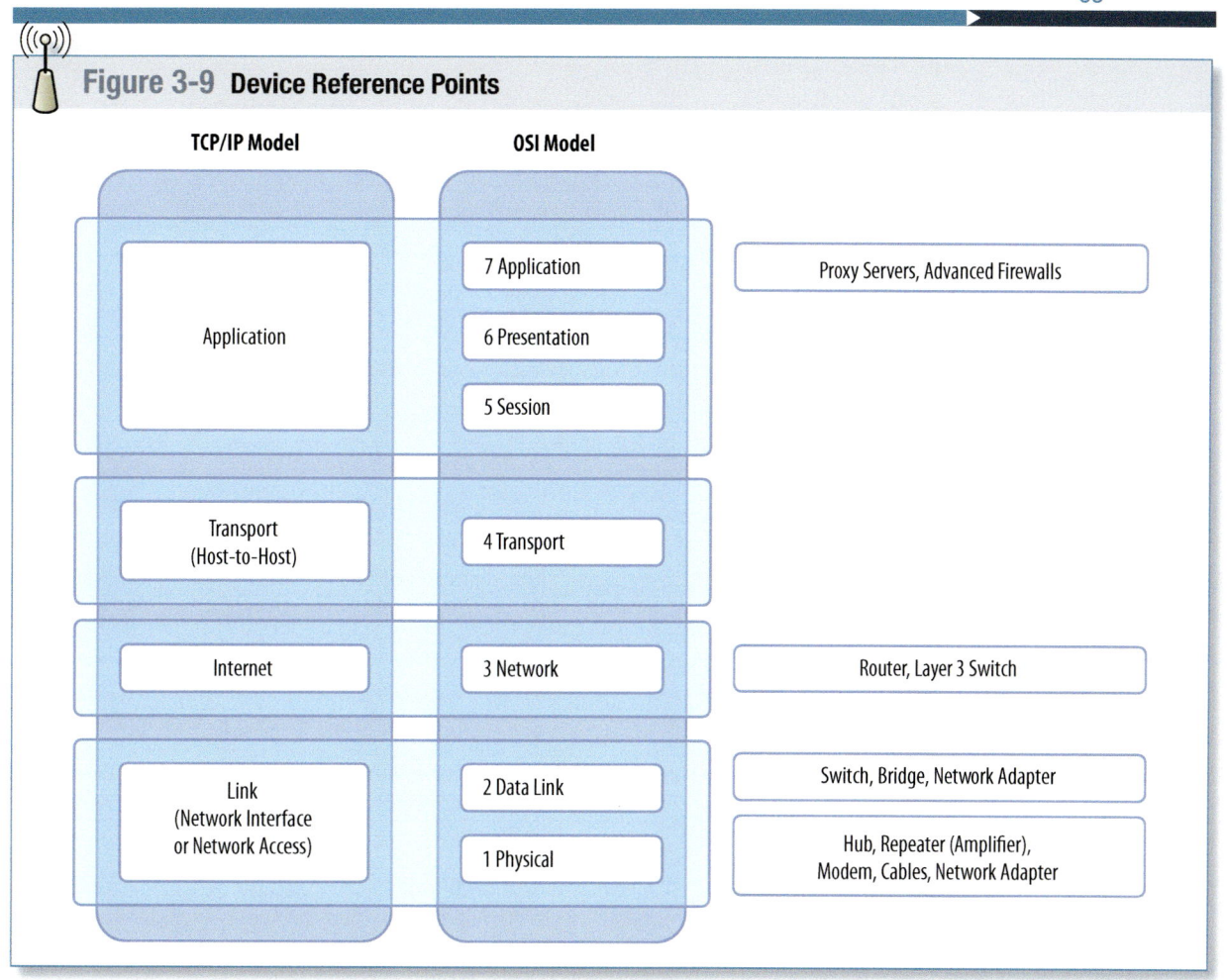

Figure 3-9. Devices have reference points on the OSI and TCP/IP models.

the data received on one port to all other ports of the hub. Switches (on layer 2) learn which port computers are connected and internally switch the traffic. Routers (on layer 3) can talk to other routers and learn the best path to any subnet within a network.

Mapping devices to specific layers of the OSI or TCP/IP models will facilitate better network troubleshooting. For example, consider a problem where a computer is not communicating on the network.

There are multiple reasons why network communication might not be working. If there are no lights on the network adapter or switch, a layer 1 problem might have caused the physical connection to fail. This could be a faulty cable or faulty network adapter. The network adapter connects to a hub or a switch, so the problem could also be a faulty network device or faulty port on the network device.

If the network adapter LED is not lit, it is important to realize that the problem is a layer 1 problem. There is no need to troubleshoot the TCP/IP configuration, the operating system, or the applications that are on different layers.

This is similar in concept to troubleshooting car problems. Imagine turning the ignition key but nothing happens. There is no sound, no clicking, nothing. Checking the oil or gas may be a waste of time. The problem is more likely with the battery or ignition system.

✓**FACT**
Some hiring managers include basic troubleshooting questions about the OSI model and/or TCP/IP model during interviews.

Physical Layer

Devices at the Physical layer are concerned only with the physical aspects of communication—actual data transmission through physical connectivity. The Physical layer does not understand logical addressing with IP addresses or physical addressing with the MAC addresses.

The Physical layer includes cables, cable connectors, network adapters, hubs, modems, and amplifiers or repeaters.

The hub is a device found at the Physical layer. It allows network expansion by enabling the plugging in of multiple devices into a central point. As a layer 1 device, the hub is not aware of any addressing, and ignores layer 2 MAC addresses and layer 3 IP addresses. It simply passes data received on one port to all other ports.

Notice that network adapters are listed on both the Physical layer and the Data Link layer. A network adapter provides simple feedback with a lit LED, indicating that the adapter is plugged in. This is a function performed at layer 1.

The network adapter can also analyze traffic to determine whether received traffic is addressed to the computer based on the MAC address. If the traffic is addressed to the computer, the adapter processes the traffic and passes it to the internal processor. This process occurs on layer 2, making an adapter both a layer 1 and layer 2 device.

The telephone modem is also found at layer 1 of the OSI model. The modem is a modulator–demodulator; it converts digital signals from the computer into analog signals used over the telephone line. Demodulation is the conversion from an analog-to-digital signal.

Repeaters and amplifiers are sometimes referred to as the same thing, but there is a subtle difference. The repeater will regenerate a digital signal, and the amplifier will regenerate an analog signal. Both boost signal strength as it travels along a cable, allowing a signal to travel further before reaching its destination. For example, if a cable is capable of carrying a signal only 100 meters, a repeater can be inserted between two 100-meter cables to extend the distance to 200 meters.

Data Link Layer

Devices on the Data Link layer include switches, bridges, and network adapters. Switches and bridges create separate collision domains.

Switch The switch is a layer 2 device that learns MAC addresses of devices from incoming traffic. These MAC addresses tell the switch which devices are connected to which port within a subnet. The switch then internally forwards traffic to create separate collision domains.

Bridge The bridge learns the MAC addresses of devices that are connected to a port, similar to how a switch learns MAC addresses. However, a bridge will typically have multiple computers connected to each bridge port via a hub. Bridges usually have only two ports.

Network Interface Adapter The network adapter is shared between the Data Link layer and Physical layer. The adapter contains the layer 2 MAC address. It analyzes traffic at this layer and determines whether the computer should process the traffic. If the traffic is addressed to the computer, it passes the traffic to the central processor.

Network Layer

The router is the primary device at the Network layer. It routes packets based on their logical IP addresses.

Routers forward IP traffic to different subnets within a network. They can communicate with other routers using

---

✓ **FACT**

Although modems are not common in urban areas, they are still popular with rural users who do not have broadband connections.

routing protocols such as RIP and OSPF. Routers use these routing protocols to learn about multiple subnets within a network.

Layer 3 switches also operate at the Network layer. They are advanced switches that have the ability to route traffic based on a layer 3 address, similar to the way in which a router routes traffic.

Application Layer

Proxy servers and advanced firewalls work at the Application layer. They have the capability to examine traffic and make decisions based on the content. For example, a proxy server can block access to specific Internet websites based on the website address.

Any firewall can block traffic based on source or destination data contained within packets. Basic firewalls do this by blocking traffic based on IP addresses or ports in each individual packet. Advanced firewalls can analyze multiple packets within a session and make decisions to block traffic or allow it to continue.

PROTOCOLS MAPPED ON THE OSI AND TCP/IP MODELS

It is important to understand where protocols operate on the OSI and TCP/IP models. **See Figure 3-10.**

Figure 3-10 Protocols

TCP/IP Model	OSI Model	Protocols
Application	7 Application	DNS, DHCP, LDAP, HTTP, FTP, TFTP, SNMP, SMTP, POP3, IMAP4, SMB
	6 Presentation	
	5 Session	
Transport (Host-to-Host)	4 Transport	TCP, UDP
Internet	3 Network	IPV4, IPV6, ARP, IGMP, ICMP, IPSec, RIP, OSPF
Link (Network Interface or Network Access)	2 Data Link	Token Ring, Frame Relay, ATM
	1 Physical	Ethernet

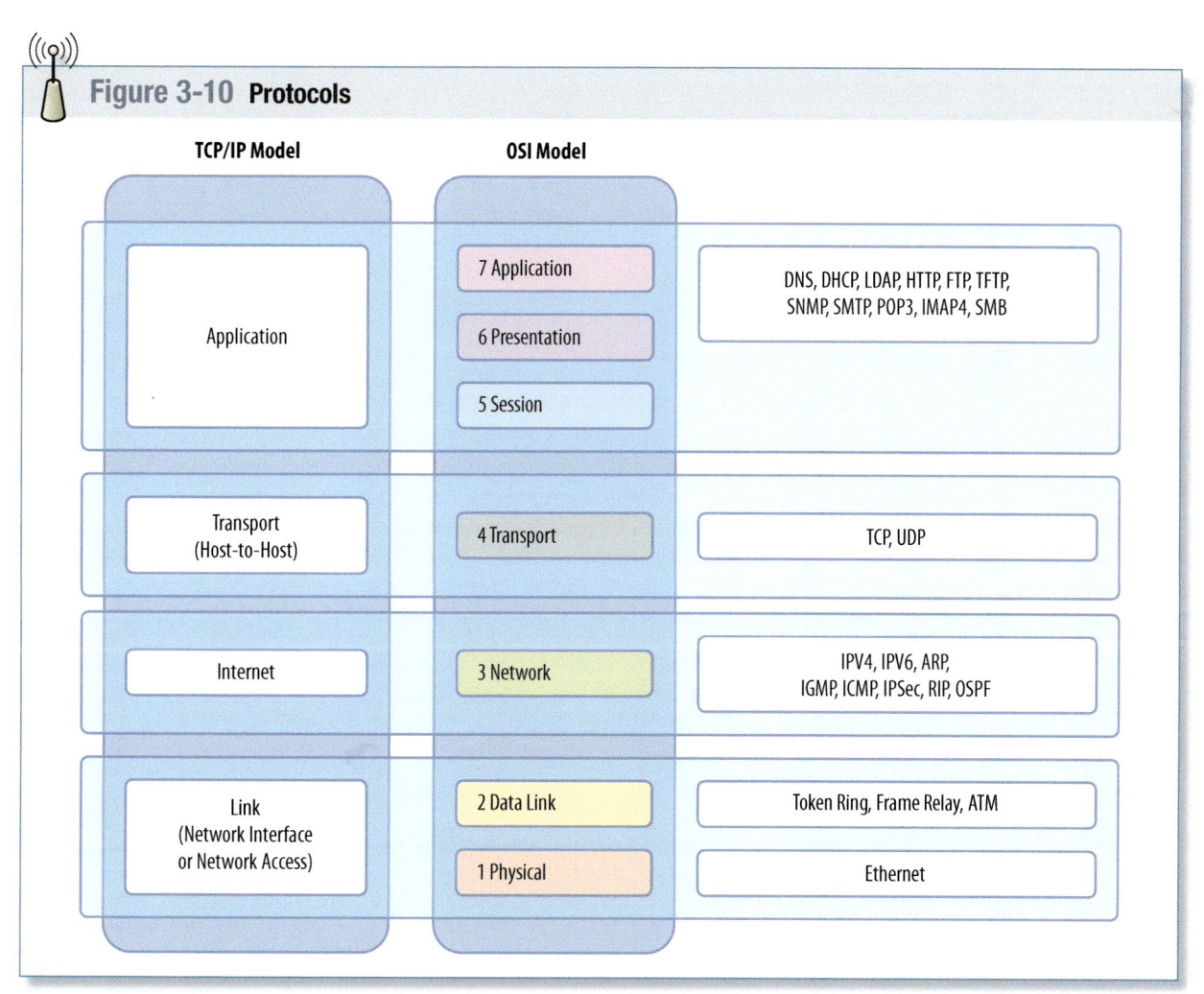

Figure 3-10. *Protocols can be mapped on the OSI model and the TCP/IP model.*

Summary

The OSI model is a framework, or set of guidelines, used to develop and standardize networking protocols. This model has seven layers known as the Application, Presentation, Session, Transport, Network, Data Link, and Physical layers. The TCP/IP model includes four layers: Applications, Transport, Internet, and Link. Protocols and devices are designed to work on specific layers of the OSI and TCP/IP models, and certain layers are associated with specific protocols and devices.

Review Questions

1. How many layers does the OSI model have?
   a. 4
   b. 5
   c. 7
   d. 8

2. Which of the following mnemonics could be used to remember the OSI model?
   a. All People Seem to Use Data Processing
   b. A Perfect Storm Never Seems Delightful
   c. Please Do Not Throw Sausage Pizza Away
   d. The Purple Dress Needs a Seamstress

3. TCP is a connectionless protocol.
   a. True
   b. False

4. What is a unit of data called at the Transport layer?
   a. Frame
   b. Packet
   c. Protocol data unit (PDU)
   d. Segment

5. Which of the following could be a valid MAC address for a server named Server 1?
   a. `192.168.1.5`
   b. `A4-BA-DB-FA-60-AD`
   c. `G4-BA-10B-FA-60-AT`
   d. `Server1`

6. IPv4 operates on which layer of the OSI model?
   a. Data Link
   b. Network
   c. Presentation
   d. Transport

7. Which of the following protocols operates on the Transport layer of the OSI model?
   a. HTTP
   b. IPv6
   c. OSPF
   d. TCP

8. Devices that operate on layer 7 of the OSI model are more intelligent than devices that operate on layer 1.
   a. True
   b. False

9. Routers operate on which of the following layers of the OSI model?
   a. Layer 1
   b. Layer 2
   c. Layer 3
   d. Layer 4

Ethernet

A complete discussion of the history of the local area network (LAN) must include the history of Ethernet. Ethernet was the first LAN protocol, originally conceived in 1973, and it has been evolving steadily ever since. Other Data Link layer protocols (such as Token Ring and ARCnet) have come and gone, but Ethernet has remained. Its only rival in the local area networking market today is the wireless LAN.

The longevity of Ethernet is primarily because of its continuous evolution. The earliest commercial Ethernet networks ran at 10 megabits per second (Mbps), and successive iterations of the protocol increased the network's transmission speed to 100, 1,000, and 10,000 Mbps. There are now even 100,000 Mbps (100 Gbps) Ethernet devices on the market. Despite recent massive changes, many of the basic elements of an Ethernet network remain the same.

OBJECTIVES

- ▶ Identify Ethernet standards
- ▶ Describe Ethernet components

CHAPTER 4

TABLE OF CONTENTS

Ethernet Standards 72
    DIX Ethernet 72
    IEEE 802.3 72

Ethernet Components 73
    The Ethernet Frame 73
    CSMA/CD. 75
    Physical Layer Specifications 77
    The 5-4-3 Rule 78
    Normal Link Pulse Signals 78

Summary 86
Review Questions 87

ETHERNET STANDARDS

The Xerox Corporation was responsible for the original development of the Ethernet networking system, and in 1980, joined with Digital Equipment Corp (DEC) and Intel to publish the first Ethernet standard, called "The Ethernet, A Local Area Network: Data Link Layer and Physical Layer Specifications." This standard (one of two upon which commercial Ethernet implementations were based) was known informally as DIX Ethernet.

DIX Ethernet

Thick Ethernet or *10Base5* was the first Ethernet standard and it described a network that used RG-8 coaxial cable in a bus topology up to 500 meters long, with a transmission speed of 10 Mbps. *Thin Ethernet* or *10Base2* is a second version of the standard published in 1982, called DIX Ethernet II. This version added a second Physical layer specification, calling for RG-58 coaxial cable. Because RG-58 cable is thinner and less expensive than RG-8, the maximum segment length is restricted to 185 meters.

Development of the DIX Ethernet standard stopped after the publication of version II.

IEEE 802.3

The IEEE began work in 1980 on an international standard defining the Ethernet network, one not privately owned, as was DIX Ethernet. The result was a document called "IEEE 802.3 Carrier Sense Multiple Access with Collision Detection (CSMA/CD) Access Method and Physical Layer Specifications," which the IEEE published in 1985.

The original 802.3 standard describes a network that is almost identical to DIX Ethernet, except for a minor change in the frame format. The main difference between IEEE 802.3 and DIX Ethernet is that the IEEE has continued to revise its standard in the years since the original publication.

The 802.3 working group updates its standards by publishing amendments that contain additional Physical layer specifications and descriptions of other new technologies. IEEE publishes the amendments with a letter and a date. For example, the first amendment, 802.3a—1988, added the 10Base2 Physical layer specification to the original standard.

At regular intervals, the working group incorporates the published amendments into the main 802.3 document. The current standard is called IEEE 802.3-2015, "EEE Standard for Ethernet." The document includes the specifications from dozens of amendments and defines Physical layer specifications ranging from 1 Mbps to 100 Gigabit Ethernet, over a variety of physical media.

The latest amendment, IEEE 802.3bw—2015, is called "IEEE Standard for Ethernet Amendment 1: Physical Layer Specifications and Management Parameters for 100 Mb/s Operation over

---

> ✓ **FACT**
>
> The terms 10Base5 and 10Base2 are shorthand designations for Ethernet Physical layer specifications. The number 10 refers to the speed of the network (10 Mbps). The word "base" refers to the use of baseband signaling on the network. The numbers 5 and 2 refer to the maximum length of a cable segment, which is 500 meters for Thick Ethernet and 200 (actually 185) meters for Thin Ethernet. Subsequent designations have used letters representing the cable type, rather than numbers indicating cable lengths. For example, the T in 10Base-T refers to the use of twisted-pair cable. The designations beginning with 10Base-T also include a hyphen to prevent people from pronouncing them "bassett."

> **Usage of the Term "Ethernet" in the Industry**
>
> Technically speaking, the only Ethernet networks are those running on coaxial cable using a bus topology. All of the Physical layer specifications defining networks using twisted-pair or fiber-optic cable are part of the IEEE 802.3 standard and should actually be called by that name. However, the name "Ethernet" is still ubiquitous in the networking industry, both on product packaging and in common usage among network administrators. It is universally understood that an Ethernet network today actually refers to one that is compliant with the IEEE 802.3 standard.

a Single Balanced Twisted Pair Cable (100BASE-T1)."

ETHERNET COMPONENTS

The Ethernet standards (both DIX Ethernet and IEEE 802.3) are made up of three components:

The Ethernet Frame The frame is the packet format that Ethernet systems use to transmit data over the network.

Carrier Sense Multiple Access with Collision Detection *Carrier Sense Multiple Access with Collision Detection (CSMA/CD)* is the media access control (MAC) mechanism that early Ethernet systems used to regulate access to the network medium.

Physical Layer Specifications The Physical layer specifications define the various types of network media that can be used to build Ethernet networks, as well as the topologies and signaling types they support.

The Ethernet Frame

The *Ethernet frame* is the mailing envelope that the protocol uses to transmit data to other systems on the LAN. The frame consists of a header and a footer (or trailer) that surround the information that the Data Link layer protocol receives from the Network layer protocol operating at the layer above (which is usually the Internet Protocol, or IP).

The frame is divided into sections of various lengths called fields, which perform different functions. **See Figure 4-1.**

Preamble (7 Bytes) The "Preamble" field contains 7 bytes of alternating 0s and 1s, which the communicating systems use to synchronize their clock signals.

Start of Frame Delimiter (1 Byte) The "Start Of Frame Delimiter" field contains 6 bits of alternating 0s and 1s, followed by two consecutive 1s, which is a signal to the receiver that the transmission of the actual frame is about to begin.

Destination Address (6 Bytes) The "Destination Address" field contains the 6-byte hexadecimal MAC address of the network interface adapter on the local network to which the frame will be transmitted. XE "fields:Destination Address"

Source Address (6 Bytes) The "Source Address" field contains the 6-byte hexadecimal MAC address of the network interface adapter in the system generating the frame.

Ethertype/Length (2 Bytes) The "Ethertype/Length" field in the DIX Ethernet frame contains a code identifying the Network layer protocol for which the data in the packet is intended. The IEEE 802.3

Figure 4-1 Ethernet Frame

Preamble (7 bytes)

Start of Frame Delimiter (1 byte)

Destination Address (6 bytes)

Source Address (6 bytes)

Ethertype/Length (2 bytes)

Data and Pad (46–1,500 bytes)

Frame Check Sequence (4 bytes)

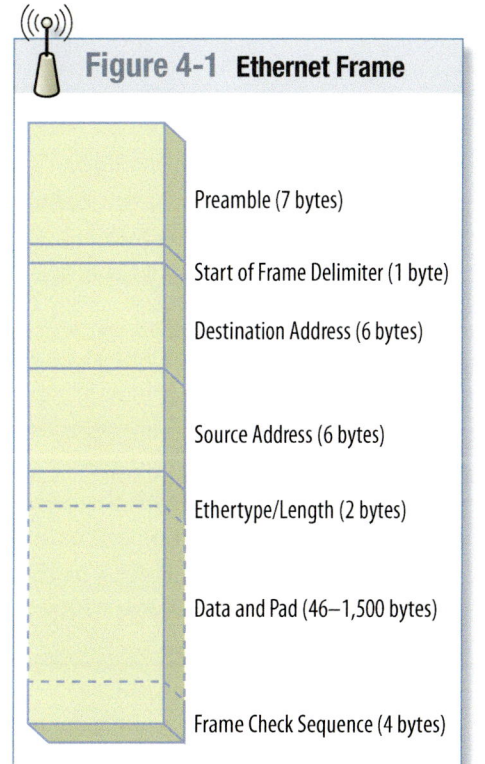

Figure 4-1. *The Ethernet frame format is divided into sections called fields.*

For additional information, visit qr.njatcdb.org
Item # 1791

frame, specifies the length of the data field (excluding the pad).

Data and Pad (46 to 1,500 Bytes) The "Data And Pad" field contains the data received from the Network layer protocol on the transmitting system, which is sent to the same protocol on the destination system. Ethernet frames must be at least 64 bytes long. Therefore, if the data received from the Network layer protocol is less than 46 bytes, the system adds padding bytes to bring it up to its minimum length.

Frame Check Sequence (4 Bytes) The frame's footer is a single field that comes after the Network layer protocol data and contains a 4-byte checksum value for the entire frame. The sending computer computes this value and places it into the field. The receiving system performs the same computation and compares it to the field to verify that the frame was received without error.

Ethernet Addressing

The primary function of the Ethernet frame—as with a mailing envelope—is to identify the addressee of the frame. Ethernet networks use 6-byte hexadecimal values called hardcoded into network interface adapters, to identify systems on the local network. The first three bytes are an organizationally unique identifier (OUI) and the last three bytes are a device identifier.

Data Link layer protocols are concerned only with communications on the LAN. Therefore, the values in the Destination Address and Source Address fields must identify systems on the local network. If a computer on the LAN is transmitting to another computer on the same LAN, then its frames contain the address of that target computer in their Destination Address fields. If a computer is transmitting to another computer on a different network, then the value in the Destination Address field must be the address of a router on the LAN. In this case, it is up to the Network layer protocol to supply the address of the packet's final destination.

Protocol Identification

The other main function of the Data Link layer protocol is to identify the protocol at the Network layer that is the destination of the data in the frame. By identifying the protocol at the Network level it allows for the system receiving the frame to pass the data up through the protocol stack. The method by which the frame identifies the Network layer protocol is the primary difference between the DIX Ethernet and the IEEE 802.3 standards.

The DIX Ethernet standard uses the 2 bytes immediately following the Source Address field to store an *Ethertype* value. An Ethertype is a hexadecimal value that identifies the protocol that generated the data in the packet. The only Ethertypes values left in common use are 0800 for the Internet Protocol version 4, 86DD for IPv6, and 0806 for the Address Resolution Protocol (ARP).

The Ethertype field is the last vestige of the DIX Ethernet standard still used on networks today. Because the TCP/IP protocols were developed in the 1970s (when DIX Ethernet was still the industry standard), most TCP/IP implementations still rely on the Ethertype value from the DIX Ethernet frame format for protocol identification.

In the IEEE 802.3 standard, those same 2 bytes following the Source Address field perform a different function. They indicate the length of the information in the Data field (excluding the Pad). Because the maximum length of data permissible in an Ethernet packet is 1,500 bytes, Ethernet systems assume that any value in this field larger than 0600 hexadecimal (1536 decimal) is an Ethertype.

For protocol identification, IEEE 802.3 relies on an outside protocol called IEEE 802.2 Logical Link Control (LLC). LLC is a separate protocol that the IEEE 802 group developed to work with a number of Data Link layer protocols that were being developed at the same time.

CSMA/CD

Carrier Sense Multiple Access with Collision Detection (CSMA/CD), the MAC mechanism used in the early implementations of Ethernet, was the single most defining characteristic of the network. For multiple computers to share a single network medium, it was critical for there to be an orderly means to arbitrate network access. Each computer was required to have an equal chance to use the network, or its performance would degrade.

Media access control is the main reason why Ethernet networks have such exacting Physical layer specifications. If cable segment lengths are too long, or if there are too many repeaters on the network, the CSMA/CD mechanism does not function properly, causing access control to break down and systems to receive corrupt data.

Carrier Sense

The name Carrier Sense Multiple Access with Collision Detection describes the successive phases of the MAC process. When a computer on an Ethernet network has data to transmit, it begins by listening to the network to see if it is in use. This is the carrier sense phase of the process. If the network is busy, the system does nothing for a given period and then checks again.

Multiple Access

The *multiple access phase* is the phase during which the station transmits its data packet when the network is free and all of the stations on the network are contending for access to the same network medium.

A *collision*, also known as a signal quality error (SQE), occurs when two systems on the LAN transmit at the same time. For example, if Computer A performs its carrier sense, and Computer B has already begun transmitting but its signal has not yet reached Computer A, a collision will occur if Computer A transmits. When a collision occurs, both systems must discard their frames and retransmit them. These collisions are a normal and expected part of Ethernet networking. They are not a problem unless there are too many of them or the computers cannot detect them.

Collision Detection

The *collision detection phase* is the part of the CSMA/CD process in which systems detect when their packets collide, avoiding a situation where corrupted data reaches a packet's destination system and is treated as valid. To avoid this, Ethernet networks are designed so that packets are large enough to fill the entire network cable with signals before the last bit leaves the transmitting computer. Ethernet packets must be at least 64 bytes long; systems pad out short packets to 64 bytes before transmission. The Ethernet Physical layer specifications also impose strict limitations on the lengths of cable segments.

The amount of time it takes for a transmission to propagate to the farthest end of the network and back again is called the network's round-trip delay time. A collision can occur only during this interval.

Once the signal arrives back at the transmitting system, that system is said to have captured the network. No other computer will transmit on the network while it is captured because it will detect the traffic during its carrier sense phase.

Ethernet computers on twisted-pair or fiber-optic networks assume that a collision has occurred if they detect signals on both their transmit and receive wires at the same time. If the network cable is too long, if the packet is too short (called a runt), or if there are too many hubs, a system might finish transmitting before the collision occurs and be unable to detect it.

When a computer detects a collision, it immediately stops transmitting data and starts sending a jam pattern instead. The jam pattern alerts the other systems on the network that a collision has taken place, that they should discard any partial packets they may have received, and that they should not attempt to transmit any data until the network has been cleared. After transmitting the jam pattern, the system waits a specified period of time before attempting to transmit again. This is called the backoff period. Both of the systems involved in a collision compute the length of their own backoff periods, using a randomized algorithm called truncated binary exponential backoff. They do this to try to avoid causing another collision by backing off for the same period of time.

With CSMA/CD, the more computers there are on a network segment or the more data the systems transmit over the network segment, the more collisions occur. Collisions are a normal part of Ethernet operation, but they cause delays because systems must retransmit the damaged packets. When the number of collisions is minimal, the delays are not noticeable. But when network traffic increases, the number of collisions increases, and the accumulated delays can begin to have a noticeable effect on network performance. Ways to reduce the traffic on the LAN include installing a bridge or switch or splitting it into two LANs and connecting them with a router.

Modern Ethernet

The IEEE 802.3 standard still includes the term "Carrier Sense Multiple Access with Collision Detection" as part of the document name, but the fact is that very few Ethernet networks actually use CSMA/CD anymore. The need for a MAC mechanism hinges on the use of a shared network medium. The early networks that used coaxial cable connected all of the computers to a single cable segment in a bus topology, and the first twisted-pair networks used hubs to create a star topology.

> **Late Collisions**
>
> It is possible for a collision to occur after the last bit of data has left the transmitting system. This is called a late collision, and it is an indication of a serious problem, such as a malfunctioning network interface adapter, or cable lengths that exceed the Physical layer specifications. Because the transmitting system has no way of detecting a late collision, it considers the packet to have been sent successfully, even though the data has actually been destroyed. As a result, there is no Data Link layer process that can retransmit the data that is lost as a result of a late transmission. It is up to the protocols operating at higher layers of the OSI model to detect the data loss and use their own mechanisms to force a retransmission. This process can take up to 100 times longer than an Ethernet retransmission, which is one reason why this type of collision is a problem. While regular collisions are normal on an Ethernet network and no cause for concern, administrators should diagnose and correct late collisions as quickly as possible.

A *hub* is a multiport repeater. When a hub receives a signal through any of its ports, it transmits that signal out through all of the other ports, resulting in a shared network medium. These networks all required a MAC mechanism to arbitrate access to the network.

The big change came when switches began to replace hubs in the marketplace. As with most new technologies, the first switches were too expensive for all but large network installations. But prices soon dropped and before long, hubs were all but obsolete. On a switched network, unicast data arriving at a switch through one of its ports leaves through only one of its other ports, the one connected to the destination system. As a result, there is no shared network medium. Each pair of computers has a dedicated connection and there is no need for media access control.

Modern Ethernet variants also use full-duplex connections between hosts, which means that computers can transmit and receive data at the same time. This also eliminates the need for media access control. Although the Ethernet standards still include CSMA/CD, for backward compatibility purposes, it is all but obsolete on today's networks.

Physical Layer Specifications

While the Ethernet frame format and the CSMA/CD MAC mechanism have remained relatively stable, the IEEE has revised the Physical layer specifications for the 802.3 standard many times throughout its history. The primary motivation for this is the continual demand for more network transmission speed, which has led the Ethernet development team to increase the speed of Ethernet transmissions by tenfold no fewer than three times in just over 30 years, with a fourth increase on the way. To support these greater speeds, other changes were required as well, both in the nature of the cable and in the signaling the systems use to transmit data.

Currently there are four primary iterations of Ethernet network, represented by their respective transmission speeds of 10, 100, 1,000, and 10,000 Mbps.

10 Mbps Ethernet

The original IEEE 802.3 standard retained the RG-8 coaxial cable specification from the DIX Ethernet document. But by 1993, the IEEE had published several amendments adding unshielded twisted-pair (UTP) and fiber-optic cable specifications, also running at 10 Mbps. **See Figure 4-2.**

✓ **FACT**

10Base-T Ethernet uses only two of the four wire pairs in the UTP cable, one pair for transmitting data and one for receiving it. The other two wire pairs are unused and must remain unused for the network to function properly. The remaining pairs may not be used for telephony, or any other application.

Figure 4-2 10 Mbps Ethernet Physical Layer Specifications

DESIGNATION	CABLE TYPE	TOPOLOGY	MAXIMUM SEGMENT LENGTH
10Base5	RG-8 coaxial	Bus	500 meters
10Base2	RG-58 coaxial	Bus	185 meters
10Base-T	CAT3 UTP	Star	100 meters
FOIRL	62.5/125 multimode fiber-optic	Star	1,000 meters
10Base-FL	62.5/125 multimode fiber-optic	Star	2,000 meters

Figure 4-2. The 10 Mbps standard had multiple amendments to include newer technologies.

The Ethernet fiber-optic specifications offered longer segment lengths and resistance to electromagnetic interference (EMI), but relatively few networks used them, mainly because another Data Link layer protocol—Fiber Distributed Data Interface (FDDI)—was available at the same time and ran over the same type of fiber-optic cable at 100 Mbps, ten times the speed of Ethernet. The Fiber-Optic Inter-Repeater Link (FOIRL) specification was designed to provide long-distance links between repeaters, and 10Base-FL expanded that capability to include fiber-optic links from repeaters to computers as well.

The 5-4-3 Rule

In addition to cable types and segment lengths, the Ethernet Physical layer specifications also limit the number of repeaters permitted in a network configuration. A repeater is a Physical layer device that enables an extension of the length of a network segment by amplifying the signals. The *5-4-3 rule* states that an Ethernet network can have as many as five cable segments, connected by four repeaters, of which three segments are mixing segments.

This rule was originally intended for coaxial networks, on which a mixing segment is defined as a length of cable with more than two devices connected to it. A length of cable with only two devices—that is, a cable connecting two repeaters together—is called a link segment. Using Thin Ethernet, such a network could span as long as 925 meters, while Thick Ethernet could span as long as 2,500 meters. **See Figure 4-3.**

Despite the presence of repeaters, two computers anywhere on this network that transmit at exactly the same time will cause a collision. Therefore, this type of network is said to consist of a single collision domain. If the segments were connected with bridges, switches, or routers instead, there would be multiple collision domains.

On networks using a star topology (such as 10Base-T networks), there are no mixing segments, because all of the cables connect only two devices, but the 5-4-3 rule still applies. A 10Base-T hub functions as a multiport repeater, so a 10Base-T network of the maximum possible size would consist of four connected hubs. Therefore, the longest possible distance between two computers would be 500 meters. **See Figure 4-4.**

The 10Base-FL specification includes some modifications to the 5-4-3 rule. When there are five cable segments present on a 10Base-FL network connected by four repeaters, the segments can be no more than 500 meters long. When four cable segments are connected by three repeaters, 10Base-FL segments can be no more than 1,000 meters long. Cable segments connecting a computer to a hub can also be no more than 400 meters for 10Base-FL.

Normal Link Pulse Signals

Normal link pulse (NLP) signals are what standard Ethernet networks use to verify the integrity of a link between two devices. Most Ethernet hubs and network interface adapters have a link pulse LED that lights when the device is connected to another active device. For example, if a UTP cable that is connected to a hub is plugged into a computer's network adapter, the LEDs on both the adapter and the hub port to which it is connected should light up when the computer is turned on. This is the result of the two devices transmitting

> ✓ FACT
> The IEEE 802.3 standard includes a number of other Physical layer specifications that were either never implemented or never caught on in the marketplace.

NLP signals to each other. When each device receives the NLP signals from the other device, it lights the link pulse LED. If the network is wired incorrectly (because of a cable fault or improper use of a crossover cable or hub uplink port), the LEDs will not light. These signals do not interfere with data communications,

Figure 4-3 5-4-3 Rule on Coaxial Ethernet

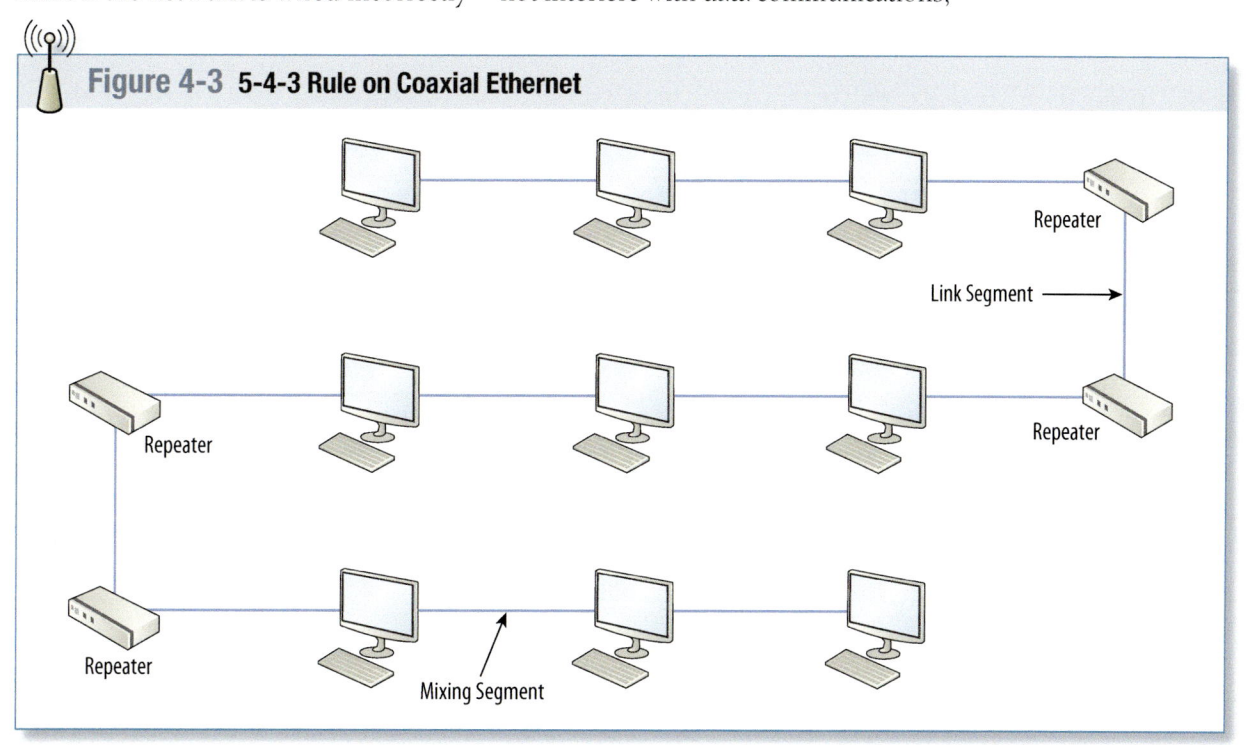

Figure 4-3. The 5-4-3 rule on a coaxial Ethernet network allows cable connecting two repeaters together.

Figure 4-4 5-4-3 Rule on 10Base-T Ethernet

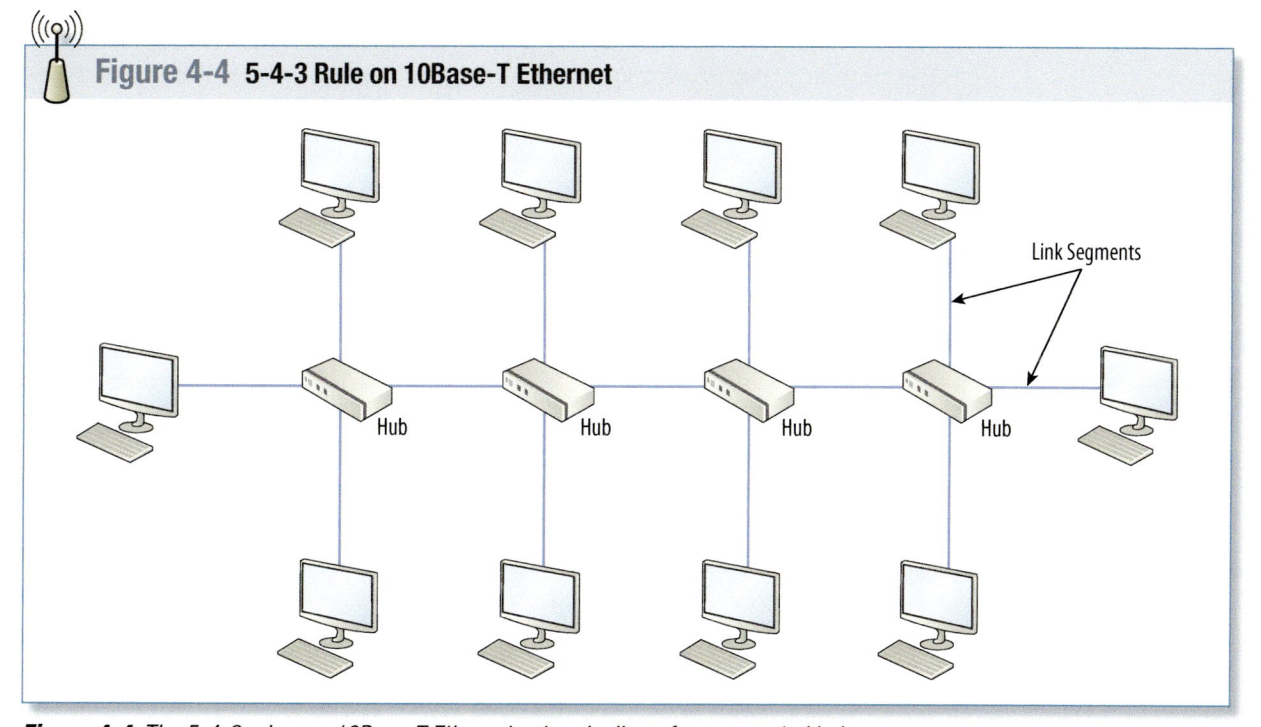

Figure 4-4. The 5-4-3 rule on a 10Base-T Ethernet network allows four connected hubs.

Figure 4-5 100 Mbps Fast Ethernet Physical Layer Specifications

DESIGNATION	CABLE TYPE	MAXIMUM SEGMENT LENGTH
100Base-TX	CAT5 UTP	100 meters
100Base-FX	62.5/125 multimode fiber-optic	412 meters (half duplex)/2,000 meters (full duplex)

Figure 4-5. The Fast Ethernet specification required the use of a higher grade of UTP cable to support the faster transmission speed.

> ✓ FACT
>
> The link pulse LED indicates only that the network is likely wired correctly, not that it is capable of carrying data. Using the wrong cable for the protocol will result in network communications problems, even though the devices might pass the link integrity test.

because the devices transmit them only when the network is idle.

Fast Ethernet

The 100Base-TX specification retains the 100-meter maximum segment length from 10Base-T, as well as the use of two wire pairs. However, to support the higher transmission speeds, the standard calls for a higher grade of cable—CAT5 instead of CAT3. **See Figure 4-5.**

On the fiber-optic side, 100Base-FX uses the same 4B/5B signaling method as 100Base-TX, and as a result, the two specifications are known collectively as 100Base-X. The maximum segment length on a 100Base-FX network depends both on the cable type and the use of full-duplex communications. Using standard multimode fiber-optic cable and full-duplex communication, a segment can be as long as 2,000 meters.

Half-duplex signaling reduces the length to 412 meters. Using singlemode cable, a full-duplex segment can be 20 kilometers long or more.

It has long been a standard practice in local area networking to connect multiple LANs together with a backbone network. Because the backbone must carry internetwork traffic from all of the LANs, many administrators run it at a higher speed than the horizontal LANs. Before the introduction of Fast Ethernet, FDDI (or its copper alternative, CDDI) was the most common high-speed backbone solution. Fast Ethernet enabled administrators to build backbones using roughly the same technology as their horizontal networks.

The 5-4-3 rule does not apply to Fast Ethernet networks. Fast Ethernet hubs are available in two classes. Class I connects different types of Fast Ethernet cable

Fast Ethernet and Gigabyte Ethernet

When Fast Ethernet first appeared, its speed seemed wondrous, and most administrators assumed that it would be used only on backbones and other networks requiring higher performance levels. However, the prices for Fast Ethernet network interface cards, hubs, and switches began to drop precipitously, and hardware manufacturers began making dual-speed equipment that could automatically sense the speed of the network and adjust itself accordingly. Before long, administrators began to realize the advantages of running Fast Ethernet to the desktop, and it quickly became the standard for the industry.

The introduction and acceptance of Gigabit Ethernet followed the same pattern. What seemed at first to be a backbone-only technology has been rapidly accepted for use on the desktop. Most new computers sold today include an integrated network interface adapter that can connect to an Ethernet network running at 10, 100, or 1,000 Mbps Fast Ethernet Cabling Limitations.

segments together, such as fiber optic to UTP or 100Base-TX to 100Base-T4. Class II connects Fast Ethernet cable segments of the same type together.

Each Fast Ethernet hub must be identified by the appropriate Roman numeral. As many as two Class II hubs can be included on a single LAN, with a total cable length (for all three segments) of 205 meters for UTP cable and 228 meters for fiber-optic cable. **See Figure 4-6.**

Because Class I hubs must perform an additional signal translation (which slows down the transmission process), only one hub is allowed on the network, with maximum cable lengths of 200 and 272 meters for UTP and fiber optic, respectively.

Full-Duplex Ethernet

The CSMA/CD MAC mechanism is the defining element of the Ethernet protocol, but it is also the source of many of its limitations. The fundamental shortcoming of the Ethernet protocol is that data can travel in only one direction at a time. This is known as half-duplex operation. With special hardware, it is also possible to run Ethernet connections in full-duplex mode, meaning that the device can transmit and receive data simultaneously. This effectively doubles the bandwidth of the network. Full-duplex capability for Ethernet networks was standardized in the 802.3x amendment to the 802.3 standard in 1997.

When operating in full-duplex mode, Ethernet systems ignore the CSMA/CD MAC mechanism. Computers do not listen to the network before transmitting; they simply send their data whenever they want to. Because both of the systems in a full-duplex link can transmit and receive data at the same time, there is no possibility of collisions occurring. Because no collisions occur, the cabling restrictions designed to support the collision detection mechanism are unnecessary. This means that, in many cases, longer cable segments can be used on a full-duplex network. The only limitation is the signal-transmitting capability (that is, the rate of attenuation) of the network medium itself.

Figure 4-6 Fast Ethernet

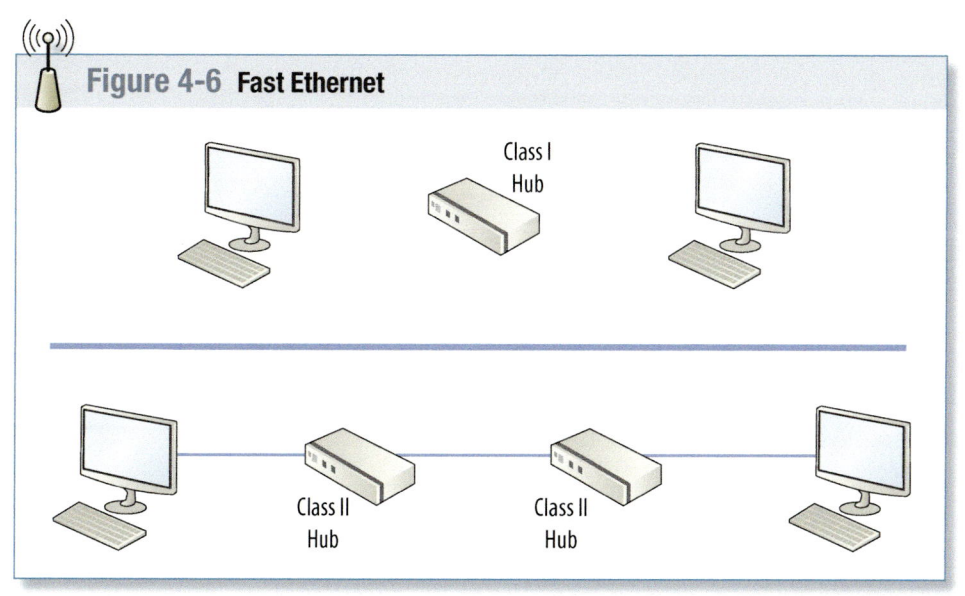

Figure 4-6. Fast Ethernet cabling guidelines allow as many as two Class II hubs to be included.

Gigabit Ethernet

The Physical layer specifications for Gigabit Ethernet, running at 1,000 Mbps, appeared in 1998 and 1999. The fiber-optic specifications (known collectively as 1000Base-X) were published as IEEE 802.3z, and the 1000Base-T UTP specification as IEEE 802.3ab. **See Figure 4-7.**

Gigabit Ethernet once again increased network speeds tenfold, while retaining the same basic UTP configuration and segment length, making upgrades possible in many cases without the need for new cable installations. As with each previous Ethernet speed increase, Gigabit Ethernet increased the requirements for the UTP cables. 1000Base-T uses all four wire pairs (unlike 100Base-TX) and is more susceptible to certain types of crosstalk. To address this issue, the Telecommunications Industry Association/Electronic Industries Alliance (TIA/EIA) created the CAT5e and CAT6 cable grades, which are designed to support Gigabit Ethernet communications.

The 1000Base-X specifications include two fiber-optic configurations, essentially long- and short-distance options, plus a unique, short-run copper alternative. 1000Base-LX is intended to be the long-distance option, supporting segment lengths up to 5 kilometers using singlemode fiber, and up to 10 kilometers using high-quality optics in a variant called 1000Base-LX10. Some specialized installations also use repeating equipment to create much longer links.

1000Base-LX is designed to be used by large carriers in a long-distance Ethernet backbone, so it is not likely that the average network administrator will ever work with it or even find 1000Base-LX network interface adapters on the shelf at a local computer store.

The 1000Base-LX specification also allows for the use of multimode fiber-optic cable at shorter distances, but 1000Base-SX is the shorter-distance specification designed for virtually any type of multimode cable. As with most fiber-optic alternatives, 1000Base-SX works well as a link between buildings and on-campus networks.

1000Base-CX is a copper specification calling for a special 150-ohm cable with two copper cores, twisted and shielded, with either an 8P8C or DE-9 connector, called a twinaxial cable. The maximum segment length is only 25 meters, making the specification good for links within data centers (such as equipment connections within server

> ✓ **FACT**
>
> The TIA also created an alternative Gigabit Ethernet Physical layer specification called 1000Base-TX. Like 100Base-TX, 1000Base-TX uses only two of the four wire pairs in a UTP cable. However, to compensate for that lack of two wire pairs, the 1000Base-TX specification requires CAT6 UTP cabling.

Figure 4-7 1,000 Mbps Gigabit Ethernet Physical Layer Specifications

DESIGNATION	CABLE TYPE	MAXIMUM SEGMENT LENGTH
1000Base-T	CAT5, CAT5e, or CAT6 UTP	100 meters
1000Base-LX	9/125 singlemode fiber-optic	5,000 meters
1000Base-LX	50/125 or 62.5/125 multimode fiber-optic	550 meters
1000Base-SX	50/125 or 62.5/125 multimode fiber-optic	500 meters/220 meters
1000Base-CX	150-ohm shielded, balanced twinaxial copper cable	25 meters

Figure 4-7. The Gigabit Ethernet specification added CAT5e and CAT6 UTP cables to support the higher speeds.

clusters) and little else. At the time the IEEE published the 802.3z document, 1000Base-CX was the only copper cable specification available, but the 1000Base-T specification appeared a year later, leaving 1000Base-CX as a marginal technology.

Gigabit Backbones

Gigabit Ethernet was virtually assured of a place in the networking market because, like Fast Ethernet before it, it uses the same frame format, frame size, and MAC mechanism as standard 10 Mbps Ethernet. Fast Ethernet quickly replaced FDDI as the dominant 100 Mbps solution because it prevented network administrators from having to use a different protocol on the backbone. In the same way, Gigabit Ethernet can prevent administrators from having to use a different protocol like Asynchronous Transfer Mode (ATM) for their backbones.

To connect an ATM or FDDI network to an Ethernet network requires that the intervening router convert the data at the Network layer from one frame format to another. Connecting two Ethernet networks together, even when they are running at different speeds, is a Data Link layer operation because the frames remain unchanged. In addition, using Ethernet throughout a network eliminates the need to train administrators to work with a new protocol and purchase new testing and diagnostic equipment. In most cases, it is possible to upgrade a Fast Ethernet backbone to Gigabit Ethernet without completely replacing the hubs, switches, and cables.

However, it is possible that some hardware upgrades will be necessary. Modular hubs and switches will need modules supporting the new protocol, and networking monitoring and testing products may also have to be upgraded to support the higher speed.

As with Fast Ethernet, Gigabit Ethernet emerged as a backbone technology, but it was not long before triple-speed 10/100/1,000 network interface adapters and switches were cheap and plentiful. Administrators who at one time scoffed at the idea of running 1,000 Mbps connections to the desktop were soon doing just that. Virtually every personal computer sold today includes a Gigabit Ethernet network interface adapter.

Gigabit Refinements

Gigabit Ethernet is designed to operate in full-duplex mode on switched networks. Full-duplex communication eliminates the need for the CSMA/CD MAC mechanism. For backward-compatibility purposes, though, Gigabit Ethernet continues to support hub-based networks and half-duplex communication.

For systems on a Gigabit Ethernet network to operate in half-duplex mode, it was necessary for the specifications to modify the CSMA/CD mechanism. Ethernet's collision-detection mechanism works properly only when collisions are detected while a packet is still being transmitted. Once the source system finishes transmitting a packet, the data is purged from its buffers, and it is no longer possible to retransmit that packet in the event of a collision.

When the speed at which systems transmit data increases, the round-trip signal delay time during which systems can detect a collision decreases. When Fast Ethernet increased the speed of an Ethernet network by ten times, the specification compensated by reducing the maximum diameter of the network.

This enabled the protocol to use the same 64-byte minimum packet size as the original Ethernet standard and still be able to detect collisions effectively.

Gigabit Ethernet increases the transmission speed another ten times, but reducing the maximum diameter of the network again was impractical because it would result in networks no longer than 20 meters or so. As a result, the 802.3z supplement increases the size of the CSMA/CD carrier signal from 64 bytes to 512 bytes. This means that while the 64-byte minimum packet size is retained, the MAC sublayer of a half-duplex Gigabit Ethernet system appends a carrier extension signal to small packets that pads them out to 512 bytes. This ensures that the minimum time required to transmit each packet is sufficient for the collision detection mechanism to operate properly, even on a network with the same diameter as Fast Ethernet.

Autonegotiation

Backward compatibility has always been a major priority with the designers of the IEEE 802.3 standards. Most of the Ethernet networking hardware on the market today enables a computer to connect to the network at Gigabit Ethernet speed or negotiate a slower-speed connection, if that is all the network supports.

The *autonegotiation system* is an optional Fast Ethernet specification that enables a dual-speed device to sense the capabilities of the network to which it is connected and then adjust its speed and duplex status accordingly. The Gigabit Ethernet specifications expand the capabilities of the autonegotiation system, enabling devices to also communicate their port type and master-slave parameters. In Gigabit Ethernet using copper cable, support for autonegotiation is mandatory.

When two Ethernet devices capable of operating at multiple speeds autonegotiate, they exchange packets to determine the best performance level they have in common and configure themselves accordingly. The systems use a list of priorities when comparing their capabilities, with full-duplex 1000Base-T providing the best performance and half-duplex 10Base-T providing the worst:

1. 1000Base-T (full-duplex)
2. 1000Base-T (half-duplex)
3. 100Base-TX (full-duplex)
4. 100Base-T4
5. 100Base-TX (half-duplex)
6. 10Base-T (full-duplex)
7. 10Base-T (half-duplex)

Autonegotiation permits administrators to upgrade a network gradually with a minimum of reconfiguration. For example, when 10/100/1,000 multispeed network adapters are used in all workstations, the network can be run at 100 Mbps using 100Base-TX switches. Later, if the switches are replaced with models supporting Gigabit Ethernet, the network adapters will automatically reconfigure themselves to operate at the higher speed during the next system reboot. No manual configuration at the workstation is necessary.

10 Gigabit Ethernet

When more Ethernet bandwidth was needed, the IEEE 802.3 working group responded in 2002 with the first standards defining an Ethernet network running at 10 Gbps (or 10,000 Mbps). With 10 Gigabit Ethernet, the developers appear to have reached a turning point, because they have abandoned their previous devotion to backward compatibility. 10 Gigabit Ethernet networks support only four-pair, full-duplex communication on switched networks. Gone is the support for half-duplex communication, hubs, and CSMA/CD. However, the standard Ethernet frame format remains, and as with all of the previous Ethernet standards, there is a copper-based UTP solution using 8P8C connectors and a 100 meter maximum segment length.

As with each of the previous Ethernet speed iterations, the 10 Gigabit Ethernet standards include a variety of Physical layer specifications. Some of them have already fallen by the wayside, but the technology is still young enough that the marketplace has not yet completed the process of winnowing out the unsuccessful ones. **See Figure 4-8**.

The designers of 10 Gigabit Ethernet intended it to be both a LAN and a WAN solution, and for many administrators, LAN means copper-based UTP cables with a 100-meter maximum segment length. It was not until 2006 that the IEEE published the 802.3 amendment, which defined the 10Gbase-T specification, but they knew it had to be done.

To support transmissions at such high speeds with copper cables, it was necessary to define a new set of cable performance standards. To support 100-meter segments, 10Gbase-T requires CAT6a cable, which has an increased resistance to alien crosstalk (that is, interference from signals on other nearby cables). With standard CAT6 cable, 10Gbase-T only supports cable segments up to 55 meters long. The standard does not support UTP cables below CAT6 at all.

The 10 Gigabit Ethernet specifications for fiber-optic cable predate the copper and provide a wide variety of options for both LAN and WAN implementations. The possibilities of 10 Gigabit Ethernet as a WAN solution led the developers to create a separate set of specifications that utilize the existing Synchronous Optical Network (SONET) infrastructure to carry Ethernet signals.

There are three pairs of fiber-optic specifications that all begin with the

Figure 4-8 10 Gigabit Ethernet Physical Layer Specifications

DESIGNATION	CABLE TYPE	WAVELENGTH IN NANOMETERS (NM)	MAXIMUM SEGMENT LENGTH
10Gbase-T	CAT6/CAT6a	N/A	55 meters/100 meters
10Gbase-SR	Multimode fiber-optic	850 nm	26 – 400 meters
10Gbase-LR	Singlemode fiber-optic	1,310 nm	10 kilometers
10Gbase-ER	Singlemode fiber-optic	1,550 nm	40 kilometers
10Gbase-SW	Multimode fiber-optic	850 nm	26 – 400 meters
10Gbase-LW	Singlemode fiber-optic	1,310 nm	10 kilometers
10Gbase-EW	Singlemode fiber-optic	1,550 nm	40 kilometers

Figure 4-8. *10 Gigabit Ethernet requires the use of CAT6/CAT6a to support the higher speeds in UTP installations.*

10Gbase abbreviation. The first letter of the two-letter code that follows specifies the type and wavelength of the fiber-optic cable. The second letter indicates whether the specification is intended for LAN use ("R") or WAN use ("W").

The "S" in the 10Gbase-SR and 10Gbase-SW specifications describes the short wavelength (850 nanometers) of the lasers used to generate the signals on the cable. As with most short-range fiber-optic solutions, these specifications call for multimode cable. The maximum segment length depends on the exact cable the network uses. For example, the 62.5 micron multimode fiber commonly used on FDDI networks (OM1) can only support segments 26 meters long. With the newly ratified OM4 cable, 10Gbase-SR segments can be as long as 400 meters.

The 10Gbase-LR and 10Gbase-LW specifications use a long wavelength laser (1,310 nm) and singlemode cables to achieve segment lengths of 10 kilometers. The extra-long wavelength of the 10Gbase-ER and 10Gbase-EW specifications can support segments up to 40 kilometers long.

None of the 10 Gigabit Ethernet Physical layer specifications indicate the types of connectors the cables should use. The actual implementation is left up to the equipment manufacturers.

10 Gigabit Ethernet Physical Implementations

With its many supported cable types and no standard connector specifications, designing, marketing, and manufacturing 10 Gigabit Ethernet equipment might seem to be a daunting task. To address this problem, networking equipment manufacturers have devised a new solution. Instead of building a specific Physical layer interface into a networking device (thus requiring multiple versions to support many different media), manufacturers have started building devices with a standard socket, into which consumers can plug a Physical layer (PHY) module that contains a transceiver and supports the desired cable and connector.

The sockets are defined by multi-source agreements (MSAs) that are not independently standardized, but which are agreed upon by groups of manufacturers. Some of the MSAs that manufacturers of 10 Gigabit Ethernet equipment most commonly use are SFP+, XFP, and XENPAK.

Summary

Ethernet was the first and is still the most popular LAN protocol. The first Ethernet standards were based on coaxial cable, but these gave way to networks based on twisted-pair and fiber-optic cables. The basic components of the Ethernet protocol are a frame format that specifies how the data units are organized, a media access control (MAC) mechanism called Carrier Sense Multiple Access with Collision Detection, and a series of Physical layer specifications that define the construction of Ethernet networks.

Review Questions

1. Which of the fields in a DIX Ethernet frame contains a value identifying the Network layer protocol contained in the frame?
   a. Ethertype field
   b. Frame Check Sequence
   c. Preamble
   d. Source Address

2. Which of the Gigabit Ethernet Physical layer specifications calls for copper cables?
   a. 10Base-T
   b. 100Base-TX
   c. 1000Base-T
   d. 10Gbase-T

3. What is the maximum length of a 1000Base-T cable segment?
   a. 100 meters
   b. 185 meters
   c. 500 meters
   d. 1,000 meters

4. The first Ethernet networks called for use of Shielded Twisted Pair cable.
   a. True
   b. False

5. What is the term for a packet on an Ethernet network that is too short?
   a. Dwarf
   b. Half-frame
   c. PDU
   d. Runt

Understanding Wireless Networking

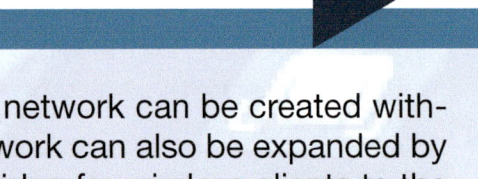

Using the concept of wireless networks, a network can be created without running cables. An existing wired network can also be expanded by adding a wireless access point (WAP) as a bridge for wireless clients to the wired network. To facilitate this requires an understanding of current wireless networking standards and an understanding of wireless security methods.

The current industry standard for wireless LANs is IEEE 802.11, which includes several different wireless standards, such as 802.11a, b, g, n, and ac. Getting the most out of a wireless network requires using compatible protocols. Some work together, but others do not. IEEE 802.11ac provides the greatest flexibility and speed.

Wireless security had a rocky start, and early wireless security methods were not secure at all. However, wireless security has increased significantly over the years, and it is possible to create a more secure wireless network today—with the right know-how.

When networks in buildings are separated by long distances, they can be connected with point-to-point wireless bridges, even if the buildings are miles away.

OBJECTIVES

- ▶ Describe the components of basic wireless networks
- ▶ Identify networking standards and their characteristics
- ▶ Explain various network security methods
- ▶ Compare wireless and wired networks
- ▶ Define point-to-point wireless

CHAPTER 5

TABLE OF CONTENTS

Basic Wireless Components 90
    Wireless Access Points 90
    Wireless Network Names. 91
    CSMA/CD Versus CSMA/CA 92

Networking Standards and
Characteristics 92
    FHSS, DSSS, and OFDM. 93
    IEEE 802.11. 94
    IEEE 802.11a. 95
    IEEE 802.11b. 95
    IEEE 802.11g. 96
    IEEE 802.11n. 96
    IEEE 802.11ac. 96

Network Security Methods 96
    Wired Equivalent Privacy 98
    Wi-Fi Protected Access 98
    WPA2 99
    IEEE 802.1x Authentication Server. . . 100

Wireless Networks. 101
    Home Wireless Networks. 101
    Wireless Networks
    in a Business 103

Point-to-Point Wireless 104

Summary. 106

Review Questions 107

BASIC WIRELESS COMPONENTS

Wireless networking is virtually everywhere today—in homes, airports, restaurants, and hotels. There are even some cities that offer citywide wireless Internet access. With newer technologies, wireless speeds are approaching that of Gigabit Ethernet.

Wireless standards and security methods are very important concepts in wireless networks. However, in order to grasp these concepts, the network technician must understand fundamental concepts related to basic wireless networks including:

- Using wireless access points and adapters
- Naming the wireless network
- Comparing CSMA/CD and CSMA/CA

Wireless Access Points

A *wireless access point (WAP)* is a device that is located between a wired LAN and wireless clients. It bridges the two networks, giving the wireless clients access to the wired network. When a WAP is used, the wireless network is functioning in infrastructure mode. **See Figure 5-1.**

Once the wireless clients connect, they are able to access resources on the wired network through the WAP. The number of clients that can be connected to the WAP depends on the available bandwidth. As more wireless clients are added, performance slows down for all of them. Just as with wired networks, high-bandwidth speeds are desirable on wireless networks. Different standards support different speeds, with 802.11ac providing the best performance today.

Wireless clients have wireless adapters that must be configured to connect to the WAP. Many laptops include a built-in adapter, but there are also Universal Serial Bus (USB)

> ✓ **FACT**
> When a WAP is not used, wireless clients connect to each other using ad hoc or peer-to-peer mode. Ad hoc wireless connections have additional security risks beyond WAP-based networks.

Figure 5-1 Wireless Access Point (WAP)

Wireless Access Point (WAP)

Wired Network

Infrastructure Mode

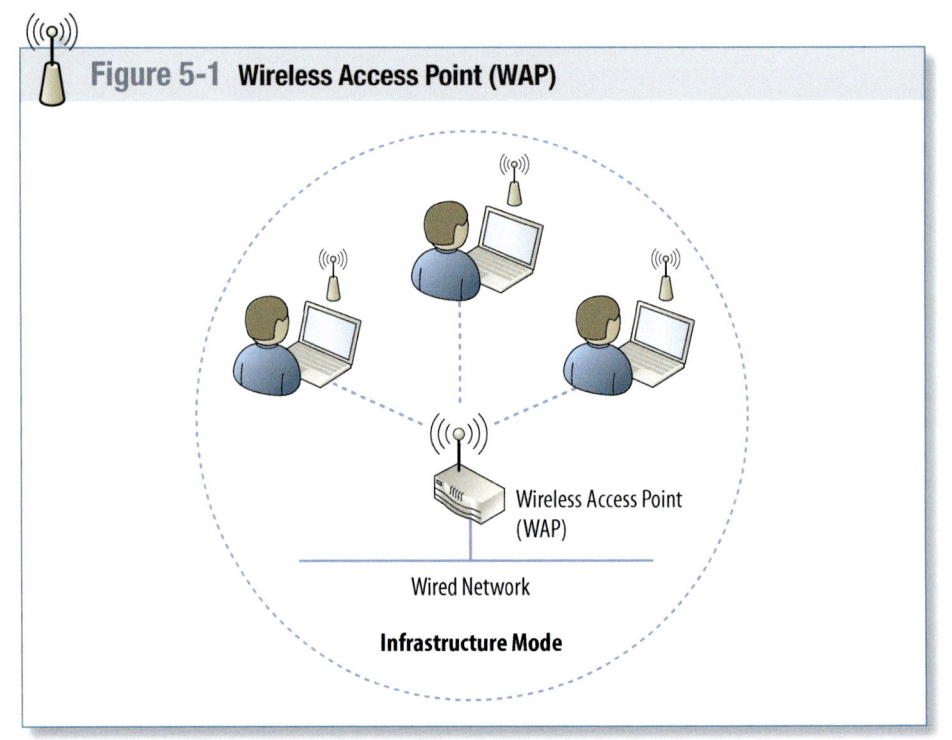

Figure 5-1. *A wireless network that uses a WAP runs in infrastructure mode.*

wireless adapters that can be plugged into a USB port, and adapter cards that can be plugged into a slot inside the computer.

Wireless Network Names

A *service set identifier (SSID)* is simply a name that someone has assigned to the network. Every wireless network includes an SSID. A wireless network name can be just about anything as long as the name does not exceed the maximum length of 32 characters.

Any wireless device that connects to the WAP uses the SSID, and the SSID is one of the primary items that must be known when configuring wireless devices. Most WAPs include a set-up screen that provides a means to name the SSID. **See Figure 5-2.**

Most WAPs provide the capability to turn SSID broadcasting off or leave it on. When SSID broadcasting is on, the WAP broadcasts the name of the wireless network. Other wireless devices can

What Is the Difference Between a WAP and a Wireless Router?

A WAP provides connectivity to a wired network for wireless clients. Think of this as a bridge between the wireless clients and the wired clients.

In contrast, a wireless router is a WAP with additional components. It includes routing components to route traffic between different networks (such as from the Internet through an ISP to a private network). The wireless router often includes a switch component to enable wired devices to plug in to the wireless router and provide connectivity for them.

When wireless connectivity is needed in an enterprise, a simple WAP (instead of a wireless router) will often be sufficient. The WAP connects the wireless devices to the wired network. Other devices on the wired network provide services, such as routing and Internet access.

In summary, a wireless router always includes the basic capability of a WAP in addition to routing capabilities. It usually includes even broader capabilities such as that of a switch and a Dynamic Host Configuration Protocol (DHCP) server. However, a WAP does not include these additional capabilities.

Figure 5-2 Service Set Identifier (SSID)

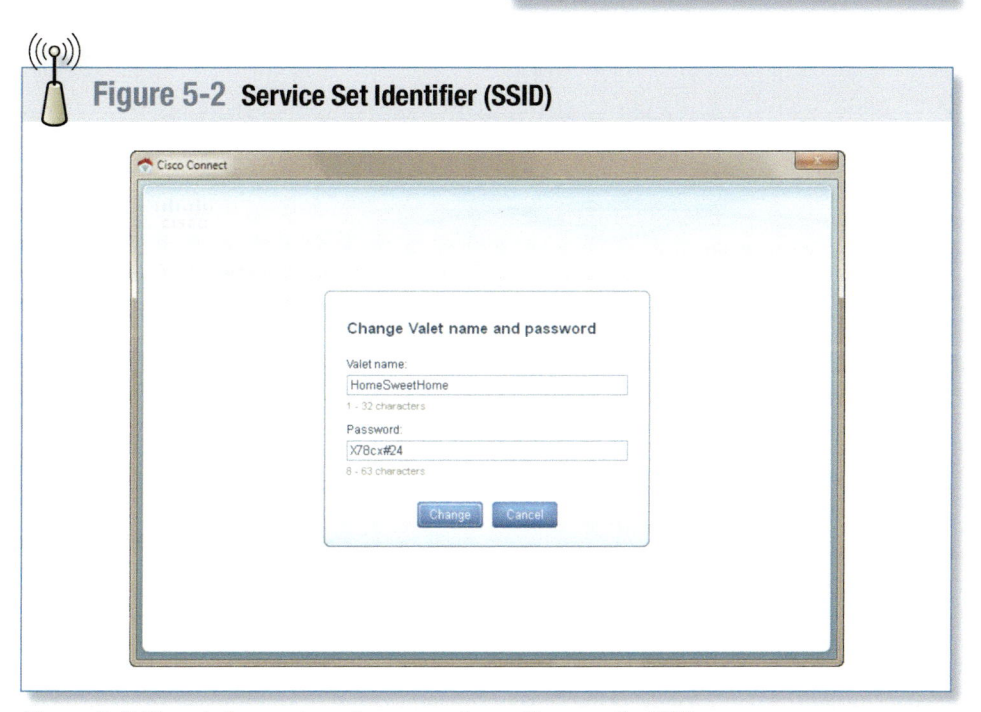

Figure 5-2. The wireless router setup screen is used to name the SSID.

easily see it and connect to it, as long as the other security settings are properly connected.

There was a time when IT professionals consistently recommended disabling SSID broadcasts. However, Microsoft recommends not disabling SSID broadcasts and configuring the WAP to broadcast its SSID.

Disabling SSID Broadcasting Does Not Enhance Security Wireless security is primarily provided by authentication and encryption. Disabling the SSID broadcasts does not help or hinder either authentication or encryption.

Disabling SSID Broadcasting Does Not Truly Hide the SSID Since the frequency ranges that different wireless protocols use are well known, any receiver can capture frames sent by the wireless devices. The SSID is included in the probe requests that clients send, and attackers can use wireless sniffers to discover the SSID.

Disabling SSID Broadcasting Requires Clients to Broadcast the SSID If SSID broadcasting is disabled on the WAP, clients must initiate the connection. Since clients do not know whether they are close to a wireless network, they must constantly send out probes looking for WAPs until they connect. When a client is away from the network (such as in a coffee shop, hotel, or airport), it sends out probes as often as every 30 seconds with the SSID name. Although it is possible to disable the client's automatic connection feature, this requires additional work on the part of the user.

Configure the wireless device to use the same security protocol that the WAP uses. This usually includes setting the passphrase and configuring it to use WPA2.

CSMA/CD Versus CSMA/CA

Ethernet uses Carrier Sense Multiple Access with Collision Detection (CSMA/CD) as its media access control (MAC) mechanism. If a collision occurs, the systems detect it, and the two parties then retransmit the data.

Wireless networks cannot detect collisions, so they use Carrier Sense Multiple Access with Collision Avoidance (CSMA/CA) instead. CSMA/CA prevents a wireless node from transmitting when another node is doing so.

In other words, if one computer wants to send data to another, it will first listen to see whether anyone else is transmitting. If no other device is transmitting, it will send data. However, if it hears data transmissions, it will wait for a random period and recheck the airwaves. This method of transmission reduces the chance of a collision in a wireless environment.

An optional method of improving this process is with Request to Send/Clear to Send (RTS/CTS) packets. **See Figure 5-3.**

PC-1 first sends an RTS frame to the other computer, asking whether it is clear. PC-2 then sends back a CTS frame indicating it is clear to send. All nodes within hearing distance (including the intended recipient) allow the sender adequate time to send the packet.

NETWORKING STANDARDS AND CHARACTERISTICS

Although wireless technologies have grown significantly in the past few years, there are not that many standards in common use. **See Figure 5-4.**

✓ **FACT**
The RTS/CTS process is not required. Most wireless devices support adding it, if required, to decrease collisions.

Figure 5-3 Request to Send/Clear to Send (RTS/CTS)

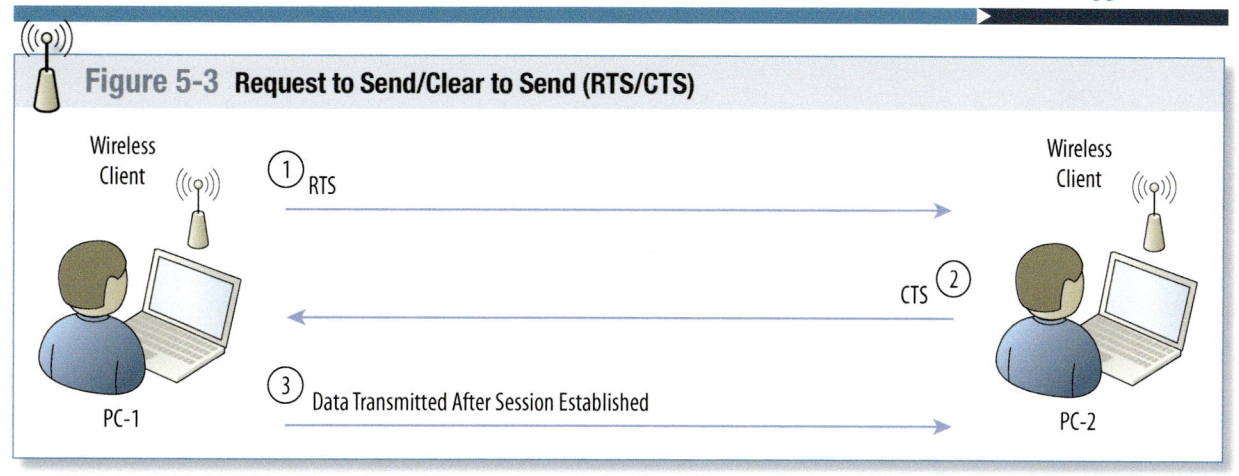

① RTS
② CTS
③ Data Transmitted After Session Established

PC-1 — PC-2

Figure 5-3. The RTS/CTS process reduces collisions on a wireless network.

Figure 5-4 Current Wireless Standards

STANDARD	SPEED IN MEGABITS PER SECOND	FREQUENCY	COMMENTS
802.11a	54 Mbps	5 GHz	Less susceptible to interference
802.11b	11 Mbps	2.4 GHz	Can configure specific channels
802.11g	54 Mbps	2.4 GHz	Widely deployed
802.11n	300 Mbps	2.4 GHz or 5 GHz	Uses MIMO antenna technology
802.11ac	500 Mbps	5 GHz .	Newer and currently most popular

Figure 5-4. As wireless technologies improve, new standards are developed.

FHSS, DSSS, and OFDM

All devices that use radio frequency (RF) signals are susceptible to interference. For example, some cordless phones use the same frequency band as many wireless networking devices. When more than one device transmits on the same frequency at the same time, it causes interference. For a wireless LAN, this interference can negatively affect performance.

To combat the interference problems, wireless technologies use different methods for transmitting data on these bands. Transmission methods include:

▶ Frequency-hopping spread spectrum (FHSS)

▶ Direct-sequence spread spectrum (DSSS)

▶ Orthogonal frequency division multiplexing (OFDM)

FHSS hops between frequencies in a pseudorandom pattern. It starts with a center frequency known to the transmitter and receiver and then quickly changes between different frequencies. The transmitter and receiver synchronize these hops so they know what frequency is next.

These random frequencies are in 1 MHz increments and do not use more than 1 MHz at any given time. FHSS was introduced with the original 802.11 specification, but it is not used with any of the current IEEE 802.11 specifications. It is used with Bluetooth wireless networking.

✓ **FACT**
FHSS makes it difficult for unintended recipients to receive the data, but it is not that effective at limiting interference problems.

✓ **FACT**
Bluetooth is a wireless technology used for short-distance personal area networks (PANs). A PAN transmits data to devices that a person is carrying or wearing.

IEEE 802.11b uses DSSS. It uses the full bandwidth (or spectrum) of the transmitted frequency and can use one of 11 possible channels in the United States. **See Figure 5-5.**

Each DSSS channel has a spectrum of 22 MHz. Therefore, devices can use channel 1, channel 6, and channel 11 without interfering with each other. This is useful when multiple WAPs for different networks are located close to each other. DSSS uses the center frequency of the channel and then modulates the signal out from the center frequency, consuming the entire 22 MHz spectrum. DSSS is resistant to interference, and it enables multiple users to share a single channel.

OFDM splits the radio frequency signal into smaller subsignals and transmits data simultaneously across these different frequencies. Each subsignal includes a separate data stream. Compare this to the multiplexing process used with cable television. A single cable includes multiple television channels, and a television can tune to any single channel. 802.11a, 802.11g, and 802.11ac use OFDM. 802.11n and 802.11ac use an enhanced OFDM by combining it with multiple antennas.

IEEE 802.11

IEEE created 802.11 as the first wireless LAN standard in 1997. It maxed out at a speed of 2 Mbps, with an actual throughput of less than 0.7 Mbps, which was simply too slow for most applications. The 802.11 standard calls for the use of FHSS.

IEEE 802.11 is also highly susceptible to radio interference from other devices using the 2.4 GHz frequency. This includes devices such as baby monitors, cordless telephones, video cameras, microwave ovens, and Bluetooth devices.

Combined with the slow speed and high susceptibility to interference, the original specification was never widely adopted.

> ✓ FACT
> WAPs used as wireless repeaters use the same center channel. WAPs physically close to other WAPs for different wireless networks use different channels.

For additional information, visit qr.njatcdb.org Item #1792

For additional information, visit qr.njatcdb.org Item #1793

Figure 5-5 Direct Sequence Spread Spectrum (DSSS)

Channel 1 Frequency Spectrum — 22 MHz

Channel 6 Frequency Spectrum — 22 MHz

Channel 11 Frequency Spectrum — 22 MHz

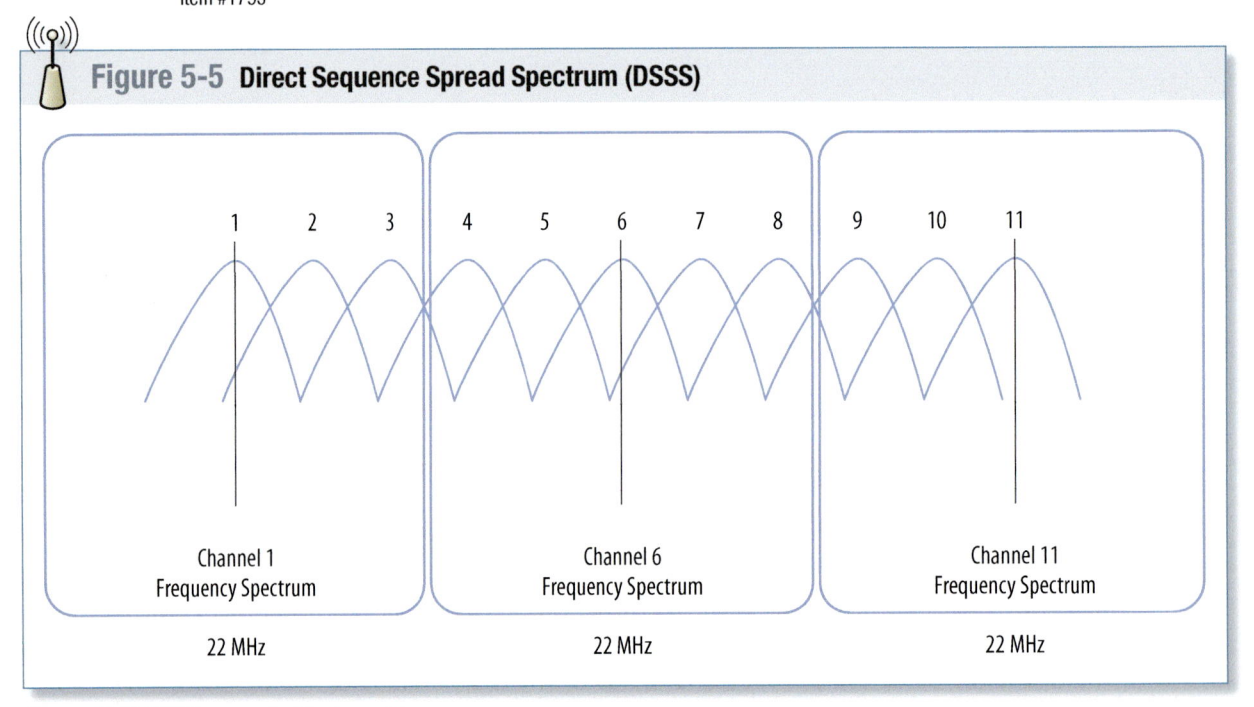

Figure 5-5. DSSS supports 11 channels.

Wireless Governing Bodies

Four governing bodies overlook wireless technology. They are the Federal Communications Commission (FCC), the Institute of Electrical and Electronics Engineers (IEEE), the International Organization for Standardization (ISO), and the Wi-Fi Alliance.

FCC The Federal Communications Commission (FCC) regulates wireless frequencies and modulation types. The FCC also regulates the use of unlicensed Instrument, Scientific, and Medical (ISM) frequency bands that are used in Wi-Fi communications.

IEEE The IEEE sets the ISO standards for the wireless IEEE 802.11 family to ensure consistency.

Wi-Fi Alliance The Wi-Fi Alliance is the trade association that promotes wireless technology. It approves products that meet their interoperability guidelines, and these products can use the Wi-Fi logo.

For additional information, visit qr.njatcdb.org Item #1794

For additional information, visit qr.njatcdb.org Item #1796

IEEE 802.11a

The IEEE designed 802.11a with a different frequency band to avoid the interference in the crowded 2.4 GHz frequency band. Instead of 2.4 GHz, IEEE uses 5 GHz. It can achieve a raw speed of 54 Mbps, which was significantly higher than the maximum of 2 Mbps for 802.11. However, the higher frequency of 5 GHz had a trade-off of a shorter range.

The advertised range of 802.11a is approximately 30 meters (100 feet). However, the full 54 Mbps speed can only be achieved at close ranges of between 50 and 100 feet.

This different frequency also caused logistical problems. The 5 GHz components were difficult to manufacture, and first-generation components often did not live up to the advertised specifications. Because of this, the release of 802.11a products was slow. IEEE 802.11a and IEEE 802.11b actually made it to market at about the same time.

IEEE 802.11b

IEEE 802.11b was another improvement over the original IEEE 802.11 specification. Like 802.11, it operates at 2.4 GHz, but increases the speed from 2 Mbps to a speed of 11 Mbps. This was a major performance increase at the time of the standard's introduction into the marketplace, and the migration to 802.11b was rapid.

Recall 802.11b uses DSSS, which improved the reliability of the signals. Also, DSSS has configurable channels. For example, if one neighbor is using channel 6 at full power, another neighbor

Wireless Speeds and Distances

The advertised speeds of different wireless devices represent maximum speeds in ideal conditions. In the real world, these speeds are rarely achievable.

For example, IEEE 802.11g advertises a speed of 54 Mbps. If the wireless access point is 5 feet away from the wireless device, a speed of 54 Mbps can probably be achieved. However, when the wireless device is moved farther and farther away, at some point, errors creep into the transmission.

Devices automatically correct for errors by slowing down the transmission speed. If there are errors at 54 Mbps, the devices try slower and slower speeds until they are able to achieve an error-free transmission. In other words, depending on the distance, interference from other transmissions, and obstructions, an advertised speed of 54 Mbps could be reduced to 6 Mbps.

✓ **FACT**
Because of the manufacturing problems with 802.11a's 5 GHz components, more users adopted 802.11b than 802.11a.

✓ **FACT**
The maximum range of wireless network devices is achievable only in ideal conditions. Obstructions such as walls and trees absorb the signal, reducing the transmission range.

> ✓ FACT
>
> 802.11b devices can work with 802.11g devices, but connections between them will operate at the slower 11 Mbps speed.

> ✓ FACT
>
> Many companies wanted to grab market share by being first to market with 802.11n products. By late 2009, there were already many 802.11n devices available.

> ✓ FACT
>
> Some devices are advertised as a/b/g compatible. They can operate at 5 GHz for 802.11a devices and operate at 2.4 GHz for 802.11b and 802.11g devices.

> ✓ FACT
>
> Actual data throughput for IEEE 802.11n is usually closer to 180 Mbps, but this is still much faster than a perfectly operating 802.11g wireless network at 54 Mbps.

can change the home network to channel 1, and neither network will interfere with the other. Changing to a less frequently used channel increases performance without any additional cost.

IEEE 802.11g

IEEE 802.11g uses OFDM, which brings a significant increase in speed up to 54 Mbps. It uses the same 2.4 GHz frequency as 802.11b, making b and g wireless devices compatible with each other.

Users were thrilled with the increased speed, and 802.11g quickly became a favorite in both homes and businesses. IEEE 802.11g is widely available, but the advances of 802.11n have overtaken 802.11g devices in market share.

802.11g increases the transmission speed to 54 Mbps. One of the reasons for this speed increase is the change in the modulation type. IEEE 802.11g uses OFDM. The average distance for maximum performance is still rated at between 80 and 100 feet, but some vendors advertise distances as great as 150 feet. The maximum distance will always vary depending obstructions, RF interference, and even atmospheric conditions.

IEEE 802.11n

The need for speed brought 802.11n to wireless networks. It advertises speeds of up to 300 Mbps and includes the possibility of reaching 450 Mbps. The improvements in speed are primarily because of equipment changes.

Multiple-input/multiple-output (MIMO) is the antenna technology IEEE 802.11n uses. MIMO includes multiple antennas at both the receiver and the transmitter to minimize errors and increase the data throughput. An intriguing improvement is the concept of smart antennas. These intelligent devices grab multiple streams of data and combine them to ensure faster speeds.

Multiple antennas also increase the transmission range of 802.11n devices. Even as far as 300 feet away, tests indicate that 802.11n networks still operate as high as 70 Mbps.

Even though the IEEE did not formally approve 802.11n until late in 2009, devices based on proposed draft versions of the standard started hitting the market in 2007. The Wi-Fi Alliance began certifying products in 2007 based on the 802.11n proposal.

Another benefit is that IEEE 802.11n is backward-compatible with 802.11a, 802.11b, and 802.11g devices. It is important to realize that being backward-compatible does not mean that the older devices will operate at the newer speeds. To achieve the 300 Mbps speed, both the WAP and the wireless device must be 802.11n.

IEEE 802.11ac

Ratified in January 2014, the IEEE 802.11ac standard enhances the innovations introduced in IEEE 802.11n to provide even greater throughput. Using the 5 GHz band only, 802.11ac increases the maximum number of MIMO streams from four to eight, and the channel bandwidth from 40 to 80 MHz, with an option for 160 MHz. The result is a throughput of at least 500 Mbps, with the potential for as much as 1 Gbps.

NETWORK SECURITY METHODS

One of the biggest concerns with wireless networking is security. Since the signals are broadcast over the air, they are easily intercepted. However, multiple security technologies are available today. Some are better than others, and some are not

Figure 5-6 Wireless Security Methods

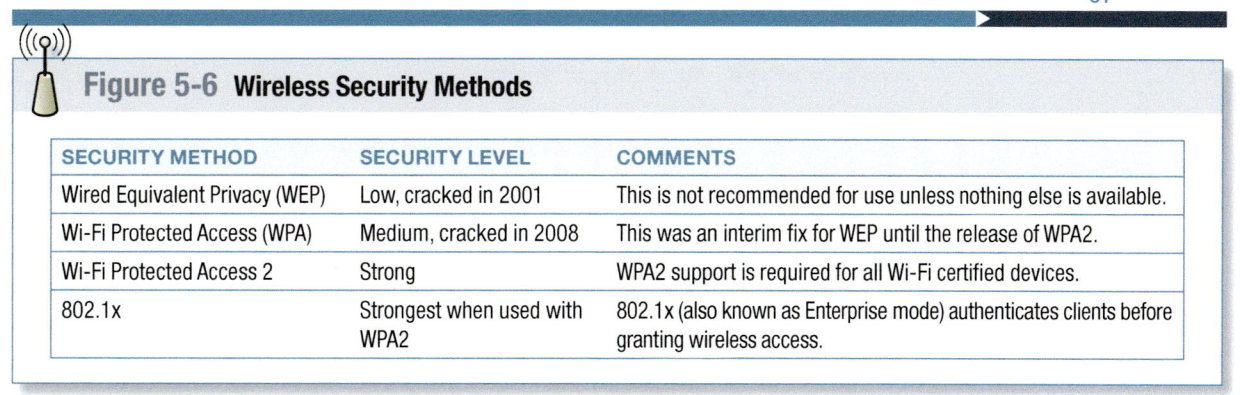

SECURITY METHOD	SECURITY LEVEL	COMMENTS
Wired Equivalent Privacy (WEP)	Low, cracked in 2001	This is not recommended for use unless nothing else is available.
Wi-Fi Protected Access (WPA)	Medium, cracked in 2008	This was an interim fix for WEP until the release of WPA2.
Wi-Fi Protected Access 2	Strong	WPA2 support is required for all Wi-Fi certified devices.
802.1x	Strongest when used with WPA2	802.1x (also known as Enterprise mode) authenticates clients before granting wireless access.

Figure 5-6. New wireless technologies have higher levels of security.

secure at all. It is important to know which security methods to implement for different wireless networks.

When the designers of wireless LAN standards first created them, their primary goal was to make them easy to use. Designers wanted to make it easy for devices to connect and transmit data between each other. The designers did a good job with this goal.

Later, they decided to add some security features. Unfortunately, their first attempt at security was not very successful. Because of this, many people still think of wireless networks as being insecure. However, it is now possible to provide strong security for wireless networks.

Knowing what security methods are available and what methods are actually secure is critical when using a wireless network. **See Figure 5-6.**

Like most brands, Cisco wireless routers support several different security modes. Note that WPA and WPA2 both support Personal and Enterprise modes. **See Figure 5-7.**

Figure 5-7 Wireless Routers

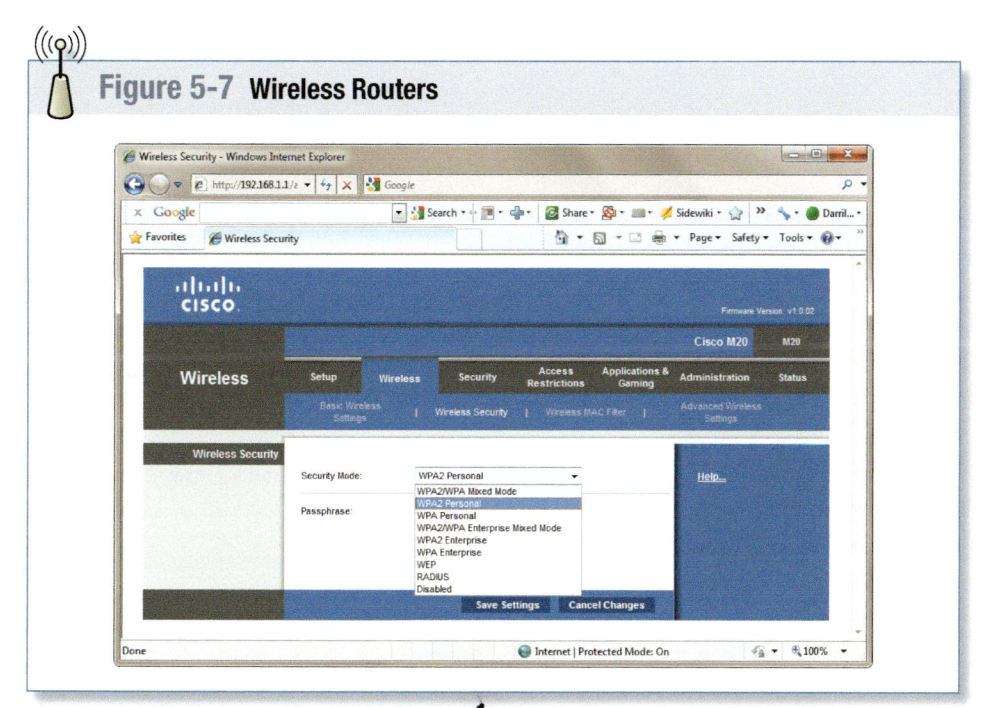

Figure 5-7. Wireless routers support multiple security modes.

In addition, a Remote Authentication Dial-in User Service (RADIUS) mode may be available on the wireless router. IEEE 802.1x can use a RADIUS server at the back end to provide authentication.

Wired Equivalent Privacy

Wired Equivalent Privacy (WEP) was the first security model used on IEEE 802.11 wireless networks. Its intent was to offer privacy equivalent to that of a wired Ethernet network.

However, attackers learned ways to listen to the data, capture it, and decrypt it. WEP had multiple faults, including:

Its Use Was Optional Instead of a "secure by default" strategy, WEP had to be enabled. Many wireless users did not understand its use and did not enable it.

Weak Encryption WEP used RC4, which is a stream cipher. Attackers can crack RC4 using freely available software downloaded from the Internet.

Poor Key Management Encryption keys are secret strings of data used to encrypt and decrypt data. They must be secret between the parties, changed often, and not repeated. However, keys used by WEP are not secure. An eavesdropping attack can determine the encryption key within a minute.

Cracking Software Widely Available Once attackers understood the cracks, they wrote and distributed tools to attack wireless networks.

Wi-Fi Protected Access

Wi-Fi Protected Access (WPA) is an improved security standard over WEP. WPA works on WEP-designed hardware without any additional cost to the consumer. This usually requires a flash upgrade to the firmware on the existing WEP hardware.

> **War Driving**
>
> War driving is the act of driving a car through an area and scanning for wireless networks. Attackers war drive to locate wireless networks and determine the security they use for protection. When attackers locate wireless networks with weak security, they sit in their car with a wireless receiver and capture the wireless transmissions.

> ✓ **FACT**
>
> The Payment Card Industry Security Standard Council sets standards for credit card processing. The use of WEP on any wireless networks processing credit cards is prohibited.

> ✓ **FACT**
>
> Flashing the wireless device is similar to flashing the BIOS on a computer. It installs new software in the programmable read-only memory (PROM).

> **Wireless Network Thefts**
>
> Wireless security has been especially problematic in the past. Early attempts to lock down wireless networks were woefully lacking, and many businesses simply did not understand the risks. They transmitted some data using insecure methods, and other transmissions did not use any security at all.
>
> In 2003 and 2004, hackers stole information from more than 45 million credit cards from two department store chains. Wireless networks transmitted all of this information. Customers who returned merchandise without receipts had to provide driver's license numbers, and it is estimated that 455,000 of these customers had their data stolen. This represents one of the biggest wireless thefts, but there have certainly been many more.
>
> Hackers were able to capture these wireless transmissions and harvest the data. Some data was sent without any security. Other data was sent using the insecure WEP. The stolen data was used to steal identities and make fraudulent charges on the credit cards.

The intent behind WPA was to improve upon WEP's weaknesses and reduce the complexity of configuration. One of WEP's weaknesses was manual key management. It was optional, but cumbersome, and oftentimes avoided. With WEP, the same key can be used for as long as desired, but when the key is changed, it must be changed on all wireless devices. It increased the administrative workload, and many users simply overlooked this step.

With WPA, rekeying the encryption keys is mandatory, and with each data frame, a new key is created automatically. Temporal Key Integrity Protocol (TKIP) manages the keys and provides several other technical improvements.

WPA can be configured in two modes:

Personal Mode, Also Known as Preshared Key (PSK) Mode Personal mode requires manual configuration similar to WEP but without the upkeep of changing the key. The initial shared key is a string of characters such as a password or passphrase. Once the initial shared key has been entered, TKIP is responsible for encryption and automatic rekeying, which eliminates the need for manual rekeying.

Enterprise Mode Enterprise mode requires authentication with a back-end server known as an 802.1x server. After authentication, the access point negotiates a separate and unique key with each client.

The significant difference between the WPA Enterprise and Personal modes is in authentication. With WPA Personal mode, every wireless client uses the same passphrase, which does not individually identify any of the clients. The WAP grants access to any client with the passphrase. With WPA Enterprise mode, authentication takes place at an authentication server, and each client requires a specific account and credentials (such as a username and password).

It may not be obvious, but the primary purpose of WPA was to provide a temporary secure solution while designers created a more secure one. Designers fully expected that researchers or attackers would crack WPA. They were right. Researchers cracked WPA in 2008.

WPA2

WPA2 is the updated version of WPA and is standardized as IEEE 802.11i. WPA2 supports encryption with the Advanced Encryption Standard (AES) algorithm. The U.S. government adopted AES as its encryption standard, and it is considered the strongest symmetric encryption available.

One drawback to WPA2 is that it requires hardware that is different from the hardware used with WEP and WPA. At this point, all new hardware is WPA2-compatible, but older hardware that is not compatible may still be in use. In those cases, use the older TKIP with WPA2.

WPA2 supports both Personal and Enterprise mode just like WPA. Personal mode uses a preshared key, and Enterprise mode requires an 802.1x server.

WPA2 Personal Mode WPA2 Personal (or WPA2-PSK for preshared key) is for home users and small businesses that are not using an authentication server. Anyone who has the passphrase and name of the network can connect. It combines the passphrase and the network name to create unique encryption keys for clients.

✓ FACT

WPA is still more secure than WEP. However, the preferred security solution for wireless networks today is WPA2.

✓ FACT

TKIP changes keys without requiring the user to change the passphrase. WEP reuses similar keys for encryption until the user changes the passphrase.

✓ FACT

AES is an extremely strong and widely respected encryption algorithm. Many different applications encrypt data with AES, including non-wireless applications.

✓ FACT

The Wi-Fi Alliance requires all new wireless devices to support WPA2 in order to be certified with the Wi-Fi logo.

> **✓ FACT**
>
> In a Microsoft environment, the authentication server will often check the credentials against an Active Directory database. Active Directory hosts accounts and their credentials.

WPA2 Enterprise Mode WPA2 Enterprise mode uses 802.1x for authentication. This requires a back-end server such as a Windows Server 2012 R2 server running Network Policy Access Services to authenticate clients. Wireless clients are not granted access to the wireless network unless they can authenticate.

Authentication means the clients must provide credentials such as a username and password. More advanced authentication procedures can require smart cards or fingerprints.

IEEE 802.1x Authentication Server

IEEE 802.1x provides port-based security. In short, it provides an authentication mechanism for either IEEE 802.3 (wired Ethernet) or 802.11 networks. IEEE 802.1x includes three elements in the authentication process:

- Supplicant: Client
- Authenticator: Access point
- Authentication server: Running RADIUS and Extensible Authentication Protocol (EAP)

Consider an example where the wireless client acts as the supplicant, the WAP as the authenticator, and a back-end server as the authentication server. **See Figure 5-8.**

The authenticator acts like a security guard and ensures the supplicant has adequate credentials before providing access to other networks. When the supplicant first connects, the access point sends the credentials to the authentication server, and the authentication server checks the credentials against its database. If the credentials are valid, the authentication server confirms them to the authenticator, and the authenticator grants access to the client.

When adding the Network Policy and Access Services (NPAS) role to a server running Windows Server 2012 R2, configure it as an 802.1x server. Any clients that are not authenticated will not be allowed access to the network. NPAS can also be configured to grant nonauthenticated clients access to isolated networks. **See Figure 5-9.**

Figure 5-8 Enterprise Authentication

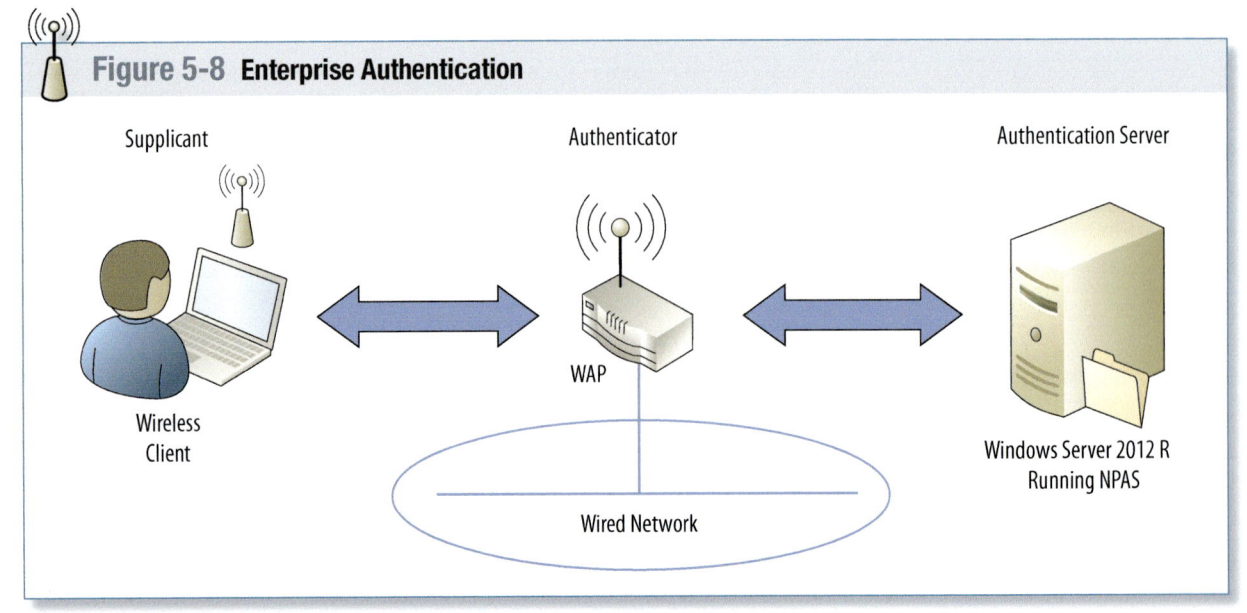

Figure 5-8. WPA2 enterprise authentication uses a RADIUS server at the back end.

Figure 5-9 Network Policy and Access Server (NPAS)

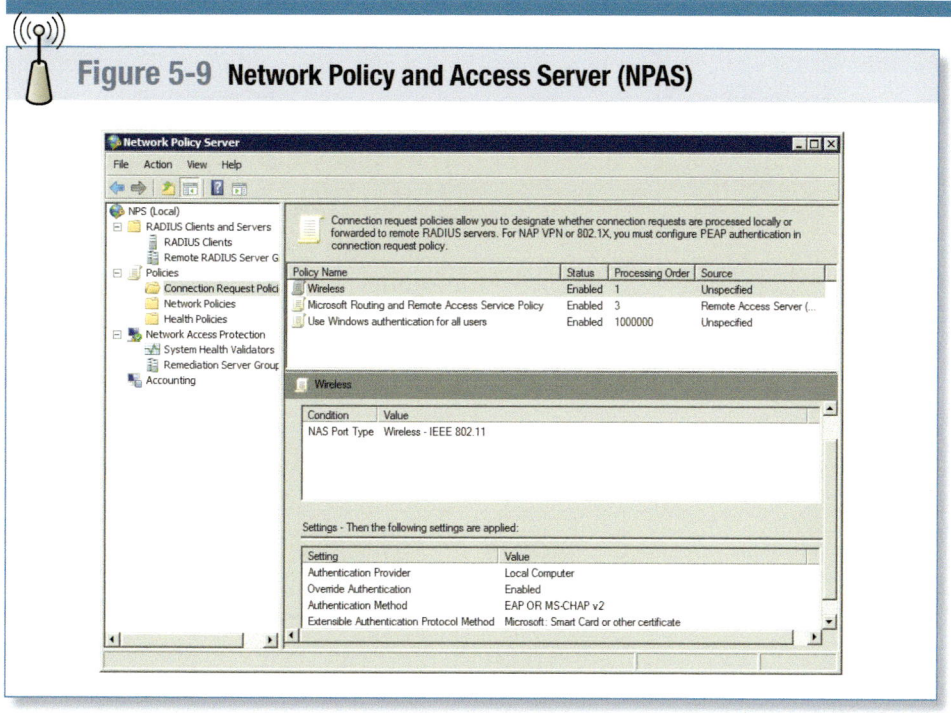

Figure 5-9. *The Network Policy and Access Server (NPAS) role on a computer running Windows Server 2012 R2 requires that the server be configured as an 802.1x server.*

Figure 5-9 shows a wireless policy (named "Wireless"). It is configured with a value of "Wireless – IEEE 802.11" to authenticate 802.11 wireless clients. Additionally, it is configured to authenticate the clients using Extensible Authentication Protocol (EAP) or Microsoft Challenge Handshake Authentication Protocol - Version 2 (MS-CHAPv2). EAP supports the use of smart cards, and MS-CHAPv2 is a secure method of authenticating username and passwords.

WIRELESS NETWORKS

Wireless networking offers many advantages over wired networks, such as its ease of installation and the elimination of wires. Once the wireless network has been configured, users can share resources such as files, folders, printers, and more, just as they can with a wired network.

Home Wireless Networks

The primary piece of equipment used to create a wireless network in a home (or small business) is a wireless router. It is important to realize that a wireless router has many different components.

MAC Filtering Does Not Provide Realistic Security

On many wireless routers, media access control (MAC) address filtering can be configured. The goal is to ensure that only computers with the specified MAC address can access the wireless router. On the surface, this sounds very secure, since MAC addresses are theoretically unique.

However, it is very easy for an attacker to spoof a MAC address. In other words, the attacker can modify packets so that it looks like the attacker's packets are coming from an approved MAC address. Although using MAC filtering does not cause any harm, it will not stop an experienced attacker from entering the network.

Introduction to Network Technologies

Wireless Access Point (WAP) The WAP provides connectivity for the wireless devices. It includes a bridge to join the wired and wireless devices.

Switch The switch provides connectivity between wired and wireless devices. All of the devices connected to the switch ports have the same network ID and share the same broadcast domain.

Router In a home network, the router is usually connected directly to the cable modem or to another Internet connection. It routes traffic from the internal switch to the Internet (and back).

DHCP DHCP provides IP addresses and other TCP/IP configuration information to all the devices on the switch's network.

Consider the rear view of a wireless router with its extra components. The first four ports are typical wired ports that connect with the wireless router's switch component. The device also connects wireless connections through the switch. DHCP provides TCP/IP information to all the ports connected to these switch ports and can assign addresses to a wired computer, a wireless printer, and a wireless laptop. **See Figure 5-10.**

> ✓ **FACT**
> Wired and wireless devices can be connected using the wireless router's WAP and switch components without connecting it to a WAN. This creates a private network.

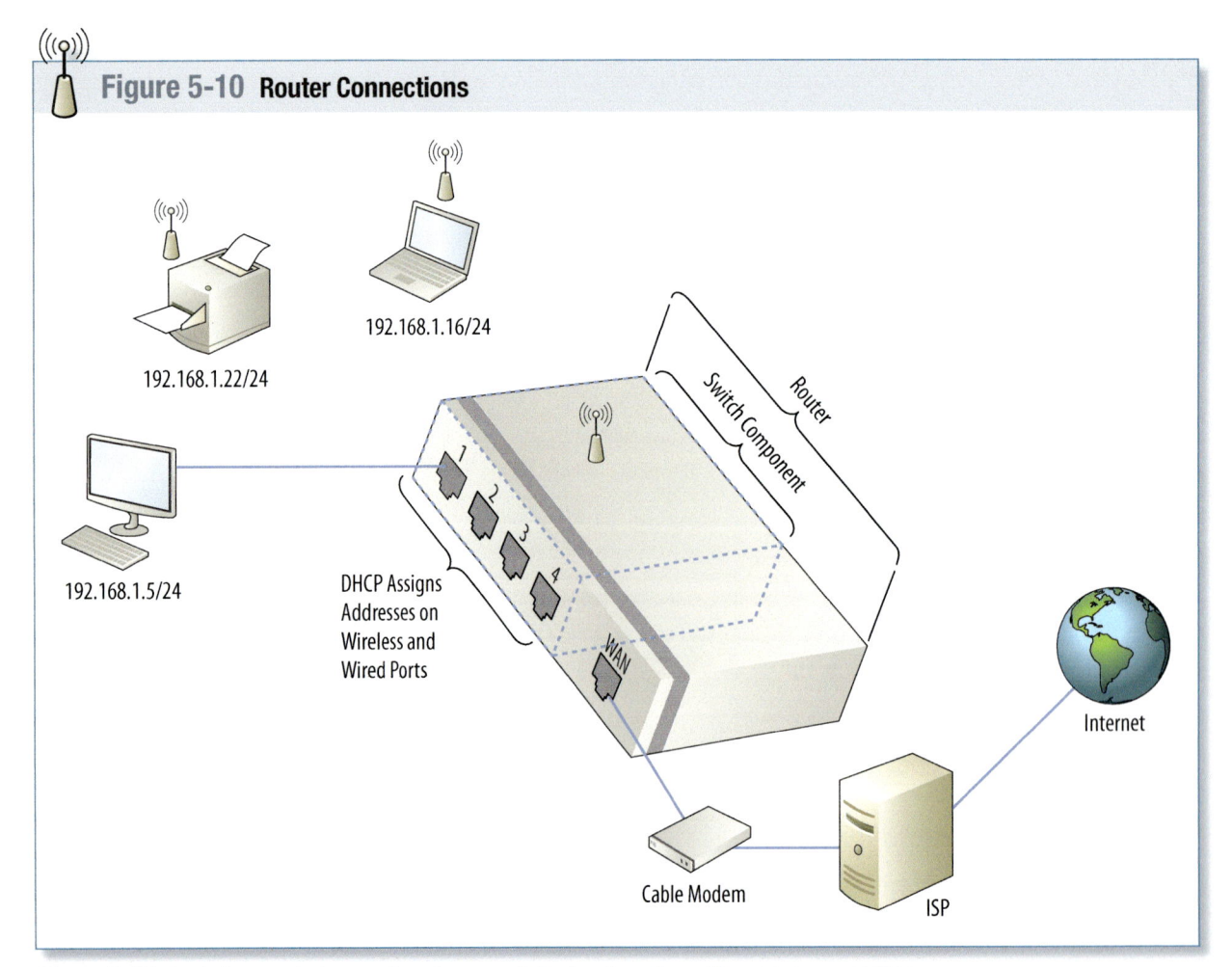

Figure 5-10. A wireless router's rear view shows switched ports and a cable modem connection.

The routing component forwards any traffic destined for the Internet through the WAN port. This usually connects—through a digital subscriber line (DSL) modem, cable modem, or other broadband device—to an Internet service provider (ISP).

Many wireless routers use the network IP address 192.168.1.0/24. The wireless router itself uses the address 192.168.1.1 and is the default gateway for computers connected via the wireless router. The router forwards traffic from the switch to the WAN port, and the WAN port provides connectivity to the Internet via an ISP.

Most wireless routers can be configured through a web interface. The web interface may consist of several pages. **See Figure 5-11.**

Most wireless routers today are extremely simple to set up. In most cases, all that must be done is to enter or change the network name (SSID), change the administrator password from the default, and choose the security method (usually WPA2 Personal). Many have installation wizards that provide guidance through the process.

Wireless Networks in a Business

Wireless networking in the business environment has grown exponentially in recent years, with no end in sight.

> ✓ **FACT**
> It is important to change the default password of the administrator account. If not, an attacker can access the network and make changes, even locking out the owner.

> ✓ **FACT**
> Many wireless routers have additional capabilities. For example, some include a virtual private network (VPN) that enables accessing a home network from a remote location.

Figure 5-11 Router Web Interface

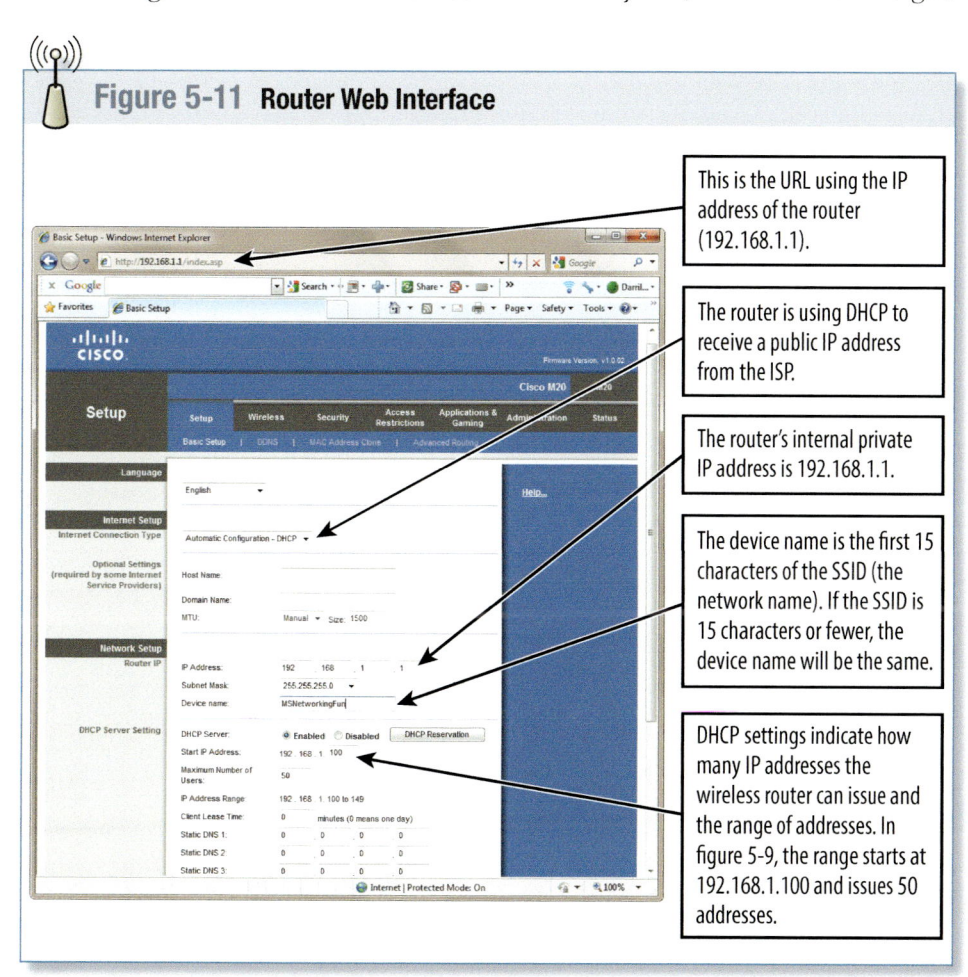

- This is the URL using the IP address of the router (192.168.1.1).
- The router is using DHCP to receive a public IP address from the ISP.
- The router's internal private IP address is 192.168.1.1.
- The device name is the first 15 characters of the SSID (the network name). If the SSID is 15 characters or fewer, the device name will be the same.
- DHCP settings indicate how many IP addresses the wireless router can issue and the range of addresses. In figure 5-9, the range starts at 192.168.1.100 and issues 50 addresses.

Figure 5-11. Most wireless routers use a web interface.

> **✓ FACT**
> Computers connected in ad hoc mode (a wireless network without a WAP) form an Independent Basic Service Set (IBSS).

> **✓ FACT**
> The placement of WAPs in business networks also becomes a network security issue, because the signals can easily bleed outdoors and across parking lots and streets.

> **✓ FACT**
> P2P wireless connections are also called wireless bridges or P2P wireless bridges. They bridge two or more wired networks with a network connection.

> **✓ FACT**
> Line of sight indicates that there is a clear path between the two bridges. The curvature of the earth limits the line of sight.

Advantages to wireless networks in a business include:

- Reduced costs compared to wired networks
- Better flexibility over wired networks
- Greater mobility between offices
- Improved scalability

The primary difference between using a wireless network in a home or small office and using one in a business is that businesses will typically use WAPs only, and not wireless routers.

Configuring a wireless network in a business is very similar to configuring a wireless network at home. The network still must be supplied with an SSID and its security must be configured. The most secure protocol for business wireless networks is WPA Enterprise, which uses 802.1x for authentication.

One difference in business wireless is the use of repeaters. Since a business can be considerably larger than a home, a single WAP may not cover the entire business. Instead, additional WAPs can serve as repeaters.

Basic Service Set (BSS) A BSS is a wireless network composed of one WAP and one or more wireless devices. For many wireless networks, a single WAP is enough.

Extended Service Set (ESS) An ESS is a wireless network with more than one WAP, with each WAP supporting one or more wireless devices. Additional WAPs act as repeaters and extend the range of the wireless network. All devices in the ESS use the same SSID and the same broadcast channel.

The key to success when adding repeaters is device placement. If a repeater is placed too far from the root WAP, a device may get dropped while roaming. If repeater is placed too close, the two WAPs may interfere with each other. As an example, say that a repeater has been added approximately 100 feet (about 30 meters) from the root access point, which is typically a reasonable distance. But testing may dictate that the distance may need to be adjusted. **See Figure 5-12.**

The repeater ensures that users who are out of range of the root WAP will still have adequate signal strength to stay connected. As users roam between repeaters, wireless devices connect with the strongest signal automatically, providing optimal performance.

POINT-TO-POINT WIRELESS

Point-to-Point (P2P) wireless is a mechanism that connects two networks using wireless technologies instead of traditional wired connections. It is sometimes cost-prohibitive to run cable between two points. The distance between points may be a few hundred feet, or a few miles, with 25 miles typically being the maximum.

Consider an example in which a main office has a newly leased building that is 10 miles down the road. Each building includes an internal network, but they need connectivity between each other. **See Figure 5-13**.

Many technologies support the P2P wireless bridge. These include microwave, infrared, and laser-optics radio transmission. Most of these are limited to line of sight, but there are other considerations.

The first step in configuring a wireless bridge is performing a site survey. This is to ensure that the area is free from radio frequency interference and

Figure 5-12 Extended Service Set (ESS)

WAP 1

100 Feet

WAP 2

Wired LAN

Extended Service Set

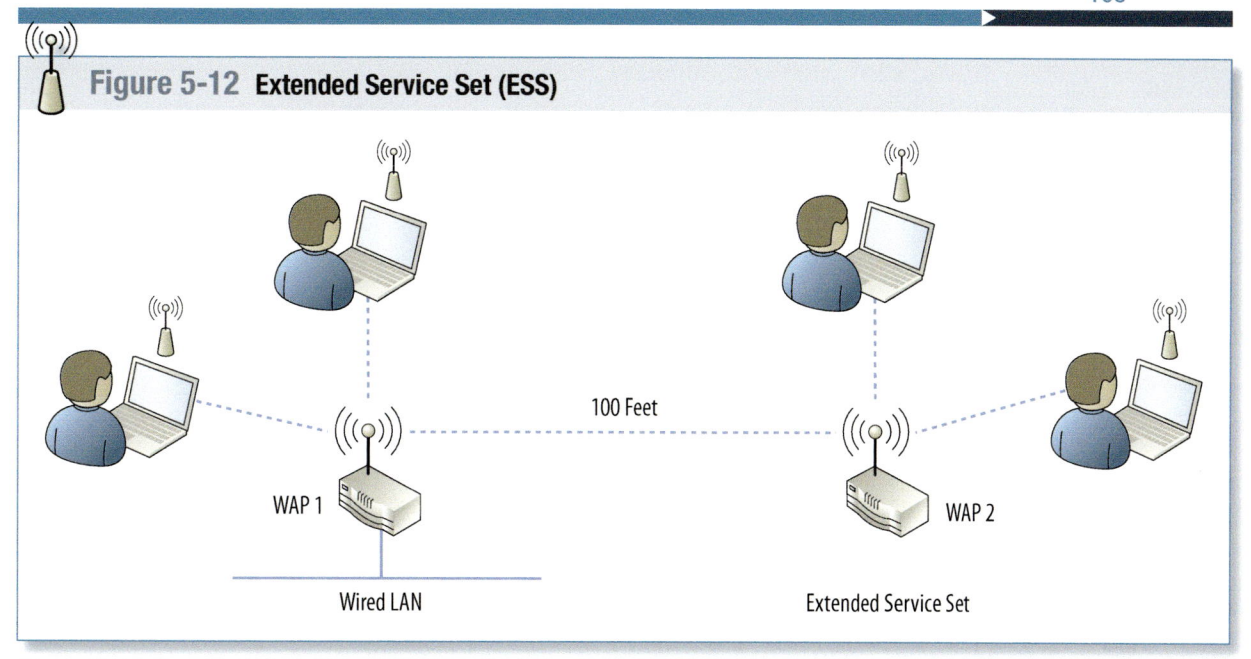

Figure 5-12. An extended service set (ESS) uses repeaters to support longer distances.

line-of-sight obstructions. Wireless radio waves in the 2.4 to 5.0 GHz range do not penetrate building structures or trees very well. The height of the transceivers on both buildings also affects the distance. Higher buildings permit longer distances.

Even if the area is clear of radio interference and physical obstructions, this is still not enough. The area underneath

Figure 5-13 P2P Bridge

Organization Main Office

Wireless Link Spanning 10 Miles

Newly Leased Building

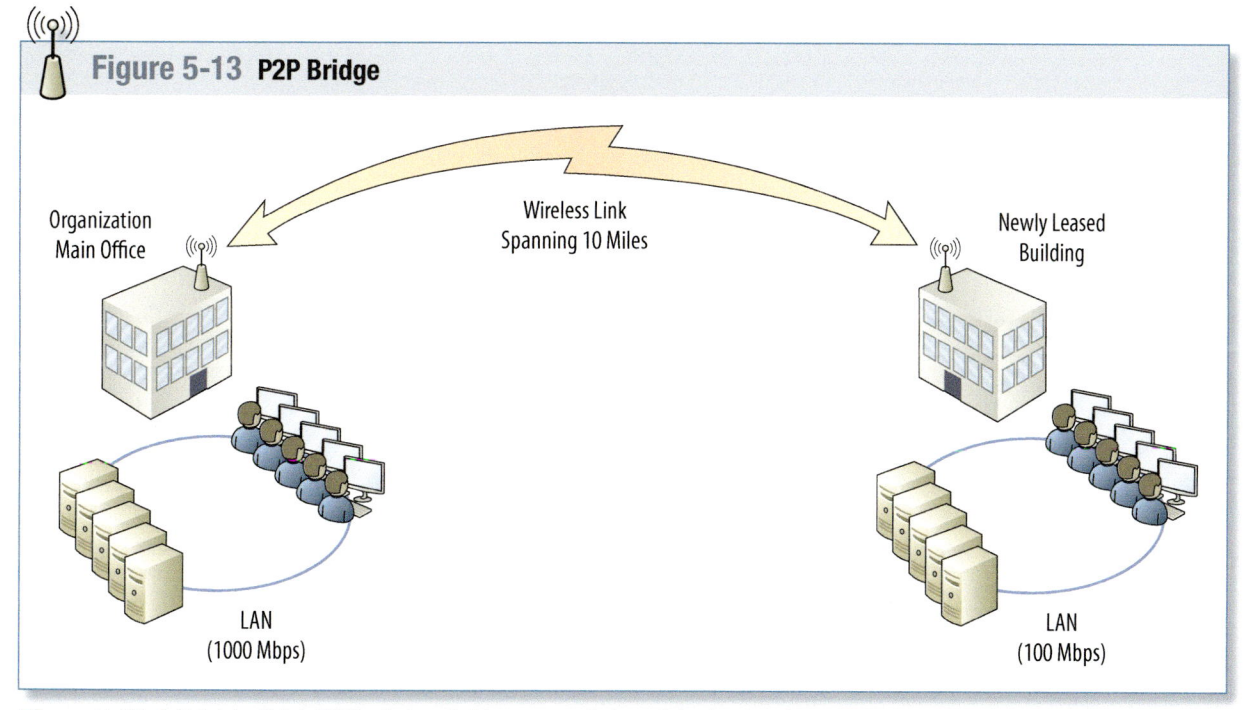

LAN (1000 Mbps)

LAN (100 Mbps)

Figure 5-13. A Point-to-Point (P2P) wireless bridge connects two distant networks.

> **✓ FACT**
>
> When the distances are farther than the eye can see, engineers use riflescopes (without the rifles) to focus and align the wireless bridges with each other.

> **✓ FACT**
>
> Performing a site survey and determining the Fresnel zone can be complex. The point to remember is that there is more to it than just adding two wireless bridges.

and above the line of direct sight must be considered. This area is the Fresnel zone. **See Figure 5-14.**

The *Fresnel zone* is the area underneath and above the direct line of sight between the two points. A house may penetrate the lower boundary of the Fresnel zone. Structures within the Fresnel zone have a tendency to absorb radio waves, and blockage greater than 40% will render the wireless connection unreliable.

Another consideration with the bridge is alignment of the directional antennas between the two points. It will require special equipment and expertise to have these two antennas focused toward each other.

Bridge antennas are typically the dish type (parabolic) directional antennas or Yagi directional antennas. A directional antenna has the best performance in a specific direction and can be pointed or directed at specific locations. In contrast, omnidirectional antennas receive signals from all directions.

Figure 5-14 Fresnel Zone

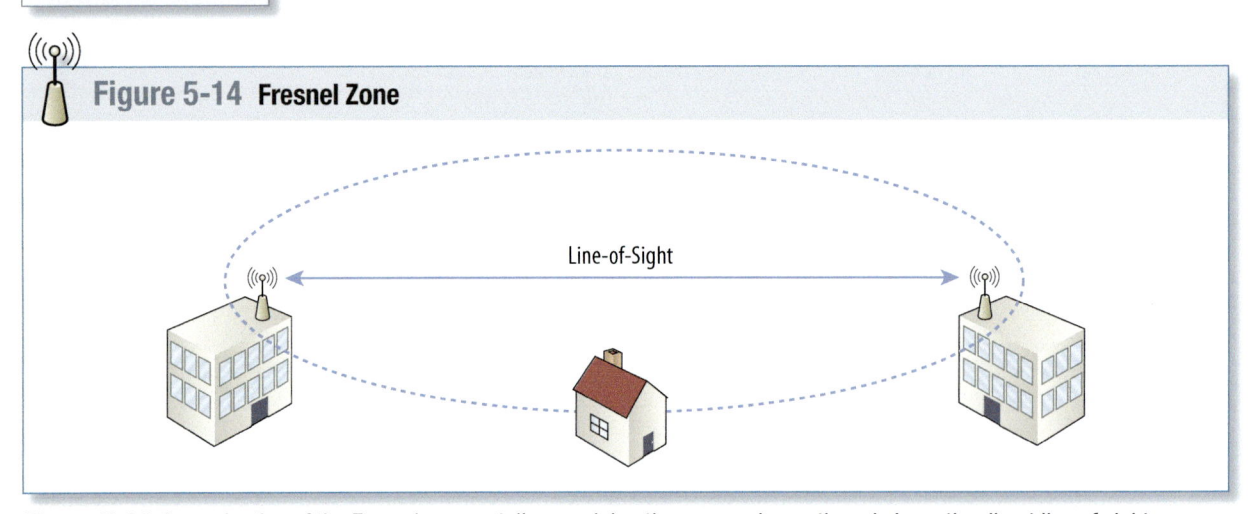

Figure 5-14. An evaluation of the Fresnel zone entails examining the area underneath and above the direct line of sight.

Summary

There are many wireless components, standards, and security methods. A WAP bridges wireless devices to a wired network. A wireless network has an SSID, which is simply the name of a wireless network. Wireless networks use several different wireless standards, including 802.11a, b, g, and n. 802.11a uses a frequency of 5 GHz. 802.11b and g use 2.4 GHz, and 802.11n uses either 2.4 GHz or 5 GHz. Security standards include WEP (old and insecure), WPA (cracked in 2008), and WPA2 (strong). Using WPA2 with 802.1x will increase security by adding authentication. For business use, a wireless network can be extended by adding repeaters. Two buildings that are miles apart can be connected by using Point-to-Point (P2P) wireless bridges.

Review Questions

1. 802.11 networks use CSMA/CD.
   a. True
   b. False

2. What frequency does an 802.11a network use?
   a. 2.4 MHz
   b. 5 GHz
   c. 11 Mbps
   d. 54 Mbps

3. Which of the following frequency ranges does 802.11b use?
   a. 2.4 GHz
   b. 2.4 MHz
   c. 4.1 GHz
   d. 5 GHz

4. What is the maximum speed of an IEEE 802.11b network?
   a. 2 Mbps
   b. 11 Mbps
   c. 54 Mbps
   d. 300 Mbps

5. IEEE 802.11n networks can operate at speeds as high as 300 Mbps.
   a. True
   b. False

6. Of the following security methods, which one is the most secure?
   a. WEP Enterprise Mode
   b. WEP Personal Mode
   c. WPA2 Enterprise Mode
   d. WPA2 Personal Mode

7. A WAP and a wireless router are the same thing.
   a. True
   b. False

8. A company is planning to lease a second building, which is about two miles away. How can the networks between the two buildings be connected?
   a. Add roaming WAPs.
   b. Connect the buildings with a P2P bridge.
   c. Extend the network by adding additional WAPs.
   d. Run a twisted-pair cable between the buildings.

IPv4

IPv4 addresses are the most common type of address used on the Internet and on internal networks today. It is important to understand the components of an IPv4 address in order to troubleshoot basic problems when a computer has been misconfigured.

Large organizations often divide the network into subnets, and one of the rites of passage for network administrators is learning how subnetting works. It is not necessary to be a master at subnetting, but understanding the basics is important.

Most organizations also use the Dynamic Host Configuration Protocol (DHCP) to automatically assign IP addresses and other TCP/IP configuration information. Although this normally works well, it occasionally fails. When a client cannot reach a DHCP server, it gives an obvious telltale sign—with the proper know-how.

OBJECTIVES

- ▶ List the components of an IPv4 address
- ▶ Describe a binary IPv4 address
- ▶ Explain the reasons for subnetting an IPv4 Address
- ▶ Compare the manual assignment and automatic assignment of an IPv4 address

CHAPTER 6

TABLE OF CONTENTS

Components of an IPv4 Address 110
    Network ID and Host ID of an
    IP Address 110
    Default Gateway. 113
    Local and Remote Addresses 114
    Classful IP Addresses 115
    Reserved IP Address Ranges 117

Binary IPv4 Address 117
    The Bits of an IP Address 117
    CIDR Notation 119
    IP Address Masking. 120
    Classless IP Addresses 121

IPv4 Address Subnetting. 121
    Number of Subnet Bits 122
    Number of Hosts in a Network 124
    Local and Remote Addresses 124
    Subnetting Usage 126

Manual Versus Automatic Assignment of
IPv4 Addresses 128
    Configuring IPv4 Configured
    Manually. 128
    DHCP 128
    APIPA 130

Summary. 132

Review Questions 133

COMPONENTS OF AN IPV4 ADDRESS

Internet Protocol version 4 (IPv4) has been the standard Network layer protocol and addressing scheme since the 1980s. The primary function of IP is to get TCP/IP traffic from one computer to another computer over a network. All computers on the Internet have unique IP addresses. As long as the IP addresses are valid, any computer can reach any other computer on the Internet using its address.

Similarly, internal networks also use IP addresses. All of the computers on each internal network have unique addresses within the network, and they use these IP addresses to get traffic from one computer to another.

Think of an IP address like the street address of a home or business. As long as the full address is valid, the post office will deliver a properly addressed letter. This also works worldwide. With a valid address, a letter will reach its destination. A valid address in the United States has a street address (or a post office box number), a city, a state, and a ZIP code.

Valid IP addresses have four decimal values separated by three dots. Additionally, the only valid decimal numbers in an IPv4 address are 0 through 255. For example, the following IP addresses are valid:

- 10.80.1.5
- 172.16.5.254
- 192.168.1.4

In comparison, the following are not valid IPv4 addresses:

- 10.80.256.5 (no number can be greater than 255)
- 172.16.254 (there must be four decimals)

On a computer running Windows, the configured IP address is obtained by using the `ipconfig` command at the command prompt. **See Figure 6-1.**

Network ID and Host ID of an IP Address

An IP address has two components: a network ID and a host ID. The network ID identifies the subnetwork, or subnet, where the computer is located. A subnet is a portion of a logically divided network that shares a range of IP addresses. The host ID uniquely identifies the computer within that subnet.

In internal networks, IP addresses are accompanied by a subnet mask. The *subnet mask* identifies the portion of the

> ✓ **FACT**
> An IPv4 address expressed with decimal numbers separated by dots is in dotted decimal format, or dot-decimal notation.

> ✓ **FACT**
> The default gateway is the address of the near side of a router. It will typically provide a path to the Internet or other subnets and is discussed later in this chapter.

> ✓ **FACT**
> Subnet masks can be more complex, but many internal networks use simple subnet masks with only the numbers 255 or 0 in dotted decimal format.

Figure 6-1 IP Address on Windows

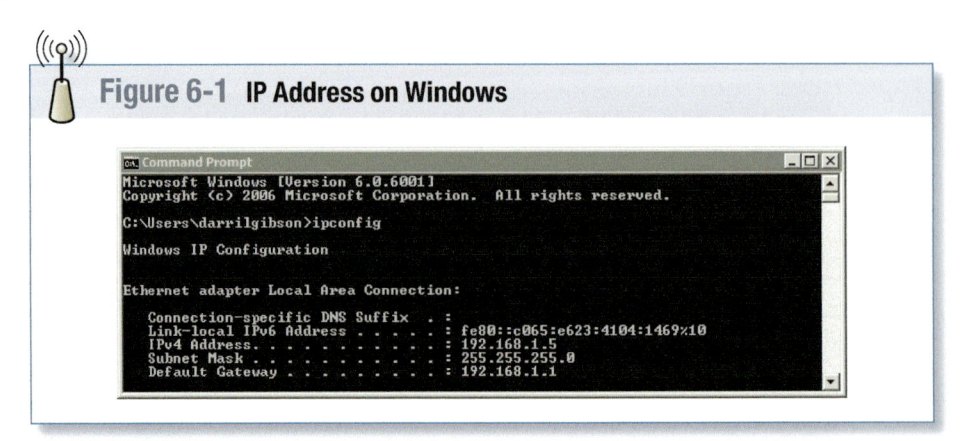

Figure 6-1. Enter the ipconfig *command at the command prompt to view the IP address on a Windows computer.*

Comparing a Network ID and a ZIP Code

A U.S. Postal Service (USPS) address includes the street address, city, state, and ZIP code. When the USPS receives a letter, it will get it to the post office in the correct city and state simply by using the ZIP code. The USPS then uses the street address to get it to the correct home or business.

Similarly, TCP/IP uses the network ID to get a packet to a router on the correct subnet. Once the packet reaches the subnet, it uses the host ID to get the packet to the correct computer on the subnet.

Throughout the United States, ZIP codes are unique. Each ZIP code represents a group of addresses relatively close to each other. Similarly, within a network, network IDs are unique. Each network ID represents a group of two or more hosts on a subnet of a network.

Within each ZIP code, each address is unique. For example, no two addresses will be 777 Success Road. If the addresses were the same, mail could not be accurately delivered to both addresses. Similarly, within a subnet, computers with the same network ID must have unique host IDs. If two computers have the same host ID and the same network ID (the same IP address), it results in an IP address conflict.

IP address that is the network ID. Common subnet masks include:

- 255.0.0.0
- 255.255.0.0
- 255.255.255.0

TCP/IP uses the subnet mask to determine which portion of the IP address is the network ID and which portion is the host ID. More specifically, when the subnet mask is configured at its maximum value (255), that indicates that the corresponding portion of the IP address is part of the network ID. The remaining portion of the IP address is the host ID. **See Figure 6-2.**

Figure 6-2 Subnet Mask

Network ID			Host ID	
192	168	1	5	IP Address

255	255	255	0	Subnet Mask

When the subnet mask is maximum (255), that portion of the IP address is the network ID.

Network ID: 192.168.1.0

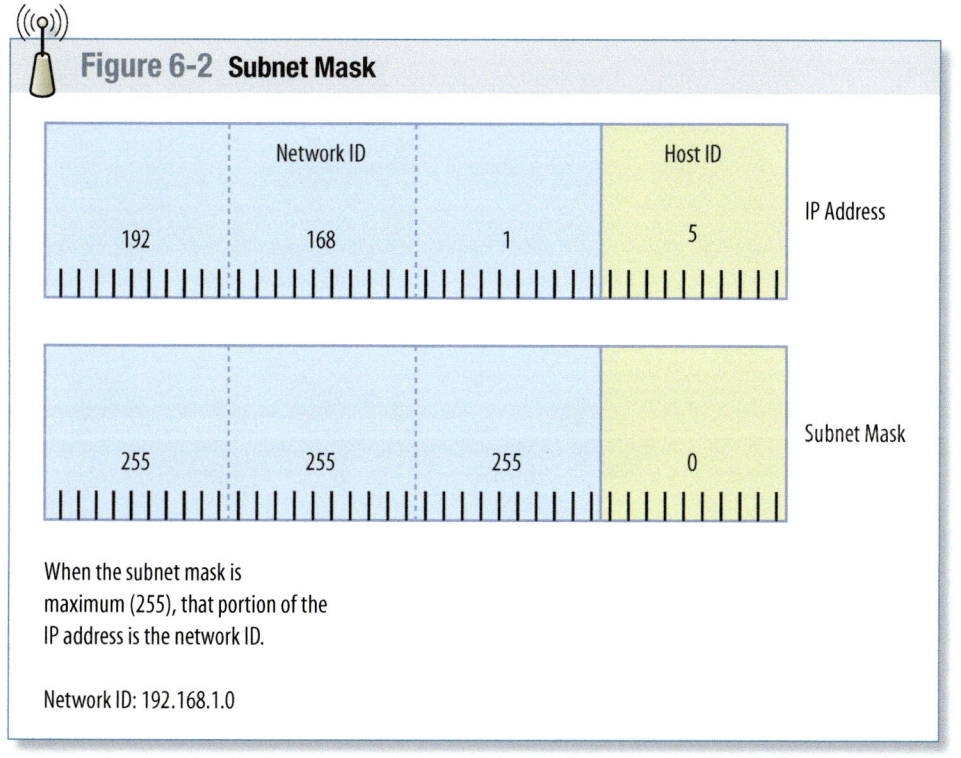

Figure 6-2. *An IP address needs a subnet mask to differentiate between the network ID and the host ID.*

112 Introduction to Network Technologies

Consider an IP address of 192.168.1.5 with a subnet mask of 255.255.255.0. Since the first three decimals of the subnet mask are 255, the first three decimals of the IP address make up the network ID. The network ID is always expressed with trailing zeros filling the place where the host ID would go. For example, the network ID in this case is 192.168.1.0. It is incorrect to express it as 192.168.1 without the trailing zeros. The host ID is simply whatever remains after identification of the network ID. In this case, the host ID is the number 5.

The subnet mask must have contiguous maximum numbers. Once the first zero is used, the remaining numbers must be zero. A subnet mask of 255.0.255.0 is not valid.

Identifying the network ID and host ID is very important. Every computer on a subnet must have the same network ID as part of the IP address, and each of these computers needs a unique IP address. When administrators assign IP addresses manually, a simple typographical error can result in a computer not being able to communicate.

Consider an example network consisting of two subnets (subnets A and B) with four computers each, connected by a router. It should be easy to identify the network ID and the host ID of each of the computers on the two subnets. Additionally, the configuration of two of the computers in subnet A and two of the computers in subnet B may be incorrect, while the addresses assigned to the router interfaces may be correct. **See Figure 6-3.**

Subnet A is on the left and subnet B is on the right. They are separated by a router. The router has two network interface adapters, and each adapter has an IP address. For all computers in subnet A, the adapter labeled A is the default gateway to subnet B. Similarly,

✓ **FACT**
When using advanced subnetting, this rule is worded a little more specifically. Once the first zero bit is used, the remaining bits must be zero.

Figure 6-3 ID Errors

192.168.1.5	192.168.1.6		192.168.4.5	192.168.14.6
255.255.255.0	255.255.255.0		255.255.255.0	255.255.255.0
①	②	Router (A)(B)	⑤	⑥
	Subnet A	192.168.1.1 192.168.4.1	Subnet B	
③	④	255.255.255.0 255.255.255.0	⑦	⑧
192.168.2.7	192.168.1.8		192.168.4.7	192.168.4.8
255.255.255.0	255.255.0.0		255.0.255.0	255.255.255.0

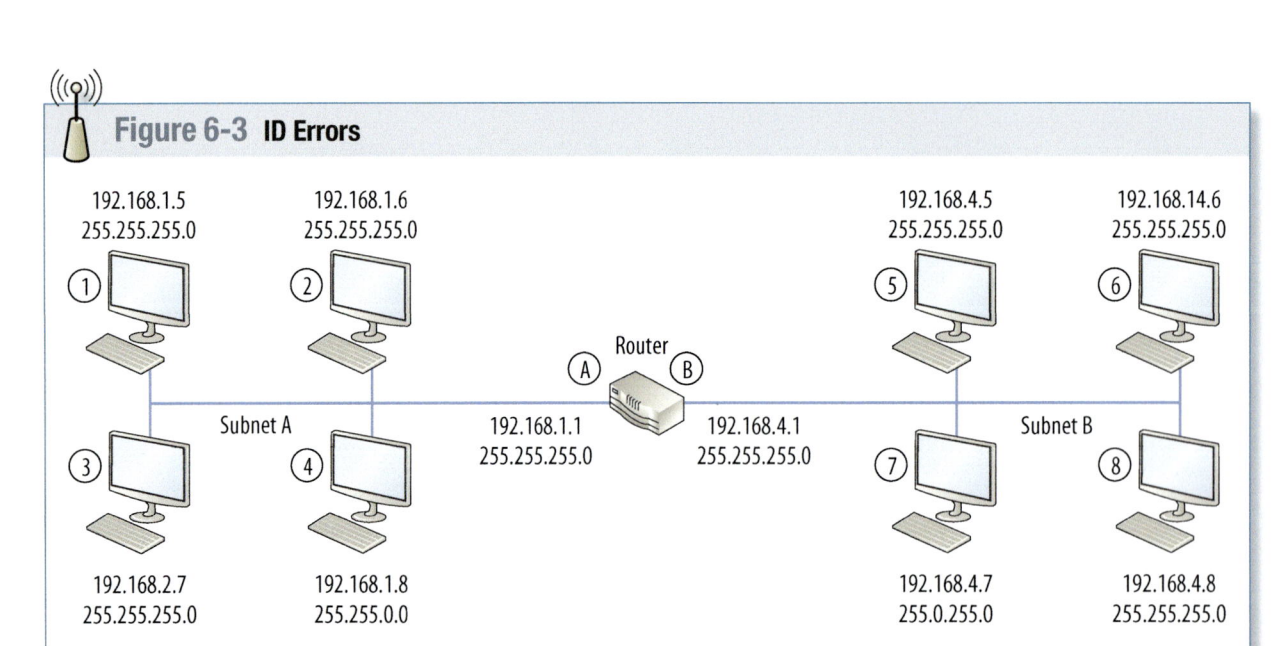

Figure 6-3. *Subnet A and Subnet B both have network ID and/or Host ID issues.*

for all computers in subnet B, the adapter labeled B is the default gateway to subnet A.

Router Connection A   The network ID is 192.168.1.0, and the host ID is 1. Since this is known to be correct, all computers on this subnet must have the same network ID (192.168.1.0). Adapter A on the router is the default gateway for computers on subnet A.

Computer 1   The network ID is 192.168.1.0, and the host ID is 5. This computer is configured correctly.

Computer 2   The network ID is 192.168.1.0, and the host ID is 6. This computer is configured correctly.

Computer 3   The network ID is 192.168.2.0, and the host ID is 7. Note that the network ID is different from the default gateway. Since the network ID is different from other computers on this subnet and also different from the default gateway, this computer will not be able to communicate on the network.

Computer 4   The network ID is 192.168.0.0, and the host ID is 1.8. Note that the subnet mask is 255.255.0.0 with only two 255s instead of three. Since the network ID is different from other computers on this subnet, and also different from the default gateway, this computer will not be able to communicate on the network.

Router Connection B   The network ID is 192.168.4.0, and the host ID is 1. Since this is known to be correct, all computers on this subnet must have the same network ID (192.168.4.0). Adapter B on the router is the default gateway for computers on subnet B.

Computer 5   The network ID is 192.168.4.0, and the host ID is 5. This computer is configured correctly.

Computer 6   The network ID is 192.168.14.0, and the host ID is 6. Note that the third number in the IP address is 14 and not 4. Since the network ID is different from other computers on this subnet, and also different from the default gateway, this computer will not be able to communicate on the network.

Computer 7   The subnet mask of 255.0.255.0 is invalid. A valid subnet mask cannot have numbers greater than zero once the first zero is used. This computer will not be able to communicate on the network.

Computer 8   The network ID is 192.168.4.0, and the host ID is 8. This computer is configured correctly.

Default Gateway

The *default gateway* is the IP address of the router's interface on the local subnet. If there is only one router, it is easy to determine the default gateway. However, if a subnet has more than one router, only one can be the default. The default gateway will usually provide a path to the Internet.

Consider an example network that includes three routers with multiple subnets. Subnets x, y, and z have computers and IP addresses assigned. All the IP addresses in subnet A have a network ID of 192.168.1.0, and all the IP addresses in subnet B have a network ID

> ✓ FACT
>
> It is common to give a default gateway the first usable IP address in the subnet, such as 192.168.1.1. However, this is not required.

of 192.168.4.0. Both subnet A and subnet B have two routers. However, only one router will provide a path to the Internet. **See Figure 6-4.**

If this is a typical network, all the computers in subnet A will be configured with a default gateway of 192.168.1.1, and all the computers in subnet B will be configured with a default gateway of 192.168.4.1.

Local and Remote Addresses

The IP protocol looks at the source and destination addresses to determine whether they are both on the same local subnet. If they are on the same subnet, it then uses the Address Resolution Protocol (ARP) to broadcast the IP, learn the physical address, and deliver the packet to the destination computer.

However, if the destination IP address has a different network ID, it is considered to be on a remote subnet. The IP protocol then sends the data to the default gateway.

It should be easy to look at two IP addresses and determine whether they are both on the same local subnet, or whether the destination IP address is on a remote subnet. To do this, follow these steps:

1. Determine the network ID of the source IP address.
2. Determine the network ID of the destination IP address.
3. Determine whether they are the same:

    If they are the same, they are local to each other. If not, the destination address is on a remote network and data must be sent through the default gateway. **See Figure 6-5.**

Example 1 The network ID of the source IP address is 192.168.1.0. The network ID of the destination IP address is 192.168.1.0. These are the same, so they are local to each other.

Figure 6-4 Default Gateways

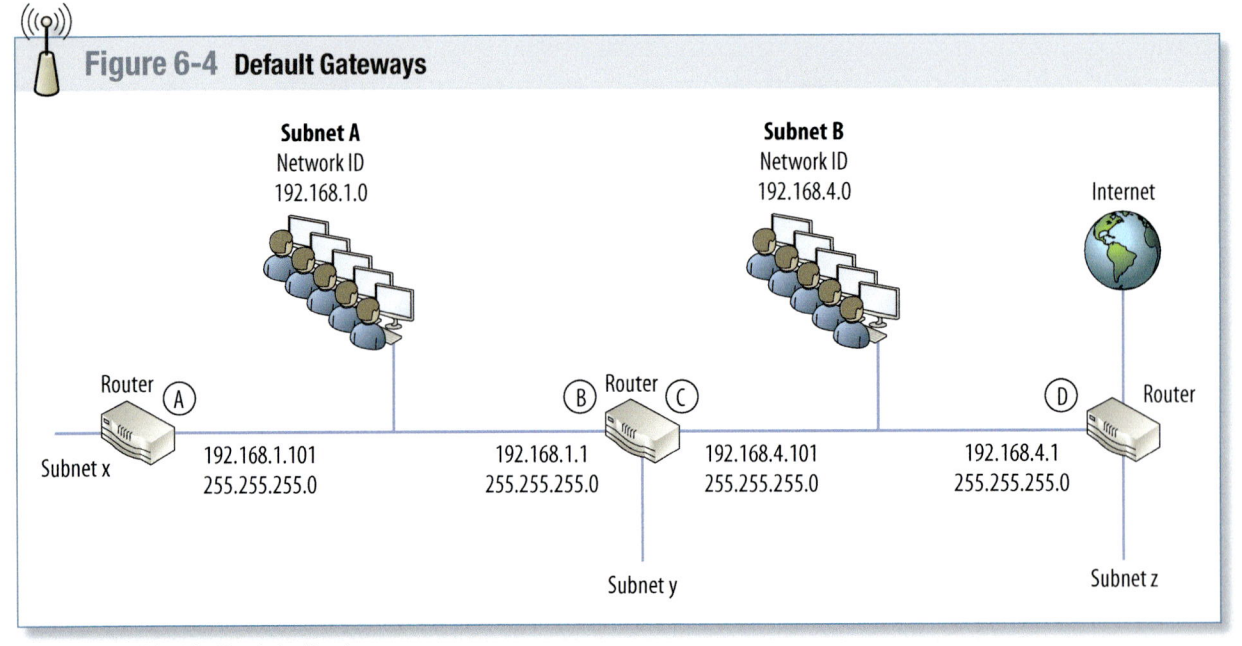

Figure 6-4. Identify the default gateways.

Figure 6-5 Comparing Two IP Addresses

EXAMPLE	SOURCE IP	DESTINATION IP	LOCAL OR REMOTE
1	192.168.1.5	192.168.1.254	
	255.255.255.0	255.255.255.0	Local
2	10.80.1.23	10.80.2.27	
	255.255.0.0	255.255.0.0	Local
3	192.168.1.17	192.168.11.23	
	255.255.255.0	255.255.255.0	Remote

Figure 6-5. The source IP address/subnet mask and the destination IP address/subnet mask will help determine if it is a local or remote network.

Example 2 The network ID of the source IP address is 10.80.0.0. The network ID of the destination IP address is 10.80.0.0. These are the same, so they are local to each other.

Example 3 The network ID of the source IP address is 192.168.1.0. The network ID of the destination IP address is 192.168.11.0. These are different, so the destination IP address is remote.

Classful IP Addresses

IPv4 is a classful logical addressing scheme using three primary address classes: Class A, Class B, and Class C. The class of the address is determined by the first number in the IP address. Additionally, the subnet mask is predetermined for each class. **See Figure 6-6.**

Note that the first example IP address (10.80.1.15) has a 10 as the first number. The number 10 is in the range 1 through 126, making this a Class A address, with a subnet mask of 255.0.0.0 and a network ID of 10.0.0.0. The second example IP address (172.16.32.15) has the number 172 first, making it a Class B address with a subnet mask of 255.255.0.0 and a network ID of 172.16.0.0. The third example (192.168.1.5) has a 192 first, making it a Class C address with a subnet mask of 255.255.255.0 and a network ID of 192.168.1.0.

Figure 6-6 Classful IP Addresses

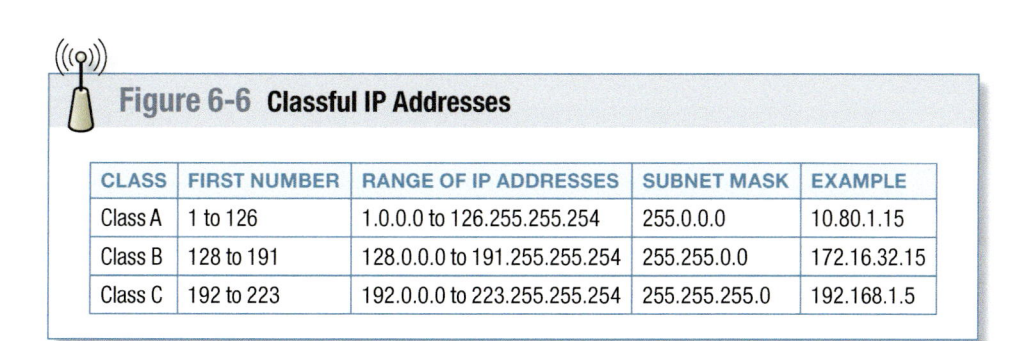

CLASS	FIRST NUMBER	RANGE OF IP ADDRESSES	SUBNET MASK	EXAMPLE
Class A	1 to 126	1.0.0.0 to 126.255.255.254	255.0.0.0	10.80.1.15
Class B	128 to 191	128.0.0.0 to 191.255.255.254	255.255.0.0	172.16.32.15
Class C	192 to 223	192.0.0.0 to 223.255.255.254	255.255.255.0	192.168.1.5

Figure 6-6. Class A, Class B, and Class C IP addresses cover the usable range of IP addresses.

Introduction to Network Technologies

Understanding Class D and Class E Addresses

Class D and Class E addresses also exist, but are not as important when understanding classful addressing. Class D is used for multicasting and includes the address range from 224.0.0.0 through 239.255.255.255. Class E is a reserved range from 240.0.0.0 through 255.255.255.255.

✓ **FACT**

To test the TCP/IPv4 stack on a local Windows computer, enter the `ping 127.0.0.1` command at the command prompt. It should return four replies.

Consider three classful IP address ranges with their network IDs and host IDs separated. It should be easy to identify the value of the high-order bits for each of the classes. **See Figure 6-7**.

One of the benefits of using classful IP addressing is being able to determine the subnet mask by looking only at the IP address. Once the subnet mask is known, then determining the network ID should be simple. For example, to determine the network IDs of the following classful IP addresses:

- 192.168.1.3
- 172.16.4.7
- 10.80.20.4

Determine the class of each IP address, identify the subnet mask of the class, and then use that information to determine the network ID. **See Figure 6-8**.

Although most of the addresses in the three address ranges are usable, there are some restrictions. For example, notice that the entire range starting with 127 is missing. This Class A address range is reserved for testing and other purposes.

Figure 6-7 High-Order Bit Identification

Class A Addresses
Range 1 to 126
SM 255.0.0.0

Network ID (8 Bits) | Host ID (24 Bits)
0 0

Class B Addresses
Range 128 to 191
SM 255.255.0.0

Network ID (16 Bits) | Host ID (16 Bits)
1 0

Class C Addresses
Range 192 to 223
SM 255.255.255.0

Network ID (24 Bits) | Host ID (8 Bits)
1 1

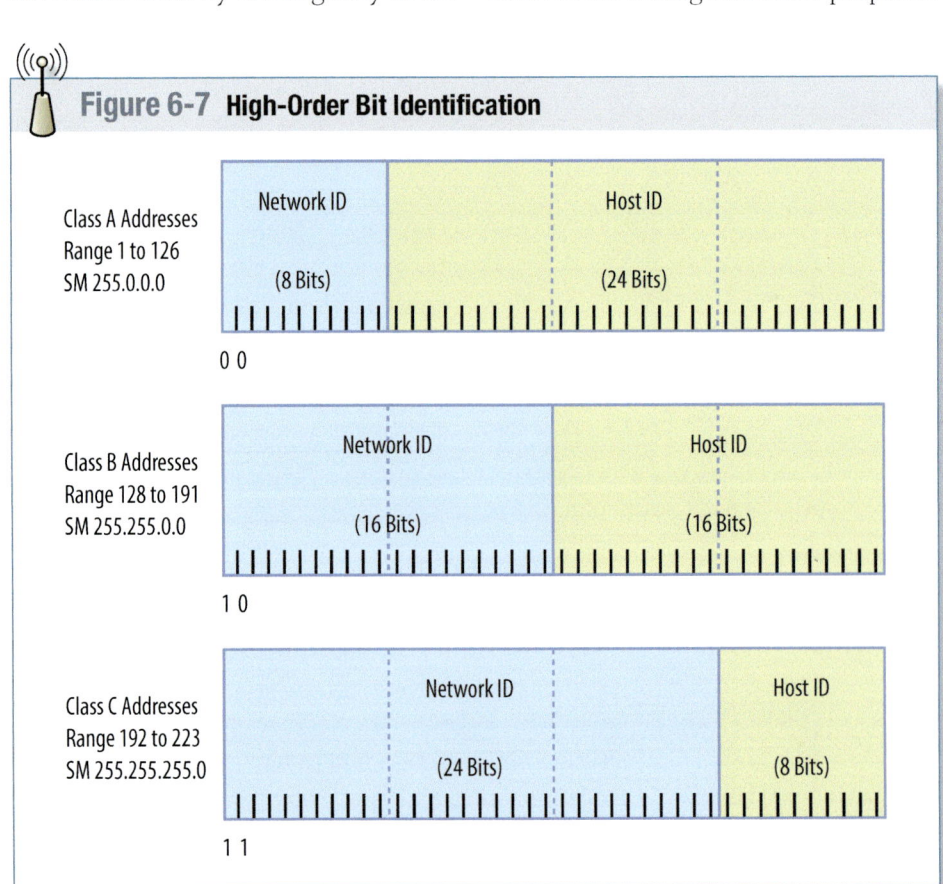

Figure 6-7. *It is easy to identify the value of high-order bits when the classful IP addresses have separate network IDs and host IDs.*

Figure 6-8 Network ID of a Classful IP Address

IP ADDRESS	CLASS	SUBNET MASK	NETWORK ID
192.168.1.3	Class C	255.255.255.0	192.168.1.0
172.16.4.7	Class B	255.255.0.0	172.16.0.0
10.80.20.4	Class A	255.0.0.0	10.0.0.0

Figure 6-8. If classful addressing is used, the default subnet mask of the class may be used to determine the Network ID.

The address of 127.0.0.1 is known as the loopback address. The *Loopback address* is used to test the TCP/IPv4 protocol stack (the software). It is important to note that the network interface card (NIC) could be damaged, yet the computer could still successfully ping 127.0.0.1. There are also several other IP address ranges reserved for use on internal networks only.

Reserved IP Address Ranges

RFC 1918 identifies several IP address ranges for use in private networks only. These addresses are not assigned to any computers on the Internet, but instead are left for assignment to computers on internal networks. These private IP ranges are as follows:

- 10.0.0.0 through 10.255.255.255
- 172.16.0.0 through 172.31.255.255
- 192.168.1.0 through 192.168.255.255

However, the first and last address in each of these ranges is not usable. The only usable addresses in these ranges are as follows:

- 10.0.0.1 through 10.255.255.254
- 172.16.0.1 through 172.31.255.254
- 192.168.1.1 through 192.168.255.254

IP addresses on the Internet must be unique. No two computers on the Internet can use the same IP address. However, since private addresses are internal to a company, different companies can use the same IP addresses on their internal networks. In other words, company A can use a range of 192.168.1.1 through 192.168.1.254 for computers in their network, and company B can use the exact same numbers.

BINARY IPV4 ADDRESS

Although the dotted decimal format can usually work with IPv4 addresses, it may occasionally be necessary to dig a little deeper.

The Bits of an IP Address

People have ten fingers and generally count using the decimal system with a base of 10. However, computers only understand ones and zeroes, and count using the binary system with a base of 2. Each binary number is a bit and can have a value of 1 or 0.

An IPv4 address has 32 bits. The IP address is commonly expressed in dotted decimal format, but it can also be expressed in four groups of eight bits. Each group of eight bits is also known as an octet in the IP address. In other words, an IP address has four decimals in dotted decimal format, which can also be expressed as four octets in binary format.

✓ FACT

A special Automatic Private IP Addressing (APIPA) range is from 169.254.0.1 through 169.254.255.254.

✓ FACT

Octet means eight, and it is accurate to say that an IP address has four octets. Eight bits is also a byte, and the address can be referred to as four bytes.

Consider an IP address and subnet mask expressed in both decimal and binary form. **See Figure 6-9.**

A logical question is, "How does 192 in decimal equate to 1100 0000 in binary?" The answer is based on which bits are ones in the binary string. The low-order bit (2^0) is 1, since any number raised to the 0 power is 1. And 2^1 is 2, since any number raised to the first power is equal to itself. The high-order bit (2^7) has a decimal value of 128. **See Figure 6-10.**

If the first two bits are a 1 (1100 0000), it represents one decimal value of 128 and one decimal value of 64. The sum of 128 + 64 equals 192. These eight bits can represent any value between 0 and 255. If all eight bits are a 0 (0000 0000), the decimal value is 0. If all eight bits are a 1 (1111 1111), the value is 255.

To determine the decimal values of different examples, add the decimal value for each bit that has a binary 1. **See Figure 6-11.**

Figure 6-9 Comparing Dotted Decimal and Binary

IP DECIMAL	IP BINARY	SUBNET MASK DECIMAL	SUBNET MASK BINARY
192	1100 0000	255	1111 1111
168	1010 1000	255	1111 1111
1	0000 0001	255	1111 1111
5	0000 0101	0	0000 0000

Figure 6-9. Binary numbering is expressed in 1's and 0's.

Figure 6-10 Binary and Decimal Values

	2^7	2^6	2^5	2^4	2^3	2^2	2^1	2^0
Decimal value	128	64	32	16	8	4	2	1

Figure 6-10. The highest order bit is equal to 128 while the lowest order bit is equal to 1.

Figure 6-11 Binary Values by Bit

	2^7	2^6	2^5	2^4	2^3	2^2	2^1	2^0
Decimal value	128	64	32	16	8	4	2	1
Example 1	1	1	0	0	0	0	0	0
Example 2	1	0	1	0	1	0	0	0
Example 3	0	0	0	0	0	0	0	1
Example 4	0	0	0	0	0	1	0	1
Example 5	0	0	0	0	1	0	1	0
Example 6	0	0	0	0	0	0	0	0
Example 7	1	1	1	1	1	1	1	1

Figure 6-11. The decimal value is equal to the sum of each column with a 1 in it.

Following is the solution:

- Example 1 = 192 (128 + 64)
- Example 2 = 168 (128 + 32 + 8)
- Example 3 = 1 (1)
- Example 4 = 5 (4 + 1)
- Example 5 = 10 (8 + 2)
- Example 6 = 0 (none of the bits is a 1)
- Example 7 = 255 (128 + 64 + 32 + 16 + 8 + 4 + 2 + 1)

Notice that Examples 1 through 4 are the binary values of the IP address 192.168.1.5. In binary form, the full IP address is 1100 0000 . 1010 1000 . 0000 0001 . 0000 0101.

A simpler way of converting binary to decimal and decimal to binary is with the Calculator application built into the Windows operating system. From the View drop-down menu, select the Programmer View. Other operating systems provide the same capability in the Scientific View. **See Figure 6-12.**

After selecting the proper view in the Calculator, enter the decimal number, and then click Bin to convert it to binary. To convert it back to decimal, simply click Dec. It is also possible to convert numbers to base-16 hexadecimal numbers (Hex) or base-8 octal numbers (Oct).

CIDR Notation

The subnet mask is sometimes referenced in a type of shorthand called Classless Inter-Domain Routing (CIDR) notation, based on the number of bits having the maximum value (1) in the subnet mask. For example, if the subnet mask is 255.0.0.0, it has eight bits in

Digit Grouping

When using decimal numbers, it is common to group digits in threes separated by a comma for better readability. For example, the number 1,234,567 is easier to read than the number 1234567.

Similarly, binary numbers are grouped with four bits separated by a space. It is easier to read 1100 0000 than it is to read 11000000. When digit grouping is used, it is easy to see that it is two groups of four, but when digit grouping is not used, it is not always immediately apparent how many digits are in the binary string.

The actual value does not change when digit grouping is used. The value of 1,234,567 is the same as 1234567, and the value of 1100 0000 is the same as 11000000.

✓ **FACT**

Use the Calculator to check conversions. Some exams include a calculator, but a binary-to-decimal converter may not always be available.

Figure 6-12 Windows Calculator

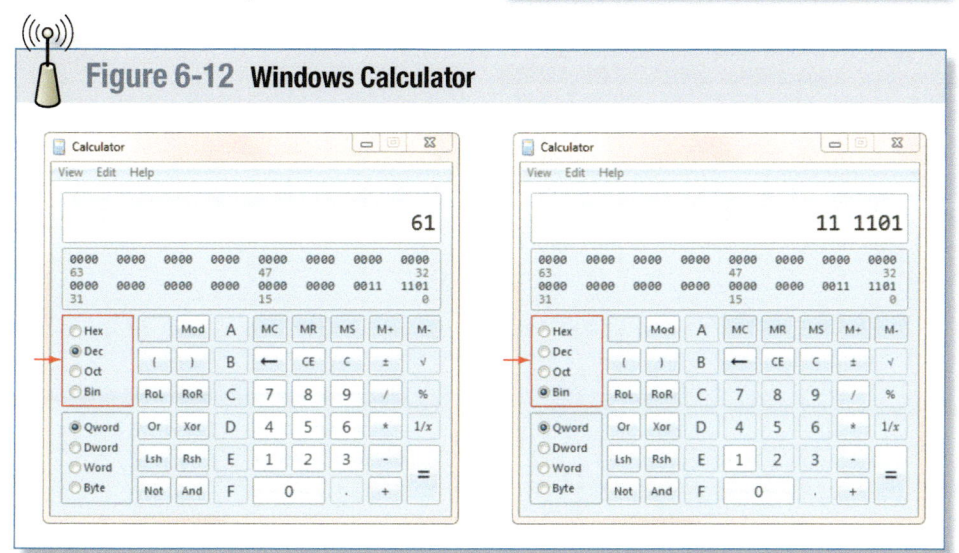

Figure 6-12. *The Windows Calculator application can assist in converting decimal to binary.*

use and can be referenced as /8. If the subnet mask is 255.255.0.0, it has 16 bits in use and can be referenced as /16. If the subnet mask is 255.255.255.0, it has 24 bits in use and can be referenced as /24. **See Figure 6-13.**

IP Address Masking

An IPv4 address includes two components: a network ID and a host ID. The subnet mask identifies which is which by masking out the network ID. When the subnet mask is configured at its maximum value (255), that portion of the IP address is the network ID.

The same point is true when using binary numbers. Consider an IP address of 192.168.1.5 with a subnet mask of 255.255.255.0. When the subnet mask is configured at its maximum (1 in binary), that portion of the IP address is the network ID. **See Figure 6-14.**

The computer looks at the first bit in the IP address (in this case, it is set to 1) and the first bit in the subnet mask (also set to 1), and then ANDs them together, which yields a 1 as the first bit in the network ID. It then looks at the second bit in the IP address (1) and the second bit in the subnet mask (also set to 1) and is set to a 1 as the second bit in the network ID. It does this with each of the bits to determine the network ID.

It is possible to do this bit by bit, but it is simpler to just look at which bits

Boolean AND Logic

Within the computer, the "masking" is done by using Boolean AND logic. Boolean AND logic compares two bits and provides a single bit as the output. If both bits are a 1, the output is a 1. However, if either of the bits is a 0, or both bits are a 0, the output is a 0. The following list shows the four possibilities when ANDing two bits:

- 0 AND 0 = 0
- 0 AND 1 = 0
- 1 AND 0 = 0
- 1 AND 1 = 1

Figure 6-13 Examples of CIDR Notation

IP ADDRESS	SUBNET MASK	CIDR NOTATION
192.168.1.5	255.255.255.0	192.168.1.5 /24
172.17.34.5	255.255.0.0	172.17.34.5 /16
10.80.4.7	255.0.0.0	10.80.4.7 /8

Figure 6-13. A "/8" after the IP address represents the number of 1s in the subnet mask when viewed in binary.

Figure 6-14 Masking an IP Address

	FIRST OCTET	SECOND OCTET	THIRD OCTET	FOURTH OCTET
192.168.1.5	1100 0000	1010 1000	0000 0001	0000 0101
255.255.255.0	1111 1111	1111 1111	1111 1111	0000 0000
Network ID	1100 0000	1010 1000	0000 0001	0000 0000

Figure 6-14. The network ID is determined by the number of 1s in the subnet mask.

are 1s in the subnet mask and recognize that the corresponding bits in the IP address make up the network ID.

Classless IP Addresses

Classful IP addresses include the Class A, Class B, and Class C addresses. Remember that when using a classful IP address, the subnet mask is implied and does not need to be included. Classless IP addressing can also be used.

When using a classless IP address, it is required to know both the IP address and the subnet mask to determine the network ID. As an example, an address of 10.80.1.5 is a classful IP address with a subnet mask of 255.0.0.0 and a network ID of 10.0.0.0. However, the same IP address can be used as a classless IP address with different subnet masks, resulting in different network IDs. **See Figure 6-15.**

IPV4 ADDRESS SUBNETTING

Classful IPv4 address ranges can be divided into smaller groups of addresses, thereby creating more subnets. Smaller organizations rarely need to do this, but large organizations frequently subnet the network instead of using the typical classful IP address ranges.

Administrators and IT support personnel must understand subnetting to ensure that systems have the correct IP addresses assigned to them.

Consider a single class C network of 192.168.1.0/24. It can host 254 computers on the same subnet. **See Figure 6-16.**

Figure 6-15 Examples of Classless IP Address

IP ADDRESS	SUBNET MASK	NETWORK ID
10.80.1.5	255.255.0.0	10.80.0.0
10.80.1.5	255.255.255.0	10.80.1.0

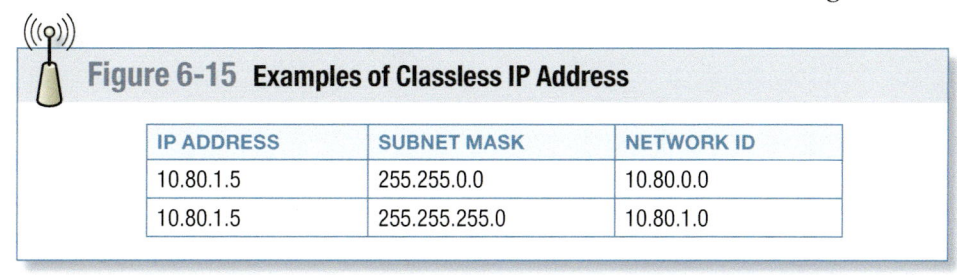

Figure 6-15. The IP address and subnet mask are needed to determine the network ID.

Figure 6-16 Class C Network

Network ID 192.168.1.0 / 24
192.168.1.1 to 192.168.1.254
Subnet Mask 255.255.255.0

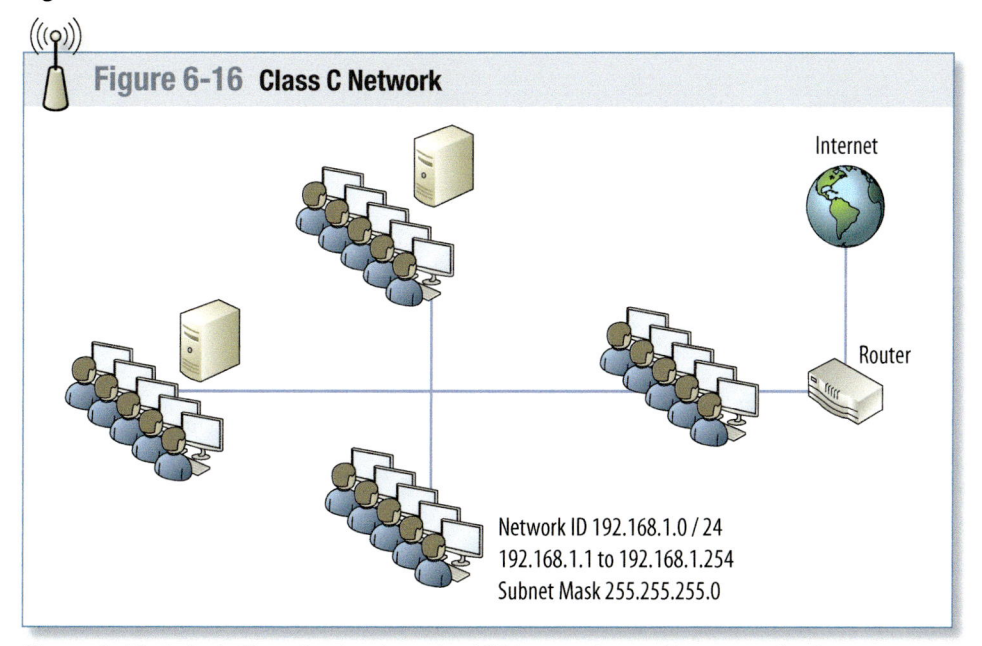

Figure 6-16. A single Class C network can host 254 computers on the same subnet.

Imagine that users on this network are in four primary groups as follows:

- One group is regularly streaming video from a server.
- Another group is regularly uploading and downloading large graphics files.
- A third group is downloading large volumes of data from the Internet.
- The last group is just a regular group of users with occasional server and Internet access.

If all four groups of users are on the same subnet, their traffic will compete with each other for network bandwidth. The overall performance of the subnet may be slow. However, if the four groups are divided into different subnets, each subnet will enjoy better performance.

Number of Subnet Bits

With this example, it makes sense to create four separate subnets. This is accomplished by borrowing bits from the host ID portion of the IP address and adding them to the network ID. They create a new portion of the network ID referred to as the subnet ID. By borrowing one bit, two subnets can be created. A single bit has two states, either a 1 or a 0, but four subnets are needed, not two. By borrowing two bits, four subnets can be created.

The single Class C address is subdivided by borrowing the two high-order bits from the host ID portion of the address. Say that the Class C address has a subnet mask of 255.255.255.0, with 24 bits in the network ID portion of the IP address and 8 bits in the host ID portion. **See Figure 6-17.**

Figure 6-17 Creating Subnets

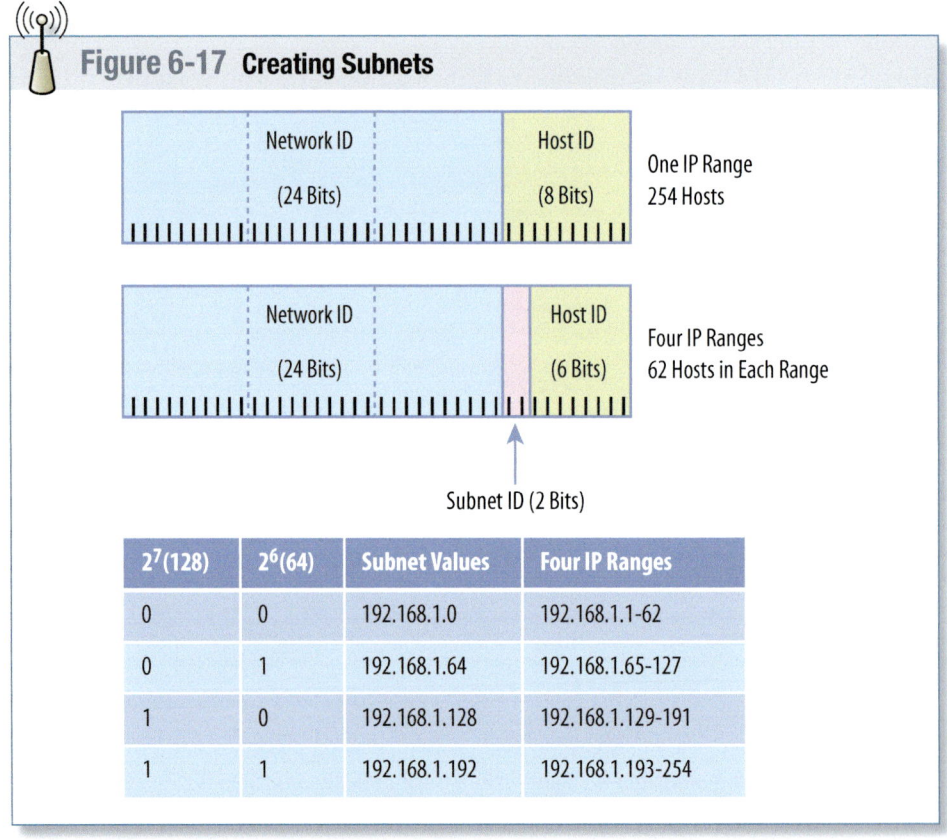

2^7(128)	2^6(64)	Subnet Values	Four IP Ranges
0	0	192.168.1.0	192.168.1.1-62
0	1	192.168.1.64	192.168.1.65-127
1	0	192.168.1.128	192.168.1.129-191
1	1	192.168.1.192	192.168.1.193-254

Figure 6-17. Borrowing bits from the host ID portion of the IP address and adding them to the network ID facilitates the creation of subnets.

After borrowing two bits from the original eight bits in the host ID portion, six bits are left for the host ID. The 24 bits of the original network ID and the two bits of the subnet ID are combined to give a total of 26 bits for the network ID.

It is important to realize that the two borrowed bits are the high-order bits in the host ID. They have the values of 128 and 64. These two bits have four possible combinations of 0 0, 0 1, 1 0, and 1 1, which will be used within the host ID to create four separate subnets. **See Figure 6-18.**

The network was a single large subnet of 254 hosts. By subnetting the Class C address, the four subnets can be created, each with 62 hosts. **See Figure 6-19.**

Note that each subnet has a specific non-overlapping range of IP addresses. This is an extremely important point. If an IP address of 192.168.1.200 was assigned to a computer in subnet A, it would not be able to communicate with

Figure 6-18 Subnetting with the Two High-Order Bits

2^7 (DECIMAL 128)	2^6 (DECIMAL 64)	DECIMAL VALUE
0	0	0
0	1	64
1	0	128
1	1	192

Figure 6-18. Subnetting is accomplished by borrowing from the high-order bits.

Figure 6-19 Class C Subnets

Subnet A
Network ID 192.168.1.0 / 26
Subnet Mask 255.255.255.192
192.168.1.1 to 192.168.1.62

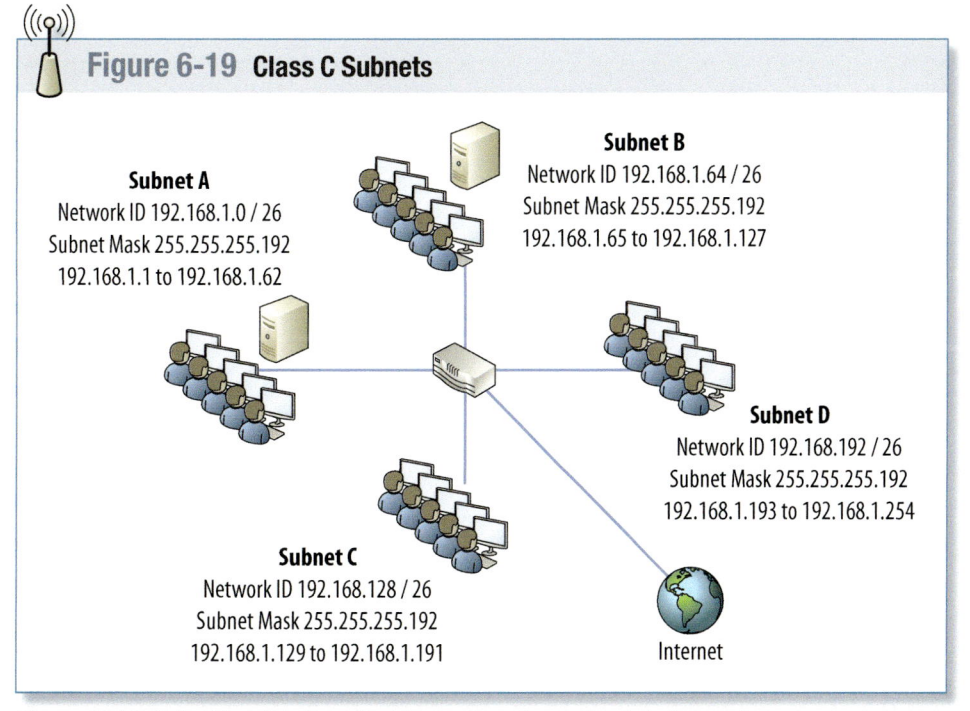

Subnet B
Network ID 192.168.1.64 / 26
Subnet Mask 255.255.255.192
192.168.1.65 to 192.168.1.127

Subnet C
Network ID 192.168.128 / 26
Subnet Mask 255.255.255.192
192.168.1.129 to 192.168.1.191

Subnet D
Network ID 192.168.192 / 26
Subnet Mask 255.255.255.192
192.168.1.193 to 192.168.1.254

Internet

Figure 6-19. Subnetting the Class C address allows four subnets to be created.

any other computer, since it has an incorrect network ID for the subnet.

To create more subnets, borrow more bits. The following formula helps to determine how many subnets can be created based on how many bits are borrowed: 2^n (where n is the number of bits borrowed).

For example, by borrowing two bits (2^2), four subnets can be created. By borrowing three bits (2^3), eight subnets can be created. Of course, the more bits that are borrowed from the host ID, the fewer hosts that can be created in a network.

Number of Hosts in a Network

Valid IP addresses cannot have all 0s in the host ID because that represents the network ID. Also, they cannot have all 1s in the host ID because that represents a broadcast address within the subnet. This eliminates two possible IP addresses in the range of IP addresses for any subnet.

For example, consider a typical Class C network address of 192.168.10 with a subnet mask of 255.255.255.0. The following two IP addresses are not possible:

- 192.168.1.0 (since this is the network ID)
- 192.168.1.255 (since this is the broadcast address for the network ID)

This gives a valid range of 192.168.1.1 through 192.168.1.254 for a total of 254 possible hosts. Use the formula $2^h - 2$ (where h is the number of bits in the host ID) to determine how many hosts are supported in any subnet.

A Class C address uses 24 bits in the network ID and 8 bits in the host ID, so the formula is $2^8 - 2$. This gives a value of 254 ($2^8 = 256$, and $256 - 2 = 254$). **See Figure 6-20.**

The same concepts can be applied to subnet a Class B or Class A network. Remember, a Class B network starts with 16 bits for the network ID and 16 bits for the host ID. **See Figure 6-21.**

A Class A network starts with 8 bits for the network ID and 24 bits for the host ID. **See Figure 6-22.**

Local and Remote Addresses

Recall how to determine whether simple IP address and subnet mask combinations were on the same subnet (local to

Figure 6-20 Determining the Number of Subnets and Hosts in a Class C Network

BORROWED BITS FROM A CLASS C ADDRESS	SUBNET MASK VALUE	NUMBER OF SUBNETS (2^n)	NUMBER OF HOSTS ($2^h - 2$)
1 0	192.168.1.128	2 (2^1)	126 ($2^7 - 2$)
2 0 0	192.168.1.192	4 (2^2)	62 ($2^6 - 2$)
3 0 0 0	192.168.1.224	8 (2^3)	30 ($2^5 - 2$)
4 0 0 0 0	192.168.1.240	16 (2^4)	14 ($2^4 - 2$)
5 0 0 0 0 0	192.168.1.248	32 (2^5)	6 ($2^3 - 2$)
6 0 0 0 0 0 0	192.168.1.252	64 (2^6)	2 ($2^2 - 2$)
7 (not valid since zero hosts are supported)	192.168.1.254	126 (2^7)	0 ($2^2 - 2$)

Figure 6-20. *A Class C network with 1 borrowed bit creates two subnets with 126 hosts each.*

Figure 6-21. Determining the Number of Subnets and Hosts in a Class B Network

BORROWED BITS FROM A CLASS B ADDRESS	SUBNET MASK VALUE	NUMBER OF SUBNETS (2^n)	NUMBER OF HOSTS ($2^h - 2$)
1 0	172.16.1.128	2 (2^1)	32,766 ($2^{15} - 2$)
2 0 0	172.16.1.192	4 (2^2)	16,384 ($2^{14} - 2$)
3 0 0 0	172.16.1.224	8 (2^3)	8,190 ($2^{13} - 2$)
4 0 0 0 0	172.16.1.240	16 (2^4)	4094 ($2^{12} - 2$)
5 0 0 0 0 0	172.16.1.248	32 (2^5)	2046 ($2^{11} - 2$)
6 0 0 0 0 0 0	172.16.1.252	64 (2^6)	1022 ($2^{10} - 2$)
7 0 0 0 0 0 0 0	172.16.1.254	126 (2^7)	510 ($2^9 - 2$)

Figure 6-21. A Class B network with 1 borrowed bit creates two subnets with 32,766 hosts each.

each other), or whether the destination IP address was on a remote subnet. It should be simple to make the same determination even when advanced subnetting techniques are in use on the network.

However, when using advanced subnetting techniques, it is a little more difficult to determine the network ID. For example, it is not readily apparent what the network ID is in the following IP address and subnet mask combinations:

▶ Source IP: 192.168.1.61, 255.255.255.192

▶ Destination IP: 192.168.1.65, 255.255.255.192

What Is Supernetting?

The term "supernetting" may crop up from time to time. Although the process is rather advanced, it is worthwhile understanding the big picture of supernetting. In short, supernetting is the opposite of subnetting.

Recall that subnetting divides a larger network into multiple smaller networks by taking bits from the host ID. Supernetting combines multiple smaller networks into a single larger network by taking bits from the network ID. This is a useful function when optimizing routing devices on a network.

Figure 6-22. Determining the Number of Subnets and Hosts in a Class A Network

BORROWED BITS FROM A CLASS A ADDRESS	SUBNET MASK VALUE	NUMBER OF SUBNETS (2^n)	NUMBER OF HOSTS ($2h - 2$)
1 0	192.168.1.128	2 (2^1)	8,388,606 ($2^{23} - 2$)
2 0 0	192.168.1.192	4 (2^2)	4,194,302 ($2^{22} - 2$)
3 0 0 0	192.168.1.224	8 (2^3)	2,097,150 ($2^{21} - 2$)
4 0 0 0 0	192.168.1.240	16 (2^4)	1,048,574 ($2^{20} - 2$)
5 0 0 0 0 0	192.168.1.248	32 (2^5)	524,286 ($2^{19} - 2$)
6 0 0 0 0 0 0	192.168.1.252	64 (2^6)	262,142 ($2^{18} - 2$)
7 0 0 0 0 0 0 0	192.168.1.254	126 (2^7)	131,070 ($2^{17} - 2$)

Figure 6-22. A Class A network with 1 borrowed bit creates two subnets with 8,388,606 hosts each.

Simplify the process of determining the network ID by following these five steps:

1. Convert the IP address to binary. Use a calculator to do this, if necessary.
2. Convert the subnet mask to binary.
3. Draw a vertical line after the last one value in the subnet mask:
    ▶ Everything to the left of the line is the network ID.
    ▶ Everything to the right of the line is the host ID.
4. Determine the network ID in binary. This is as simple as copying the IP address in binary to the left of the line and writing 0s to the right of the line.
5. Convert the network ID to decimal.

Now, identify the steps for the 192.168.1.61 IP address. **See Figure 6-23.**

Identify the steps for 192.168.1.65. See **Figure 6-24.**

Completing these steps reveals that the network ID of the source IP address is 192.168.1.0, and the network ID of the destination IP address is 192.168.1.64. These network IDs are not the same, so the destination IP address is remote.

Subnetting Usage

On the job, knowledge of subnetting is important because a misconfigured computer might not communicate with other systems. It should be possible to look at the IP address and determine whether it is correct.

Recall basic troubleshooting by evaluating the IP addresses and subnet masks of several computers, some of which were configured incorrectly. However, those examples used subnet masks of 255.255.255.0 and 255.255.0.0 only. An actual network might have advanced subnet masks. **See Figure 6-25.**

Here, the default gateways are configured correctly. The configured IP

Figure 6-23 192.168.1.61 Network ID

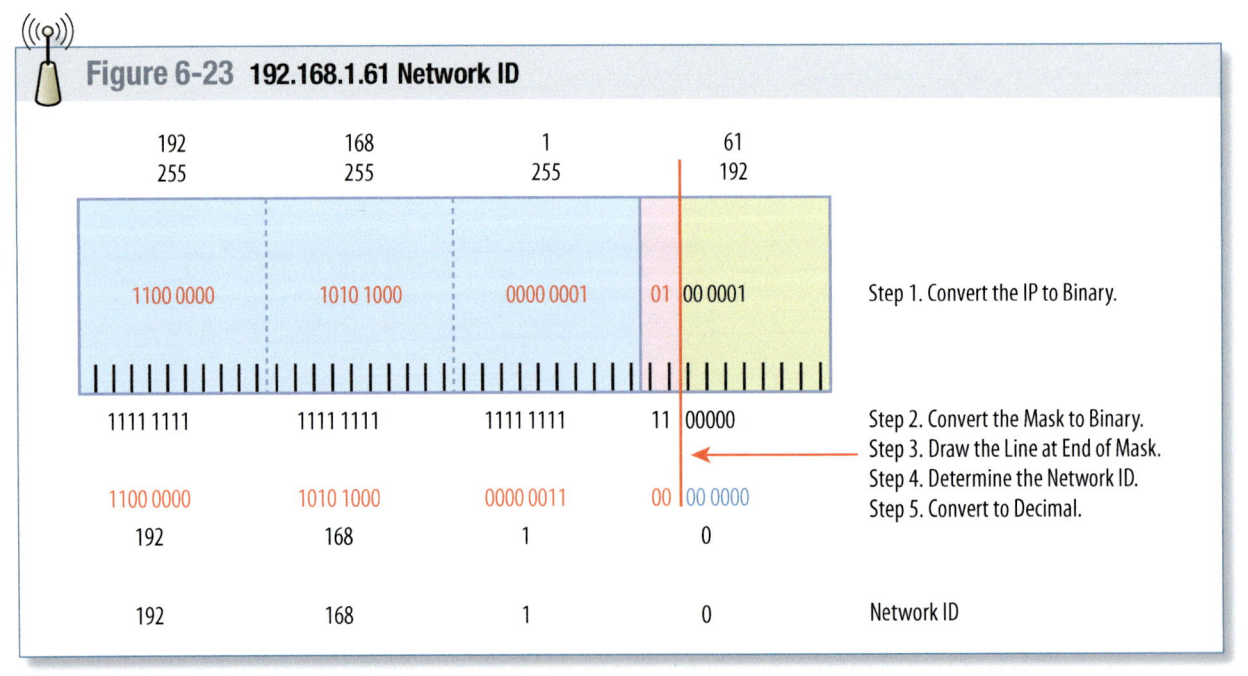

Figure 6-23. Determine the network ID of 192.168.1.61.

Figure 6-24 198.168.1.65 ID

192	168	1	65
255	255	255	192

| 1100 0000 | 1010 1000 | 0000 0001 | 01 | 00 0001 |

Step 1. Convert the IP to Binary.

| 1111 1111 | 1111 1111 | 1111 1111 | 11 | 00000 |

Step 2. Convert the Mask to Binary.
Step 3. Draw the Line at End of Mask.
Step 4. Determine the Network ID.
Step 5. Convert to Decimal.

| 1100 0000 | 1010 1000 | 0000 0011 | 01 | 00 0000 |
| 192 | 168 | 1 | 64 |

| 192 | 168 | 1 | 64 | Network ID |

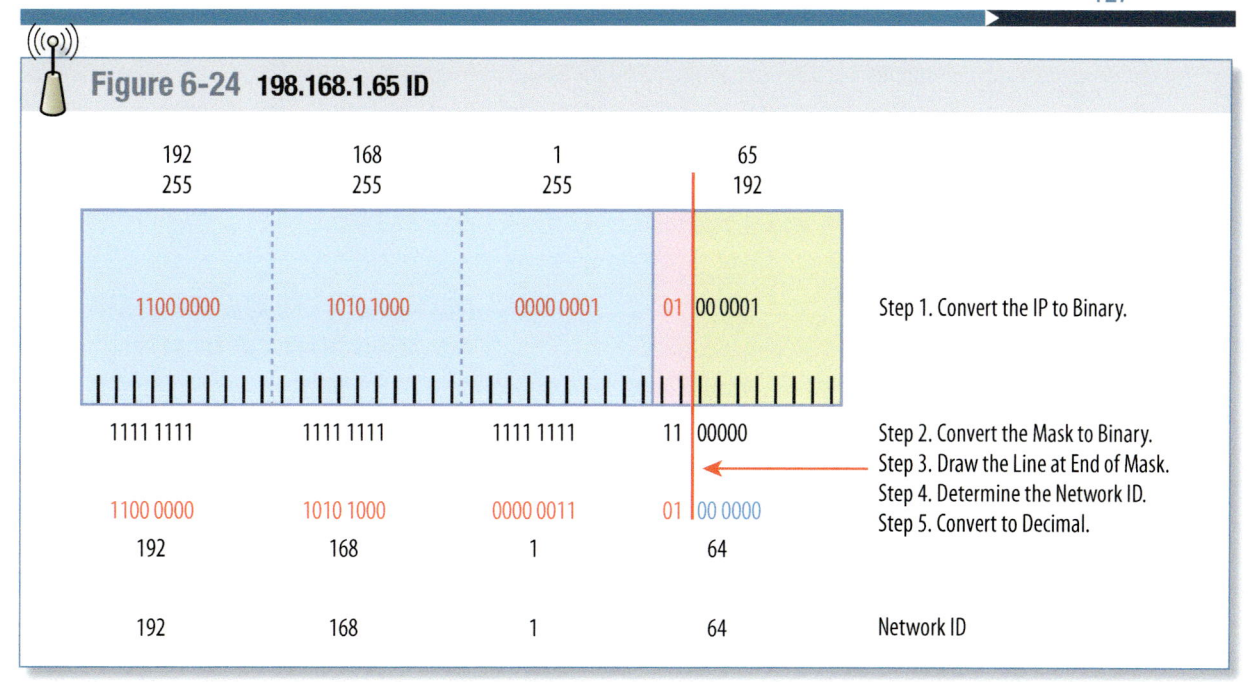

Figure 6-24. *Determine the network ID of 192.168.1.65.*

Figure 6-25 Advanced Subnet Masks

IP: 192.168.129.63
SM: 255.255.255.192
DG: 192.168.129.65

IP: 192.168.129.66
SM: 255.255.255.192
DG: 192.168.129.65

IP: 10.80.162.2
SM: 255.255.224.0
DG: 10.80.160.1

IP: 10.80.159.3
SM: 255.255.224.0
DG: 10.80.160.1

Subnet A

Router
A — 192.168.129.65 / 255.255.255.192
B — 10.80.160.1 / 255.255.224.0

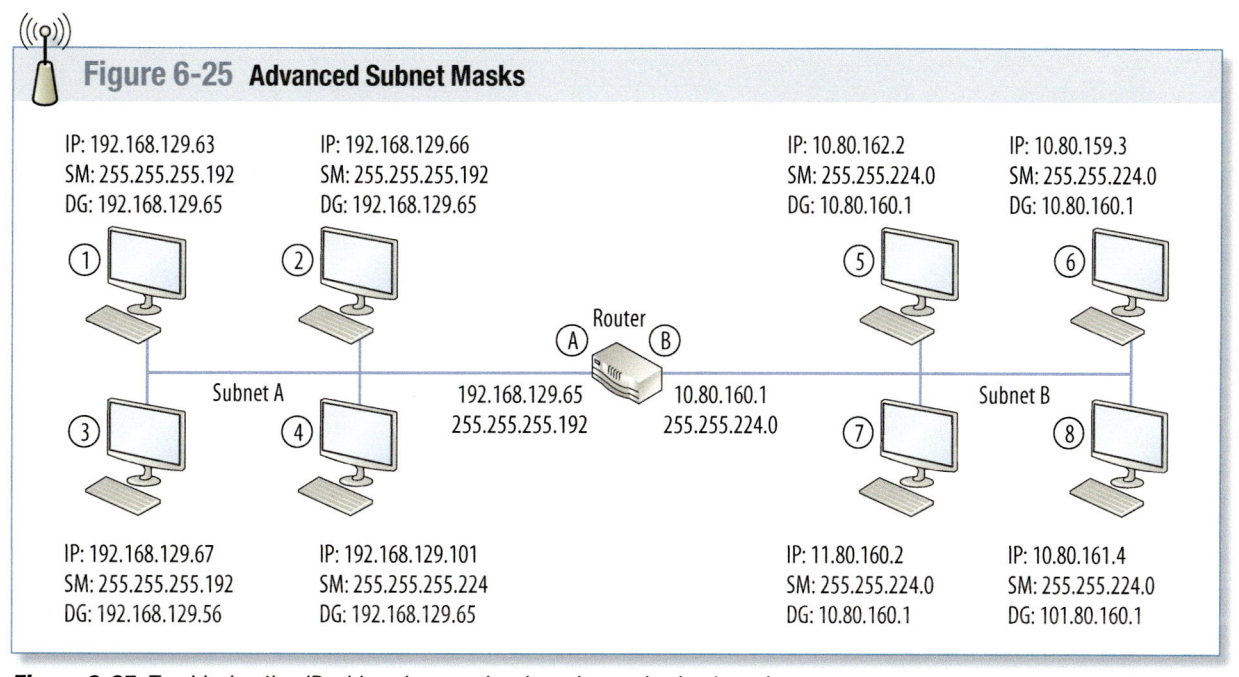

Subnet B

IP: 192.168.129.67
SM: 255.255.255.192
DG: 192.168.129.56

IP: 192.168.129.101
SM: 255.255.255.224
DG: 192.168.129.65

IP: 11.80.160.2
SM: 255.255.224.0
DG: 10.80.160.1

IP: 10.80.161.4
SM: 255.255.224.0
DG: 101.80.160.1

Figure 6-25. *Troubleshooting IP addressing may involve advanced subnet masks.*

address, subnet mask, and default gateway are shown for each of these computers, but they are not necessarily configured correctly. Instead, they show common typographical errors that occur when administrators configure a system manually.

Recall, all computers on the same subnet must have the same network ID, and they must be configured with the correct IP address of the default gateway. To identify errors in the network it may help to draw the network layout to determine the problem areas.

Computer 1 The network ID of the computer is 192.168.129.0. However, the network ID of the default gateway is 192.168.129.64. Since the network IDs are different and the default gateway is known to be correct, the computer is configured with an incorrect IP address.

Computer 2 The network ID is 192.168.129.64, which is the same as the default gateway. This computer is configured correctly.

Computer 3 The network ID is 192.168.129.64, which is the same as the default gateway. However, the default gateway is configured incorrectly as 192.168.129.56 instead of 192.168.129.65.

Computer 4 The subnet mask is incorrect on this computer. All of the computers on the same subnet must have the same subnet mask. The network ID is 192.168.129.100, which is different from the network ID of the default gateway (192.168.129.64).

Computer 5 The network ID of the computer is 10.80.160.0, which is the same as the network ID of the default gateway. This computer is configured correctly.

Computer 6 The network ID of the computer is 10.80.128.0, which is different from the network ID of the default gateway (10.80.160.0). This computer is not configured correctly.

Computer 7 The network ID of the computer is 11.80.128.0 (notice the first octet is 11 instead of 10), which is different from the network ID of the default gateway (10.80.160.0). This computer is not configured correctly.

Computer 8 The network ID of the computer is 10.80.160.0, which is the same as the network ID of the default gateway. However, the default gateway is configured with an incorrect IP address. It should be 10.80.160.1 instead of 101.80.160.1.

MANUAL VERSUS AUTOMATIC ASSIGNMENT OF IPV4 ADDRESSES

IPv4 information can be assigned manually or automatically. "Manually" requires actually typing the IP address, subnet mask, default gateway, DNS server address, and other TCP/IP information into the configuration screens on each computer. "Automatic" assignment uses a Dynamic Host Configuration Protocol (DHCP) server to assign the information to the computers without user intervention.

In most networks, it is much easier to use DHCP. The majority of the clients automatically get their TCP/IP configuration from a DHCP server. However, some clients might need to have a manually configured IP address. For example, the DHCP server itself must have a manually-assigned address. Additionally, some clients do not support DHCP, and devices such as routers must be manually configured.

Configuring IPv4 Configured Manually

To manually view or configure the IPv4 information on a Windows system, use an interface shown in the Properties dialog. **See Figure 6-26.**

DHCP

If the system is configured to obtain an IP address automatically, a server on the network must be running DHCP. Windows Server 2012 R2 servers include

Figure 6-26 Properties Dialog

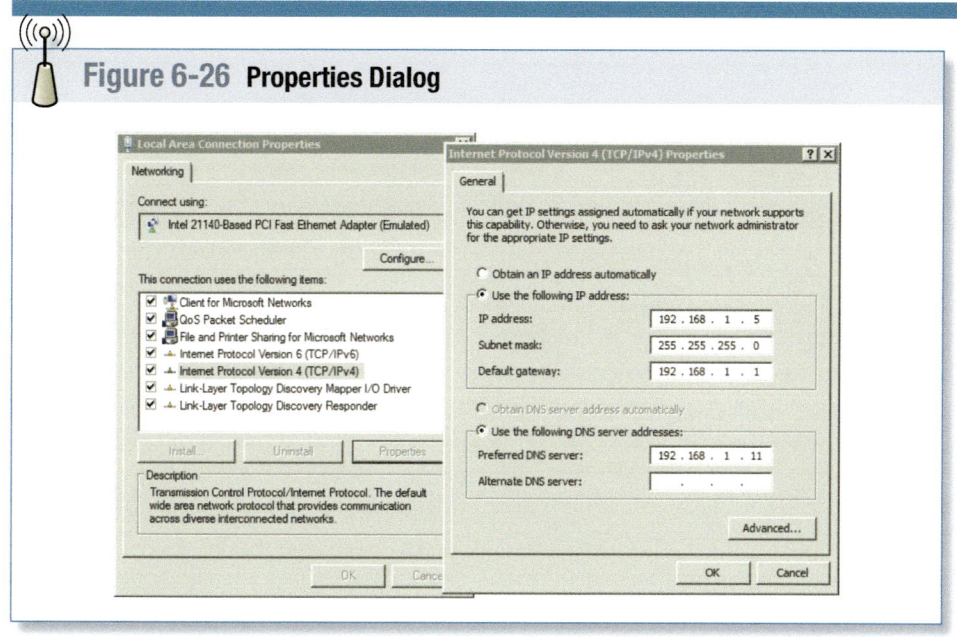

Figure 6-26. The Properties dialog provides information on the IPv4 configuration of the NIC.

the DHCP role and can be configured to run DHCP.

Consider the process a DHCP client uses to obtain an IP address and other IP information from a DHCP server. This process is commonly called the DORA process, referring to the first letter in each of the packets (Discover, Offer, Request, and Acknowledge). **See Figure 6-27.**

1. When the DHCP client turns on, it sends a broadcast looking for a DHCP server. This is the Discover packet.

2. The DHCP server answers with an Offer packet. The offer includes an IP address, subnet mask, and other information such as the address of a DNS server. This offer is also referred to as a lease offer. A lease, in this context, like an agreement to rent a piece of property for a certain length of time, is an agreement between the server and the client to exclusive use of an IP address for a period of time.

3. The DHCP client replies with a Request packet to request the lease.

Figure 6-27 DORA Process

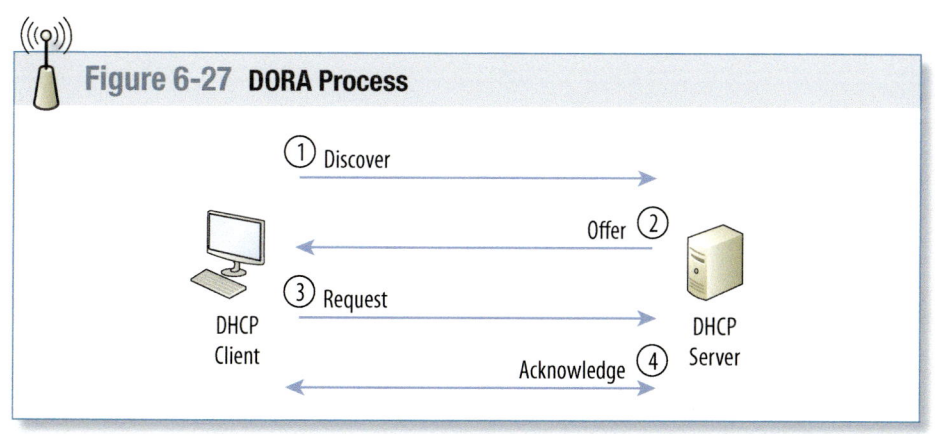

Figure 6-27. The DORA process enables a DHCP server to obtain an IP address and other information.

If the DHCP client receives offers from multiple DHCP servers, it requests a lease only from the first DHCP that offers a lease.

4. The DHCP server responds with an Acknowledge packet. The DHCP server assigns this IP address to this client and removes the IP address from the list of available IP addresses to lease to other clients.

The DHCP broadcast packets are special bootp broadcast messages defined in RFC 1542. A regular broadcast message would not pass through a router, but a bootp broadcast uses UDP ports 67 and 68. Routers on the network can be configured to pass these bootp broadcasts. Without bootp broadcasts, it would require placing a DHCP server on each subnet, or utilizing a DHCP proxy service to act as a liaison between the DHCP client and server.

APIPA

If a Windows DHCP client is unable to reach a DHCP server, it will automatically assign itself an IP address using Automatic Private Internet Protocol Addressing (APIPA). The APIPA address always starts with 169.254 in the IP address and always has a subnet mask of 255.255.0.0. The host ID is randomly generated by the client computer and then broadcast on the network to check for IP address conflicts. If no conflicts are found, the client will assume the generated IP address.

The APIPA address provides limited connectivity for clients on the network. If other clients also have an APIPA address, then they have a network ID of 169.254.0.0, and they can communicate with each other. However, APIPA does not provide a default gateway, so clients will not be able to access any resources outside of the subnet, including the Internet.

Typing ipconfig /all at the command prompt reveals whether a client has been assigned an IPv4 address. **See Figure 6-28.**

In the code listing, three lines are in boldface type:

DHCP Enabled   When set to Yes, it shows this system is a DHCP client.

Autoconfiguration Enabled   When set to Yes, this indicates that APIPA is enabled and an APIPA address will be assigned if a DHCP server cannot be reached.

Autoconfiguration IPv4 Address   An address starting with 169.254 shows this is an APIPA address. Moreover, this shows that the DHCP client could not receive an address from a DHCP server.

✓ FACT

Routers that can pass bootp broadcasts on UDP ports 67 and 68 are RFC 1542 compatible.

Figure 6-28 ipconfig /all *output*

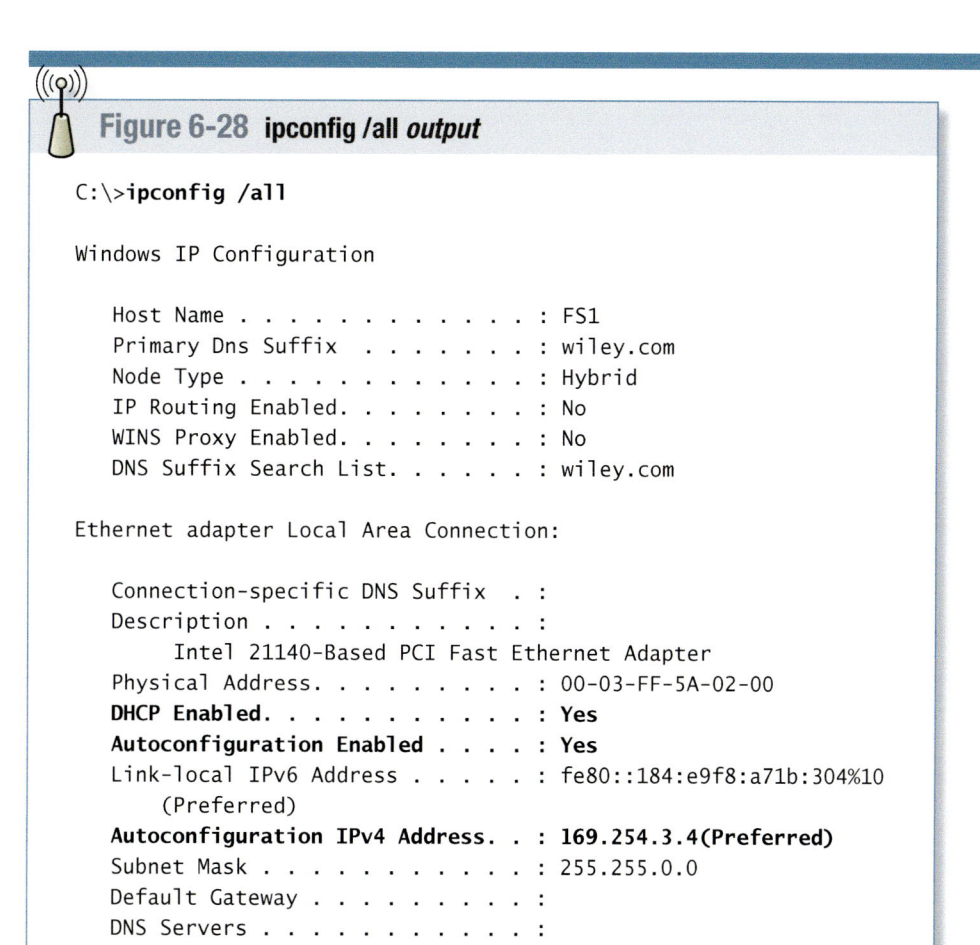

```
C:\>ipconfig /all

Windows IP Configuration

    Host Name . . . . . . . . . . . . : FS1
    Primary Dns Suffix  . . . . . . . : wiley.com
    Node Type . . . . . . . . . . . . : Hybrid
    IP Routing Enabled. . . . . . . . : No
    WINS Proxy Enabled. . . . . . . . : No
    DNS Suffix Search List. . . . . . : wiley.com

Ethernet adapter Local Area Connection:

    Connection-specific DNS Suffix  . :
    Description . . . . . . . . . . . :
        Intel 21140-Based PCI Fast Ethernet Adapter
    Physical Address. . . . . . . . . : 00-03-FF-5A-02-00
    DHCP Enabled. . . . . . . . . . . : Yes
    Autoconfiguration Enabled . . . . : Yes
    Link-local IPv6 Address . . . . . : fe80::184:e9f8:a71b:304%10
        (Preferred)
    Autoconfiguration IPv4 Address. . : 169.254.3.4(Preferred)
    Subnet Mask . . . . . . . . . . . : 255.255.0.0
    Default Gateway . . . . . . . . . :
    DNS Servers . . . . . . . . . . . :
    NetBIOS over Tcpip. . . . . . . . : Enabled
```

Figure 6-28. *The* ipconfig/all *command can be used to view the current network settings for a Windows computer.*

Summary

The two primary components of an IPv4 address are the network ID and the host ID. All computers on the same subnet must have the same network ID, and all these computers must have unique host IDs. IP addresses can be expressed in dotted decimal format or by using binary. Classful IP addresses are identified by the value in the first octet of the IP address and have known subnet masks. Classless IP addresses are accompanied by a subnet mask, which is used to determine the network ID. Classful IP addresses can be subnetted to create multiple subnetworks, or subnets. IP addresses and other TCP/IP configuration can be assigned manually or automatically using a DHCP server.

Review Questions

1. Which of the following addresses is a valid IPv4 address?
   a. 10.1.25.2
   b. 192.168.1.256
   c. 2001:0000:4137:9e76:3c2b:05ad:3f57:fe98
   d. 2001:0000:4137:9g76:3c2b:05zd:3x57:gh98

2. What class is the following IP address: 192.168.1.5?
   a. Class A
   b. Class B
   c. Class C
   d. Class D

3. The following two classful IP addresses have the same network ID: 192.168.1.5 and 192.168.2.6.
   a. True
   b. False

4. The following two classful IP addresses have the same network ID: 10.80.4.2 and 10.81.15.2.
   a. True
   b. False

5. Look at the following graphic. Which letter would represent the default gateway for subnet B?

   a. A
   b. B
   c. C
   d. D

6. What is the subnet mask for the following IP address: 192.168.1.5 /26?
   a. 192.168.1.5
   b. 255.255.255.0
   c. 255.255.255.192
   d. 255.255.255.240

7. Which of the following IP addresses is in one of the reserved IP address ranges defined by RFC 1918?
   a. 10.80.256.1
   b. 172.17.34.14
   c. 192.169.4.5
   d. 224.17.2.5

8. How many hosts are supported in a subnet with a network ID of 192.168.1.128 /26?
   a. 30
   b. 32
   c. 62
   d. 64

9. The following two classless IP addresses have the same network ID: 192.168.1.105 /26 and 192.168.1.136 /26.
   a. True
   b. False

10. A computer is unable to communicate with other computers on the network. Using `ipconfig` reveals the following information:

    IP address: 169.254.5.7
    Subnet mask: 255.255.0.0
    Default gateway: Blank
    DNS server: Blank

    Which of the following could possibly be the problem?
    a. A DHCP server cannot be reached.
    b. The default gateway needs to be manually configured.
    c. The DNS server IP address needs to be manually configured.
    d. None of the above.

IPv6

Traditionally, IPv4 has been the addressing scheme used on the Internet and on internal TCP/IP-based networks. More and more, though, IPv4 is being replaced by IPv6, and the differences between the two are fairly significant. Even technicians who have mastered IPv4 have a lot to learn to be able to properly implement IPv6. Topics they need to understand include the basics of an IPv6 address, its different components, how IPv4 and IPv6 coexist, and how to assign IPv6 addresses.

OBJECTIVES

- ▶ Explain the use of IPv6 addresses
- ▶ List the components of an IPv6 address
- ▶ Describe the dual IP stack
- ▶ Compare the manual assignment and automatic assignment of IPv6

CHAPTER 7

TABLE OF CONTENTS

IPv6 Addresses 136
    IPv4 Classes and IPv6 Prefixes 136
    Hexadecimal 136
    IPv6 Address Display 138
    IPv6 Transmission Types 138
    The Need for IPv6 139
    Neighbor Discovery 139

Components of an IPv6 Address 139
    Global Unicast Addresses 140
    Link-Local Addresses 141
    Unique Local Addresses 142

The Dual IP Stack 143
    IPv4-Mapped IPv6 Addresses 143
    IPv4 to IPv6 Tunneling Protocols 143

Manual Versus Automatic
Assignment of IPv6 144
    Manually Configuring IPv6 144
    DHCPv6 144

Summary 146

Review Questions 147

IPV6 ADDRESSES

IPv4 uses 32 bits, making it possible to have about 4 billion addresses. When the Internet was in its infancy, those 4 billion addresses seemed like they would last forever, but that is not the case today. The astronomical growth of the Internet resulted in a concern that it may run out of IPv4 addresses, and IPv6 was created to help avert that crisis as well as add a few additional capabilities. In fact, of the 5 registries that govern allocation of IP addresses on the Internet, only the one in Africa has any IPv4 addresses left. North America's registry (ARIN) ran out of addresses in July of 2015.

IPv6 changed the way the addresses are expressed from binary to hexadecimal and uses 128 bits, which gives it the capacity to have more than 340 undecillion IP addresses. That is more than 340,000,000,000,000,000,000,000,000,000,000,000,000 addresses, or more than 340 trillion, trillion, trillion addresses. While IPv4 and its 32 bits lasted a few decades before needing an upgrade, IPv6 should be able to provide plenty of addresses to last for quite a while into the future.

The change from one addressing protocol to another has been slow but steady, and IPv6 is gradually replacing IPv4 on the Internet and on internal networks. It is easy to tell the difference between the two address types by looking at them—gone is the dotted decimal format of IPv4. As an example, consider the IPv4 and IPv6 addresses of a computer running Windows Server 2012 R2. The IPv6 address is `fe80::184:e9f8:a71b:304`. The `%10` at the end of the IPv6 address is a zone index and identifies the network interface card. **See Figure 7-1.**

IPv4 Classes and IPv6 Prefixes

An *IPv6 prefix* (also called a prefix length) identifies the type of IPv6 address. As an example, an IP address starting with `fe80`, is a link-local address.

Prefixes are written in a prefix notation that begins with / followed by a number—such as /3 or /32—indicating how many bits are in the prefix. For example, /3 indicates that only the first three bits are in the prefix, but /32 indicates that the first 32 bits are in the prefix. Given the size of the IPv6 address, this prefix number can theoretically range from 0 to 128. The space between the address and the / is considered optional and usually not used. **See Figure 7-2.**

Hexadecimal

IPv6 addresses are displayed in hexadecimal, so it is important to understand the

> ✓ FACT
> The IPv6 address starts with fe80, indicating that it is a link-local address. Other types include global and unique local addresses.

> ✓ FACT
> The IPv6 prefix notation is similar to the Classless Inter-domain Routing (CIDR) notation in IPv4. It indicates how many bits are in a 1 in the subnet mask.

> ✓ FACT
> There was an IPv5. It used 64 bits and was never adopted since designers realized the Internet would quickly run out of IP addresses again if it were adopted.

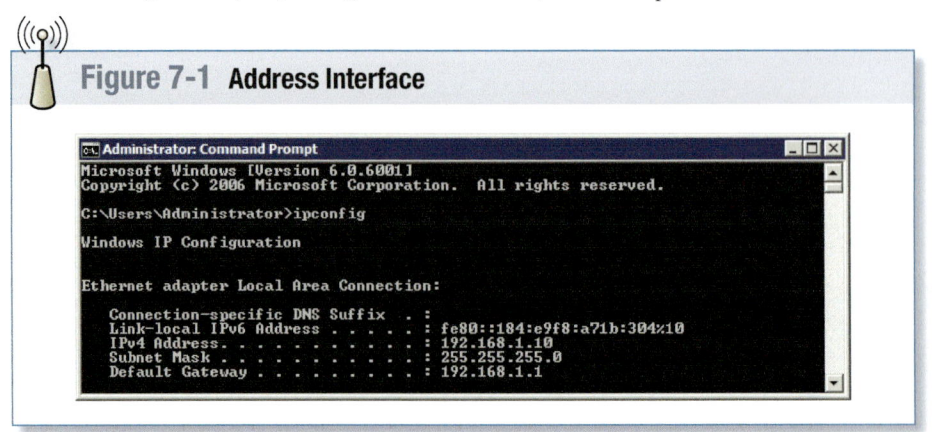

Figure 7-1 Address Interface

Figure 7-1. IPv4 and IPv6 addresses can be viewed via an interface.

basics of hexadecimal numbers. Remember, decimal uses a base of 10 with numbers from 0 through 9, while binary uses a base of 2 with 0 and 1 as the only two values.

Hexadecimal notation uses a base of 16 with the numbers 0 through 9 followed by a through f. Each hexadecimal number can be represented using four binary bits. **See Figure 7-3.**

Figure 7-2 Common IPv6 Prefixes

IPV6 PREFIX	ADDRESS TYPE	DESCRIPTION
2000::/3	Global unicast addresses	Other prefixes starting with 2 are also possible, but this is the most common.
2001::/32	Teredo tunneling protocol address	This is used for IPv4 and IPv6 compatibility.
fe80::/10	Link-local addresses	Unicast address representing an interface on a single data link. Used for communications that do not need to leave the local link.
fc00::/7 (fd /8)	Unique local unicast addresses	These are IPv6 addresses assigned in an internal network, similar to IPv4 private addresses. The prefix is only 7 bits, which is literally identified as fc in hexadecimal, but the eighth bit is always a 1, so this is always seen as fd in an IPv6 address.
::1	Loopback address	A prefix of 127 zeros followed by a single 1 as the 128th bit is the loopback address. This is similar to the IPv4 loopback address of 127.0.0.1.

Figure 7-2. A slash followed by a number identifies how many bits are used in a prefix.

Figure 7-3 Hexadecimal Values

HEXADECIMAL NUMBER	BINARY VALUE				DECIMAL VALUE
	2^3	2^2	2^1	2^0	
	8	4	2	1	
0	0	0	0	0	0
1	0	0	0	1	1
2	0	0	1	0	2
3	0	0	1	1	3
4	0	1	0	0	4
5	0	1	0	1	5
6	0	1	1	0	6
7	0	1	1	1	7
8	1	0	0	0	8
9	1	0	0	1	9
a	1	0	1	0	10
b	1	0	1	1	11
c	1	1	0	0	12
d	1	1	0	1	13
e	1	1	1	0	14
f	1	1	1	1	15

✓ FACT

Hexadecimal values can be uppercase or lowercase. For example, f is the same as F. IPv6 addresses are commonly displayed in lowercase.

Figure 7-3. Hexadecimal numbering is used in IPv6.

IPv6 Address Display

A full IPv6 address includes eight groups of four hexadecimal numbers separated by colons, similar to this:

`fe80:0000:0000:0000:0184:e9f8:a71b:0304`

To keep from writing so much, the IPv6 address can be shortened by using two techniques. First, *zero compression* is used to identify a contiguous group of zeros. Zero compression replaces a group of zeros with two colons. Second, leading zeros can be dropped in any hexadecimal grouping. For example, there are three ways to display the same IPv6 address. **See Figure 7-4.**

It is important to note that only one set of double colons can be used. With (`fe80::0184:e9f8:a71b:0304`), there are five groups of hex numbers. Since an IPv6 address has eight groups of hex numbers, the double colon (`::`) takes the place of three groups of zeros (`0000:0000:0000`).

However, if a number had two groups of double colons (for example, `fe80::0184:e9f8::0304`), there would be no way to determine how many groups of zeros each double colon represented. In other words, only one set of double colons can be used in an IPv6 address. Using two sets of double colons is not valid.

IPv6 Transmission Types

The three transmission methods for IPv4 are unicast, multicast, and broadcast.

Unicast Traffic sent from one computer to one other computer.

Broadcast Traffic sent from one computer to all other computers on the same subnet.

Multicast Traffic sent from one computer to multiple other computers using the Internet Group Management Protocol (IGMP) protocol.

IPv6 uses three types of transmission known as unicast, multicast, and anycast. These have some similarities to IPv4's methods.

Unicast An IPv6 unicast transmission is traffic sent from one computer to one other computer, just as it works in IPv4.

Multicast An IPv6 multicast transmission is traffic sent from one computer to multiple other computers, similar to multicast in IPv4. IPv6 provides some improvements in multicasting.

Anycast An IPv6 anycast transmission is traffic sent from one host to one other

Figure 7-4 Displaying an IPv6 Address

EXAMPLE	DESCRIPTION
fe80:0000:0000:0000:0184:e9f8:a71b:0304	Full IPv6 address
fe80::0184:e9f8:a71b:0304	IPv6 address using zero compression
fe80::184:e9f8:a71b:304	IPv6 address using zero compression and dropping leading zeros

Figure 7-4. *IPv6 can be expressed in many different ways.*

host from a list of multiple hosts. It is typically used to locate the nearest router or to locate services on the network.

The Need for IPv6

While there were other capabilities gained, the primary driving force behind creating IPv6 was to provide more IP addresses. At one time, experts predicted that the Internet would run out of IPv4 addresses sometime in 2011. Some earlier predictions had even indicated that the Internet was on track to run out of addresses during the late 1990s or early 2000s, but such things as CIDR and Network Address Translation (NAT) helped reduce the number of public IP addresses needed, although there are still problems because of all the reserved ranges.

In addition to providing exponentially more IP addresses, IPv6 also provides several improvements over its predecessor, and is thus a much better Layer 3 protocol choice:

Native Support for IPsec IPv6 supports Internet Protocol Security (IPsec) without any additions. This enables clients to easily encrypt IPv6 data. IPv4 can use IPsec, but it takes extra effort to make it work.

More Efficient Routing IPv6 uses global addresses on the Internet. These are designed for worldwide delivery and reduce the number of routes that Internet routers need to remember. In contrast, many Internet backbone routers maintain routing lists of more than 85,000 IPv4 routes.

Easy Host Configuration IPv6 routers can automatically configure internal computers. Dynamic Host Configuration Protocol version 6 (DHCPv6) servers can also be used to provide IPv6 information. However, even if an IPv6 router or DHCPv6 server is not available, systems can configure themselves with internal IPv6 addresses.

Neighbor Discovery

Neighbor Discovery (ND) is an IPv6 protocol that uses Internet Control Message Protocol version 6 (ICMPv6) messages to discover details about the network.

It performs the following key functions:

Discovers Routers ND identifies routers on the local subnet. These routers can then be queried for IPv6 configuration.

Discovers Prefixes The prefix in IPv6 is similar in function to the subnet mask in IPv4. ND identifies the prefix used by other hosts on the subnet.

Discovers Parameters ND messages tell the computer what IPv6 parameters other hosts on the subnet are using.

Address Autoconfiguration This determines whether the host can obtain an IP address from a router or a DHCPv6 server. If not, the host automatically generates its own IPv6 address.

Detects Duplicate Addresses This prevents the computer from using an IPv6 link-local address that is already in use.

Resolves Addresses ND can resolve a neighbor's IPv6 address to its link-layer address. This is similar to how the Address Resolution Protocol (ARP) resolves IP addresses to MAC addresses in IPv4.

COMPONENTS OF AN IPV6 ADDRESS

Recall that an IPv4 address has a network ID component and a host ID component. Similarly, an IPv6 address has

> ✓ FACT
> Anycast is sometimes called "one-to-one-of-many."

a network identifier and an interface identifier. Consider the two basic components of the IPv6 address used on a Windows computer. In IPv6, the first 64 bits are typically the network identifier, and the last 64 bits are typically the interface identifier, although there are exceptions. **See Figure 7-5.**

The interface identifier is similar to the host portion of a network address used in IPv4, but it often uses the MAC address to fill out that portion of the address. Remember that the MAC is a group of 48 bits expressed as 12 hexadecimal numbers similar to `12:34:56:78:9A:BC`. The first six hex numbers are the organizational unique identifier (OUI) identifying the manufacturer, and the last six are unique on the network interface adapter.

Windows Server 2012 R2 uses an EUI 64-bit address (EUI-64) defined as part of IPv6. Gigabit network interface adapters are configured with EUI-64 addresses. IPv6 uses the 48-bit MAC addresses on older cards and adds 16 extra bits to reach 64 bits.

Systems can also create the interface identifier using other methods, resulting in:

- A randomly generated temporary identifier
- A randomly generated permanent identifier
- A manually assigned identifier

These alternative interface identifiers provide a level of privacy in network communication by hiding the actual identifier of the host.

Global Unicast Addresses

Global unicast addresses are used on the Internet. They can be compared to IPv4 public IP addresses, but they are designed for hierarchical routing, which makes them easier to route throughout the Internet.

Consider the components of a global unicast address. The first three bits are `001`. If the first number of the address is a `2` (or, less commonly, `3`), it indicates it is a global unicast address used on the Internet. **See Figure 7-6.**

The formal definition of global unicast addresses says that only the first three digits are specified as `001`, meaning that it could be `0010` (hex `2`) or `0011` (hex `3`). However, `0011` is reserved and cannot be used. In other words, the first number will always be a `2`.

Although a wide range of global unicast addresses is possible, the one in most common use is `2001`. IPv6 global unicast addresses are assigned by the Internet Assignment Numbers Authority (IANA). They have assigned several banks of IPv6 addresses starting with `2001` (such as `2001:0000`, `2001:0200`, `2001: 0400`, and so on). They have also assigned some addresses starting with

> ✓ **FACT**
> IPv4-mapped IPv6 addresses and Teredo addresses differ from this basic format, but they still have 128 bits.

> ✓ **FACT**
> EUI-48, EUI-60, and other alternatives are also available. However, Windows Server 2012 R2 uses EUI-64.

Figure 7-5 Address Components

Network Identifier	Extended Unique Identifier (EUI)
64 Bits	64 Bits

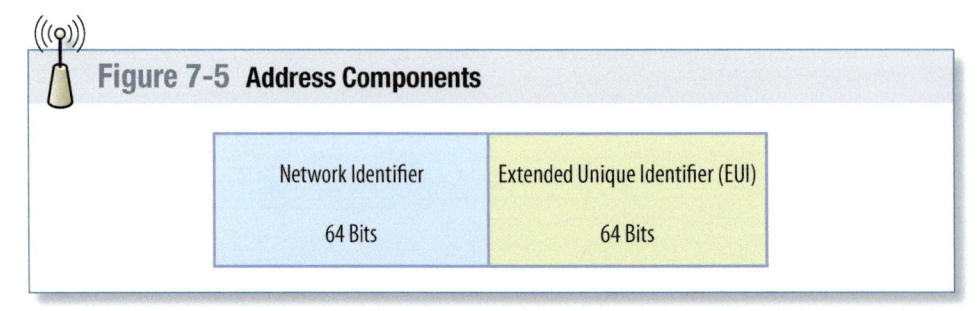

Figure 7-5. *The components of an IPv6 address include the network ID and the interface identifier.*

Figure 7-6 Unicast Address

001	Global Routing Prefix 48 Bits Public Topology	Subnet ID 16 Bits Site Topology	Interface ID 64 Bits

001 (2) – First Three Bits 001 (Hex 2) Identify a Global Unicast Address
Most Global Unicast Addresses Start with 2001:
Used on the Internet

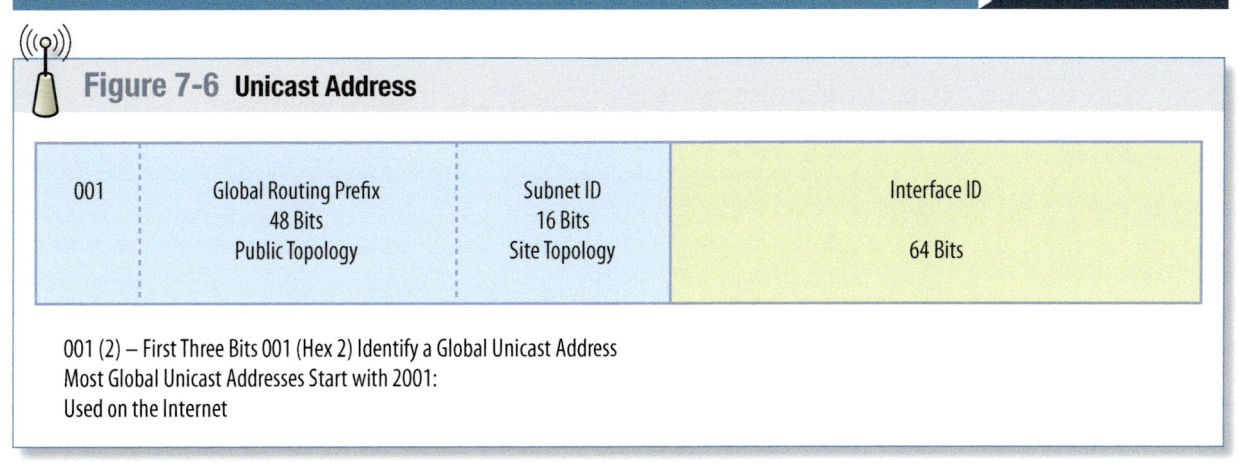

Figure 7-6. The IPv6 global unicast address is designed for hierarchical routing.

2400, 2600, and more, but 2001 remains the most popular.

The first 48 bits of the global unicast address make up the public topology, and the next 16 bits make up the site topology. Addresses in public topology are assigned to Internet Service Providers (ISPs). ISPs can then use the 16 bits in the site topology to create as many as 65,536 subnets, with each subnet having more than 18 quintillion addresses each.

Link-Local Addresses

IPv6 has a new address type that is completely different than IPv4. Every IPv6-enabled interface must have at least a link-local address to be on a subnet. But to leave the subnet, the interface must have both a link-local and a unicast address. In other words, every IPv6 interface that is connected to an internetwork must have two different addresses.

If an interface has only a link-local address, it will work on the local network, but it will not be able to access resources on any other network or the Internet. The link-local address may be manually configured or automatically configured, but manual configuration of the link-local address is only common on devices such as routers.

Every computer on every subnet in a network can actually have the exact same default gateway (such as `fe80::1`), even if the interface has a global unicast address. Remember, the point of a device having a default gateway is to have a place to send traffic that is leaving the network.

Since link-local addresses are unique only to the subnet, it is entirely possible for every router interface to be assigned the same link-local address, but those router interfaces must all have different unicast addresses. It is also possible for a computer to use the unicast address of the router interface as the default gateway, because either way, the data is sent to the router's interface.

Whether to use the unicast address or the link-local address as the default gateway is up to the preference of the network engineer. In either case, the system will work, so understand that if given an address of `2607:ea0:102:4001::-cafe:40d/64` and a default gateway of `fea0::1`, the information is likely completely correct. The default gateway must match either the link-local address or the unicast address, but again, either one will work.

For additional information, visit qr.njatcdb.org Item #1797

If a DHCPv6 server or an IPv6 router is not available to assign an IPv6 address, an IPv6 client can assign itself a link-local address to communicate with other hosts on the same network. Just as Automatic Private IP Addressing (APIPA) limits communication to the local subnet, having only a link-local address limits a client to communication only on the local subnet. This is because link-local addresses are not routable.

However, IPv6 clients can also have another IPv6 address in addition to the link-local address. For public systems, they must also have a global unicast address. For private systems, they must also have a unique local address. The link-local address is used to communicate with local nodes, and the other address is used to communicate with clients on other subnets.

Consider the format of a link-local address. The first ten bits are always 1111 1110 10. This means that the first hextet could have a value ranging from 1111 1110 1000 0000 (fe80) to 1111 1110 1011 1111 (febf). This prefix used for a link-local address is fe80::/10. On a single-link IPv6 network with no IPv6-enabled router, link-local addresses are used to communicate between devices on the link. See **Figure 7-7**.

If IPv6 is installed on a system, a link-local address will always be configured. Microsoft systems use the link-local address for Neighbor Discovery (ND) processes, and these processes will not work if IPv6 is not enabled on the system.

Autoconfiguration of IPv6 addresses is either stateless or stateful.

Stateless The configuration is performed based on router advertisements. *Stateless Address Autoconfiguration (SLAAC)* starts with a self-assigned link-local address, and then goes through a process to verify it and learn about the network by communicating with local routers.

Stateful The configuration is performed by a DHCPv6 server. DHCPv6 can be used in conjunction with SLAAC. In that case, the IPv6 address would be acquired via SLAAC and DHCPv6 would be used to "fill in the holes" (such as DNS addresses).

Unique Local Addresses

Unique local addresses are IPv6 addresses used in an internal network. They are similar to IPv4 private IP addresses in that they are assigned to computers on an internal network.

✓ **FACT**
Every IPv6 interface that wants data to leave the network must have both a link-local address and a unicast address. Understand that every host will use a link-local address for packets that do not leave the IPv6 subnet.

For additional information, visit qr.njatcdb.org Item #1798

✓ **FACT**
Link-local IPv6 addresses always start with a value ranging from fe80 to febf.

Figure 7-7 Link Local Address Format

1111 1110 10	xxx ... xxx	Interface ID
10 Bits	54 Bits	64 Bits

1111 1110 10 (fe80) – fe80 /10 Identifies a Link-Local Address

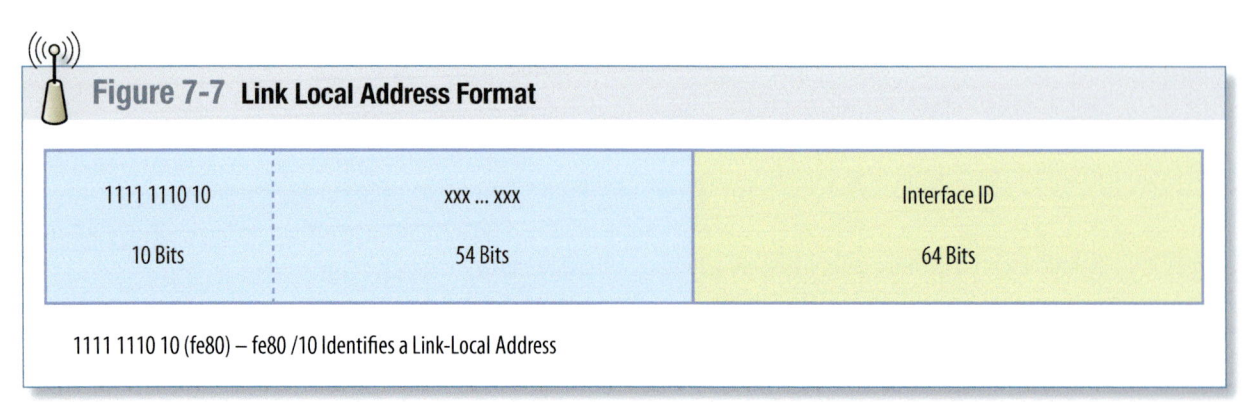

Figure 7-7. *The format of an IPv6 link-local address always includes 1111 1110 10 for the first ten bits.*

Figure 7-8 Address Identification

1111 110	1	Global ID	Subnet ID	Interface ID
7 Bits	L	40 Bits	16 Bits	64 Bits

1111 1100 – fd00: Prefix Identifies a Unique Local Address
Similar to Private IPv4 Addresses
Used on Internal Networks

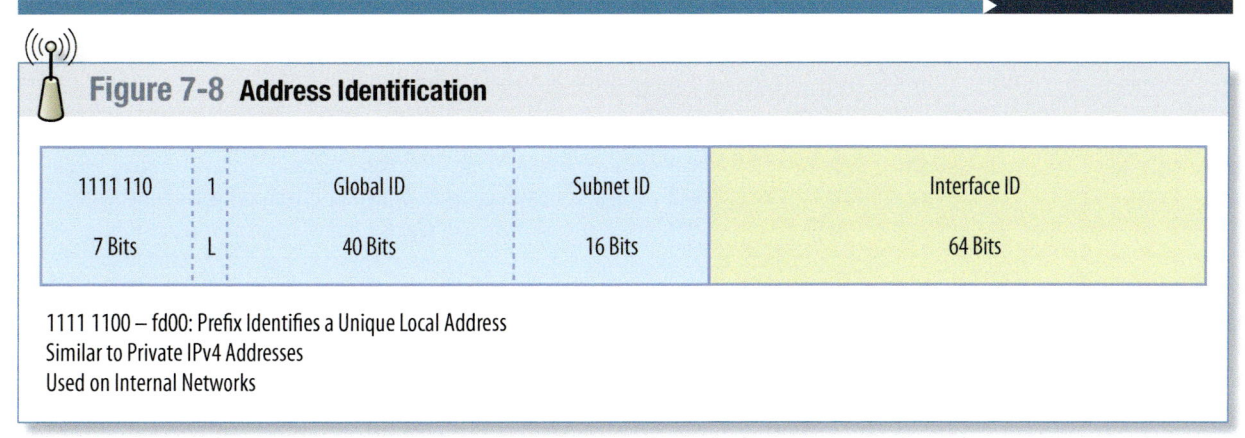

Figure 7-8. IPv6 unique local unicast addresses are identified by the first seven bits.

Consider the format of unique local addresses. Unique local addresses are identified by the first seven bits as `1111 110`. Additionally, the eighth bit is always a `1`. **See Figure 7-8.**

THE DUAL IP STACK

In a perfect world, every computer on the Internet would switch from IPv4 to IPv6 on a specific day. For example, November 9, 2010, could have been designated international IPv6 day, and everyone in the world could have magically switched each computer's configuration at midnight Greenwich Mean Time. This, however, is not realistic.

Instead, IPv4 and IPv6 must be able to operate side by side. IPv6 is currently working on the Internet and will gradually replace IPv4. In the meantime, operating systems and routers support IPv4-mapped IPv6 addresses and Teredo tunneling.

IPv4-Mapped IPv6 Addresses

One way that the IPv4/IPv6 dual IP stack works is by supporting IPv4-mapped IPv6 addresses. Consider an IPv4-mapped IPv6 address. The first 80 bits are set to `0`, the next 16 bits are set to `1`, and the last 32 bits hold the IPv4 address. **See Figure 7-9.**

The IPv4-mapped IPv6 address is expressed with the leading zeros omitted with zero compression (using a double colon, `::`), the 16 ones expressed as `ffff`, and the IPv4 address in traditional dotted decimal format.

IPv4 to IPv6 Tunneling Protocols

Some devices accessible from the Internet are not yet IPv6-enabled. If only IPv6

For additional information, visit qr.njatcdb.org Item #1799

✓ **FACT**
Unique local IPv6 addresses always start with an IP prefix of `fd`.

Figure 7-9 IPv4-Mapped IPv6 Address

000 ... 000	111 ... 111	IPv4 Address
First 80 Bits Set to Zero	16 Bits	32 Bits

::ffff:192.168.1.5

Figure 7-9. An IPv4-mapped IPv6 address has the first 80 bits set to 0.

fc or fd for Unique Local Addresses?

Some documentation indicates that the first seven bits of a unique local address are always set to 1111 110, giving it a 7-bit prefix of fc hexadecimal. Other documentation indicates that all unique local IPv6 addresses start with a prefix of fd hexadecimal. Which one is correct? Actually, both are.

RFC 4193 defines unique local addresses and specifies the eighth bit should be a 1. This "L" bit indicates the address is locally assigned. RFC 4193 states that the value of 0 for the "L" bit may be defined in the future, but for now it is always a 1.

If only the first seven bits are counted (1111 110), then the eighth bit is implied as a zero. This gives a value of binary value 1111 1100, or fc in hex. However, if the eighth bit is a 1 in the actual IPv6 address, then the first eight bits are 1111 1101, which equates to fd.

In other words, if the full IPv6 address is shown, the prefix is fd in hex. If only the first seven bits are represented, the value is fc in hex.

---

was used, data could not transit through these devices. *Teredo* is a tunneling protocol that encapsulates IPv6 packets within IPv4 datagrams. This allows the IPv6 packets to transit through these devices.

Teredo is needed for NAT devices that translate private IPv4 addresses to public IPv4 addresses and translate public IPv4 addresses back to private IPv4 addresses. Once IPv6 is fully implemented, Teredo will not be needed anymore.

If Teredo traffic needs to pass through a firewall, the firewall must be configured to allow the Teredo traffic to pass through. By default, a firewall will block traffic that uses a Teredo tunnel.

Consider the mapping of a Teredo IPv6 address. **See Figure 7-10.**

To tell whether a system is using Teredo, enter `ipconfig/all` at the command prompt. **See Figure 7-11.** Notice the prefix is 2001:0.

MANUAL VERSUS AUTOMATIC ASSIGNMENT OF IPV6

Just as it is possible to assign IPv4 addresses manually or via DHCP, IPv6 can be assigned manually or with DHCPv6.

Manually Configuring IPv6

To manually view or configure the IPv6 information on a Windows system, use an interface. **See Figure 7-12.**

DHCPv6

Just as DHCP can be used to assign TCP/IP information for IPv4 clients,

✓ **FACT**
Teredo clients have an IPv6 address that starts with 2001::/32. This is known as the Teredo prefix.

Figure 7-10 Teredo IPv6Address

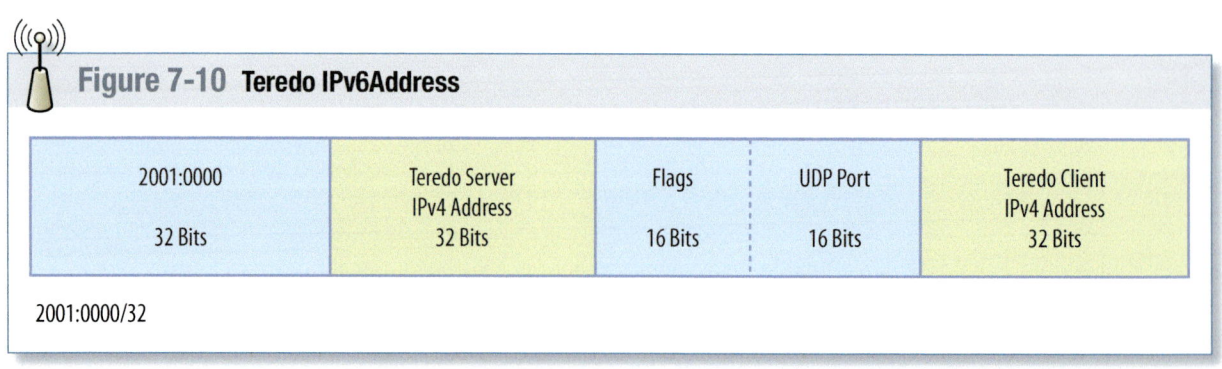

2001:0000	Teredo Server IPv4 Address	Flags	UDP Port	Teredo Client IPv4 Address
32 Bits	32 Bits	16 Bits	16 Bits	32 Bits

2001:0000/32

Figure 7-10. *A Teredo IPv6 address starts with the Teredo prefix.*

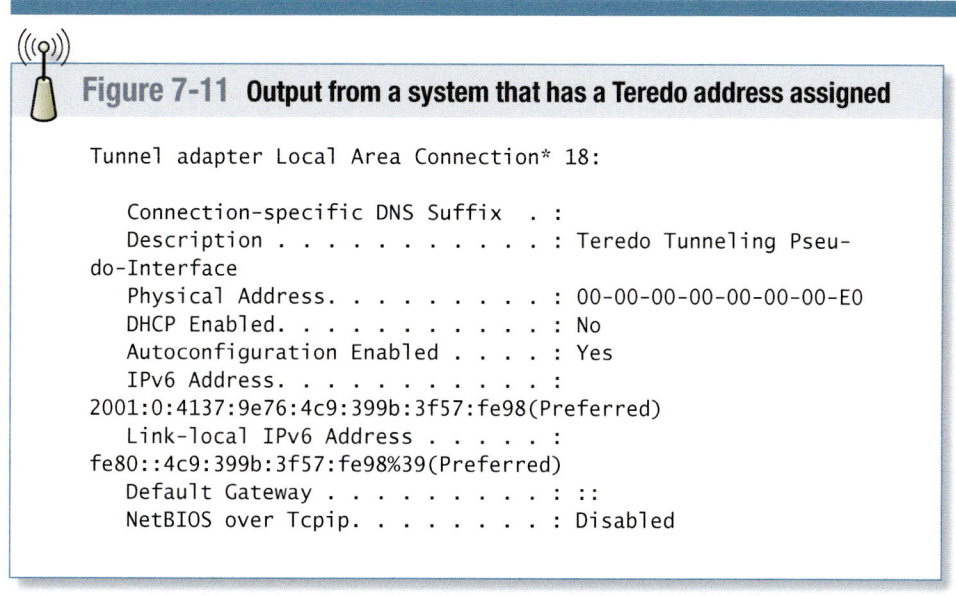

Figure 7-11 Output from a system that has a Teredo address assigned

```
Tunnel adapter Local Area Connection* 18:

   Connection-specific DNS Suffix  . :
   Description . . . . . . . . . . . : Teredo Tunneling Pseu-
do-Interface
   Physical Address. . . . . . . . . : 00-00-00-00-00-00-00-E0
   DHCP Enabled. . . . . . . . . . . : No
   Autoconfiguration Enabled . . . . : Yes
   IPv6 Address. . . . . . . . . . . :
2001:0:4137:9e76:4c9:399b:3f57:fe98(Preferred)
   Link-local IPv6 Address . . . . . :
fe80::4c9:399b:3f57:fe98%39(Preferred)
   Default Gateway . . . . . . . . . : ::
   NetBIOS over Tcpip. . . . . . . . : Disabled
```

Figure 7-11. The ipconfig/all *command shows Teredo tunneling enabled.*

✓ **FACT**
Another method that is used to help migrate between IPv4 and IPv6 is 6to4. It enables IPv6 traffic to be transmitted over an IPv4 network and is popular with routers used in small offices and home offices (SOHOs).

Figure 7-12 Adapter Configuration

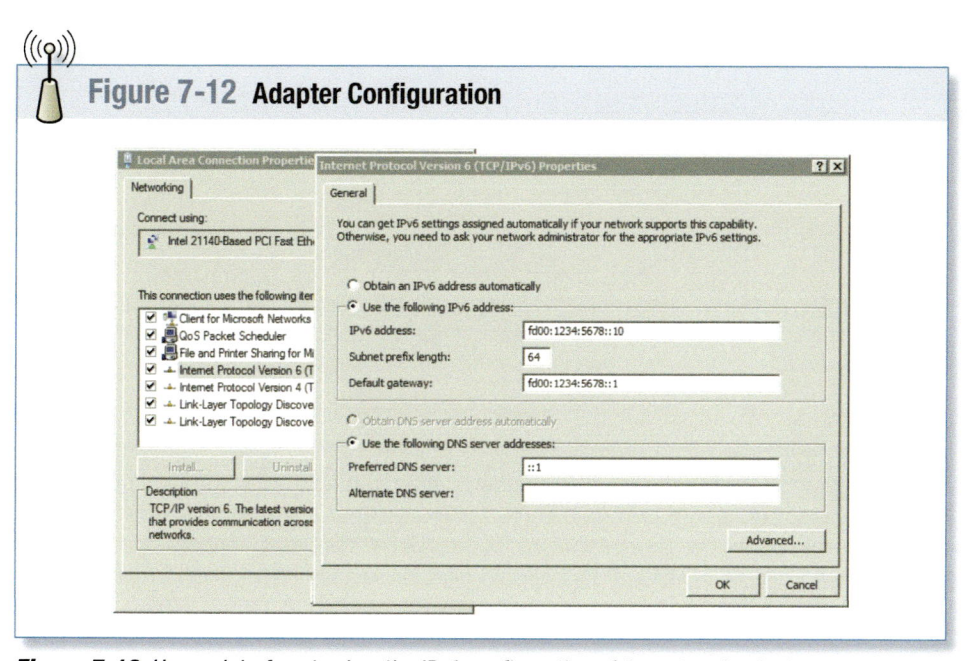

Figure 7-12. Use an interface to view the IPv6 configuration of the network adapter.

DHCPv6 can also be used to assign TCP/IP information for IPv6 clients. The process is similar. The benefit is that it is possible to assign all the TCP/IP information, including the IP addresses of DNS servers, default gateways, and more. The system can configure itself with link-local addresses, but link-local addresses will not provide access outside the subnet since they are not routable and will not include addresses of DNS servers.

Consider the process of a DHCPv6 client receiving an IPv6 address from a DHCPv6 server. **See Figure 7-13.**

1. When the DHCPv6 client turns on, it sends a Solicit message looking for a DHCPv6 server.
2. The DHCPv6 server answers with an Advertise message. This lets the client know that the server can offer IPv6 configuration information.
3. The DHCPv6 client replies with a Request message formally requesting the IPv6 information.
4. The DHCPv6 server responds with a Reply message. This message includes IPv6 information such as the IPv6 address, the default gateway, and the address of a DNS server.

DHCPv6 servers are not required on each subnet. If a network includes multiple subnets, DHCPv6 relay agents can be placed on each subnet to listen for the DHCPv6 solicit messages and then forward them to the DHCPv6 server. The relay agent then acts as the intermediary (or the proxy) for the four messages going back and forth between the DHCPv6 server and the DHCPv6 client.

Figure 7-13 DHCPv6 Process

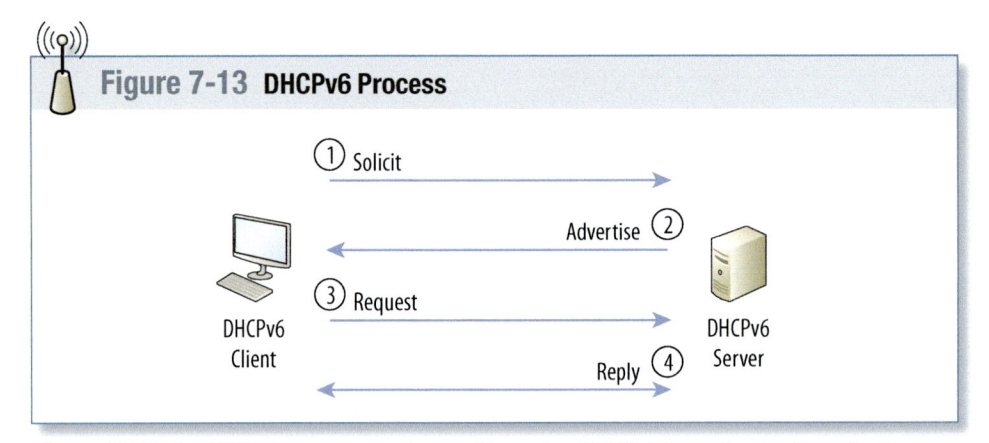

Figure 7-13. The DHCPv6 process can be used to assign TCP/IP information for IPv6 clients.

Summary

IPv6 is slowly replacing IPv4, and it uses 128 bits instead of the 32 bits used in IPv4. IPv6 addresses are displayed in hexadecimal, and the type of IPv6 address is identified by the prefix.

Review Questions

1. Which of the following is a valid IPv6 address?
   a. 10.1.25.2
   b. 192.168.1.256
   c. 2001:0000:4137:9e76:3c2b:05ad:3f57:fe98
   d. 2001:0000:4137:9g76:3c2b:05zd:3x57:gh98

2. In order to manually assign an IPv6 address to a client computer for use within a private network, which one of the following addresses should be used?
   a. 0000::a123:4567:89ab:cdef
   b. 2001:0000: fcde:ba98:7654
   c. 2001:0001::fcde:ba98:7654
   d. fd00:: a123:4567:89ab:cdef
   e. fe80:: a123:4567:89ab:cdef

3. Which of the following features is built into IPv6 to provide extra security?
   a. Global addresses
   b. IPsec
   c. Teredo tunneling
   d. Unique local addresses

4. An IPv6 address with a prefix of fd is a link-local address.
   a. True
   b. False

5. What IPv6 to IPv4 technology uses tunneling to encapsulate an IPv6 packet within an IPv4 packet?
   a. CIDR
   b. EUI-64
   c. IPsec
   d. Teredo

6. In order to assign IPv6 addresses to hosts on a private network, a link local address should be used.
   a. True
   b. False

7. What IPv6 protocol is used to identify routers on the same network?
   a. Anycast
   b. IGMP
   c. Network Discovery
   d. Teredo

8. IPv4 addresses use public address on the Internet. IPv6 uses global unicast addresses on the Internet.
   a. True
   b. False

Networking Protocols

An important part of understanding networking is the understanding of networking protocols. TCP/IP is the primary protocol suite used in networks today, including the Internet. TCP and UDP are two important protocols that are integral to most networking communications, but there are many more.

This chapter examines many of the more popular protocols with a high-level overview of these protocols and their purposes. It is important to understand the basics of TCP and UDP, such as which one is connection-oriented and which one is connectionless. However, it is not necessary to have in-depth knowledge of the common protocols other than understanding their primary purposes. Many common protocols also use specific ports identified as well-known ports. A good understanding of how ports work and the well-known ports used with specific protocols is also important.

OBJECTIVES

- ▶ List various Physical and Data Link layer protocols
- ▶ Describe various Network layer protocols
- ▶ Explain the different functions of the Transport layer
- ▶ Identify various Application layer protocols

CHAPTER 8

TABLE OF CONTENTS

Physical and Data-Link Layer
Protocols. 150
   LAN Connections. 150
   WAN Connections 150
   Address Resolution Protocol 150

Network Layer Protocols. 151
   IP Functions 151
   Internet Control Message Protocol
   (ICMP). 152
   Internet Group Management Protocol
   (IGMP) 153

Transport Layer Functions. 155
   TCP 155
   UDP 157

Ports 158
   Port Traffic and Firewalls 159
   Internally Used Ports and Protocols . 160

Application Layer Protocols 160
   Hypertext Transfer Protocol. 160
   File Transfer Protocol 162
   Trivial File Transfer Protocol 162
   Telnet 162
   Remote Desktop Services 163
   Secure Sockets Layer 163
   Transport Layer Security 163
   Secure Shell 164
   Internet Protocol Security 164
   Simple Mail Transfer Protocol 165
   Post Office Protocol v3 165
   Internet Message Access Protocol. . . 165
   Lightweight Directory Access
   Protocol 166
   Kerberos. 167
   Point-to-Point Tunneling Protocol . . . 167
   Layer 2 Tunneling Protocol. 167
   Simple Network Management
   Protocol 167

Summary. 168

Review Questions 169

PHYSICAL AND DATA-LINK LAYER PROTOCOLS

The Physical layer of the Open Systems Interconnection (OSI) reference model defines the characteristics of the hardware that connects the network interfaces in computers and other network devices together. In the case of a typical local area network (LAN), this hardware consists of a series of cables, as well as the connectors and other components needed to install them.

The second layer of the OSI model—the Data Link layer—is where the physical manifestation of the network meets with the logical one, implemented in the software running on the computer. Most of the hardware elements used to construct a local area network (LAN) are part of the Physical and Data Link layer protocol implementations.

LAN Connections

On a typical wired LAN, the industry standard Data Link layer protocol is Ethernet, installed using unshielded twisted pair (UTP) cable. Wireless LANs typically use equipment based on the IEEE 802.11 standard.

WAN Connections

Unlike LANs, the wide area network (WAN) connections that homes and small offices use to connect to the Internet and that enterprise networks use to connect remote offices together typically use a third-party service provider. There are a variety of WAN technologies providing different amounts of bandwidth at many price levels.

Address Resolution Protocol

The Address Resolution Protocol (ARP) uses broadcast transmissions to identify the Media Access Control (MAC) addresses of computers.

The Internet Protocol (IP) routes traffic to the correct subnet. When the traffic reaches the destination subnet, the router uses ARP to broadcast the destination IP address to all computers on the subnet. This ARP broadcast asks, "Who owns this IP address?" **See Figure 8-1.**

> ✓ FACT
> The MAC address is also known as the physical address. It is expressed in hexadecimal characters such as 01-23-45-AB-CD-EF.

Figure 8-1 Address Resolution Protocol

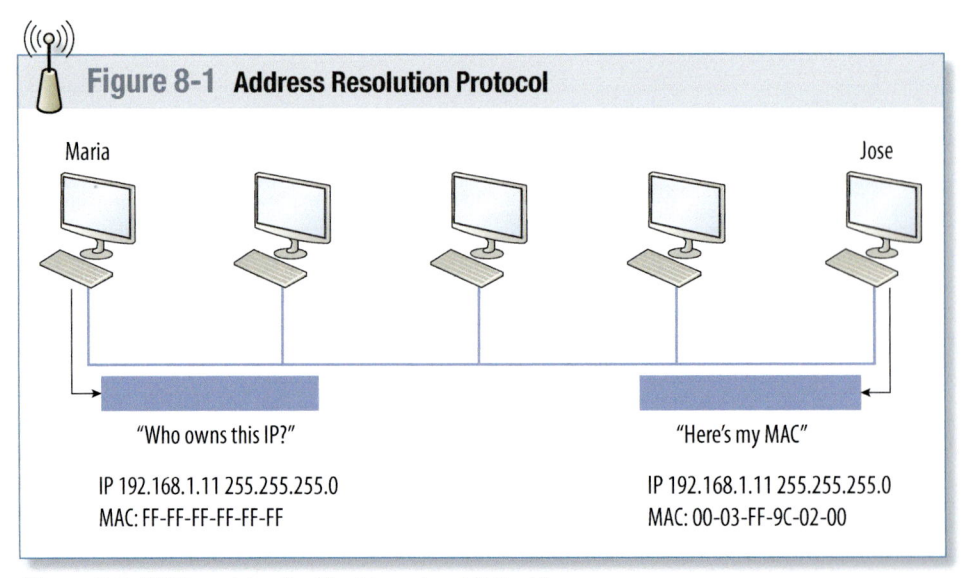

Maria — "Who owns this IP?"
IP 192.168.1.11 255.255.255.0
MAC: FF-FF-FF-FF-FF-FF

Jose — "Here's my MAC"
IP 192.168.1.11 255.255.255.0
MAC: 00-03-FF-9C-02-00

Figure 8-1. ARP translates the IP address to a MAC address.

Each computer that receives the ARP broadcast looks to see whether it has the broadcasted IP address. If so, the computer responds with its MAC address.

When a computer resolves a MAC address using ARP, it stores it in a cache for two to ten minutes, depending on the operating system. If it wants to communicate with the computer again, it does not have to send another ARP broadcast to get the MAC address, but instead retrieves it from the cache. A *cache* is an area of memory used for short-term storage. Many applications and devices use caches.

The ARP protocol is part of the TCP/IP protocol suite, but there is also a Windows command-line tool named `arp`. To view the ARP cache from the command line, run the `arp -a` command. **See Figure 8-2.**

NETWORK LAYER PROTOCOLS

The Data Link layer of the OSI reference model is about getting packets to their next destination. Data Link layer protocols are responsible for sending packets to another system on the same subnet, to a router on the subnet, or to a system at the other end of a WAN link. The Network layer, by contrast, is responsible for the delivery of data to its ultimate destination, whether that is on the same network or another network, across the hall, or halfway around the planet.

This is why it might seem as though network layer protocols have some redundant functions. For example, why should the Internet Protocol (IP) have its own addressing scheme, when Ethernet already has addresses? Ethernet addresses are only good for local delivery. They tell computers where the local post office is, something IP cannot know. IP addresses tell computers the final destination of a packet. Therefore, a packet might have addresses at the Data Link and Network layers that identify two different systems.

IP Functions

On a TCP/IP internetwork, IP is the connectionless protocol responsible for transmitting data from its source to its final destination. A *connectionless protocol* transmits messages to a destination without first establishing a connection to the receiving system. IP is also considered to be an unreliable protocol, because it does not require the receiving system to respond with acknowledgments.

The IP service is connectionless and unreliable because it carries data generated by other protocols, only some of which require connection-oriented service. The TCP/IP suite includes both connection-oriented and connectionless services at the Transport layer, making it possible for applications to select one or the other, depending on

Figure 8-2 `arp -a` **Command-Line Tool**

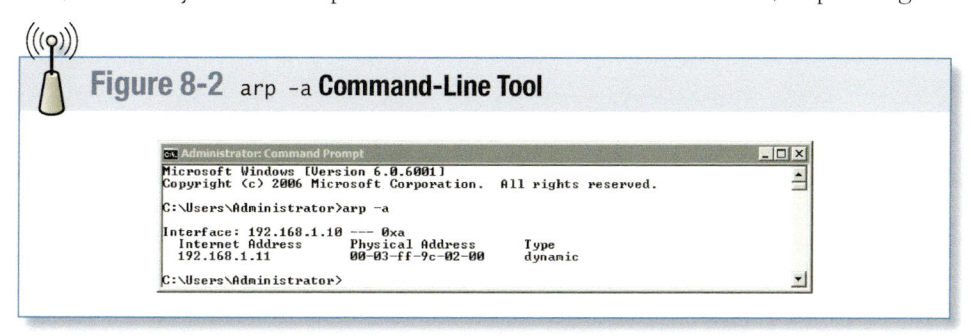

```
Administrator: Command Prompt
Microsoft Windows [Version 6.0.6001]
Copyright (c) 2006 Microsoft Corporation.  All rights reserved.

C:\Users\Administrator>arp -a

Interface: 192.168.1.10 --- 0xa
  Internet Address      Physical Address      Type
  192.168.1.11          00-03-ff-9c-02-00     dynamic

C:\Users\Administrator>
```

Figure 8-2. *Technicians can use the arp -a command to view the ARP cache.*

the quality of service they need. Because TCP provides connection-oriented service at the Transport layer, there is no need to implement a connection-oriented service at the Network layer. The Network layer can remain connectionless, reducing the amount of control overhead generated by the protocol stack.

IP performs several functions that are essential to the internetworking process.

Data Encapsulation

Data encapsulation packages the Transport layer data into an IP datagram (packet). Just as Ethernet packages Network layer data for transmission over a LAN, IP encapsulates data that it receives from the Transport layer protocols—such as the Transmission Control Protocol (TCP) and the User Datagram Protocol (UDP)—for transmission to a destination. At the Network layer, IP adds a header, thus creating a packet.

The IP packet is addressed to the computer that will ultimately use the data, whether that computer is located on the local network or on another network far away. Except for a few minor modifications, the packet remains intact throughout its journey to the destination. Once IP has created the packet, it passes it down to a Data Link layer protocol for transmission over the network.

During the transportation process, various routers might encapsulate a packet in different Data Link layer protocol frames, but the packet itself remains intact. The process is similar to the delivery of a letter by the post office, with IP functioning as the envelope. The letter might be placed into different mailbags and transported by various trucks and planes during the course of its journey, but the envelope remains sealed. Only the addressee is permitted to open it and use the contents.

IPv4 Addressing

IPv4 addressing identifies systems on the network by using unique addresses. The self-contained IP addressing system is one of the most important elements of the TCP/IP protocol suite. IP addresses enable computers running any operating system on any platform to communicate by providing unique identifiers for the computer itself and for the network on which it is located. Understanding how IP addresses are constructed and how they should be assigned is an essential part of TCP/IP network administration.

IP Routing

IP routing selects the most efficient path through the internetwork to the destination system. Routing is a critical function of IP. It is the means by which packets find their way around the vastness of the Internet and locate a single destination system among billions.

Internet Control Message Protocol (ICMP)

The Internet Control Message Protocol (ICMP) is a Network layer protocol that does not carry user data, although its messages are in IP packets. ICMP fills two roles in the TCP/IP suite. It provides error-reporting functions (informing the sending system when a transmission cannot reach its destination, for example) and it carries query and response messages for diagnostic programs. The PING utility, for example, which is included in every TCP/IP implementation, uses ICMP echo messages to determine if another system on the network is reachable and capable of receiving and sending data.

ICMPv4

As defined in RFC 792, the ICMP protocol consists of messages that are carried in IPv4 packets. Since the introduction of ICMPv6 (a new version of the protocol designed for use with IPv6 traffic), the original version of ICMP has unofficially become known as ICMPv4.

ICMPv4 Error Messages

Because of the way that TCP/IP networks distribute routing chores among various systems, there is no way for either of the end systems involved in a transmission to know what has happened during a packet's journey. IP is a connectionless protocol, so there are no acknowledgment messages returned to the sender at that level. When using a connection-oriented protocol at the Transport layer (like TCP), the destination system acknowledges transmissions, but only for the packets that it receives. If something happens during the transmission process that prevents the packet from reaching the destination, there is no way for IP or TCP to inform the sender what happened.

ICMP error messages are designed to fill this void. When an intermediate system such as a router has trouble processing a packet, it typically discards it, leaving it to the upper-layer protocols to detect its absence and arrange for a retransmission. ICMP messages enable the router to inform the sender of the exact nature of the problem. Destination systems can also generate ICMP messages when a packet arrives successfully but cannot be processed.

ICMPv4 error messages are informational only. The system receiving them does not respond, nor does it necessarily take any action to correct the situation. It may be left up to the user or administrator to address the problem that is causing the failure.

ICMPv4 Query Messages

ICMPv4 query messages are not generated in response to other activities, as are the error messages. Systems use them for self-contained request/reply transactions in which one computer requests information from another, which responds with a reply containing that information.

ICMPv6

IPv6 has its own version of ICMP. The newer version of the protocol uses the same message format and functions in much the same way. The primary difference is in the values for the various messages. There are also some important new message types.

Chief among the additions in ICMPv6 is the NDP. *Neighbor Discovery Protocol (NDP)* is a new Data Link layer protocol that performs multiple functions, including local network system discovery, hardware address resolution, duplicate address detection, router discovery, DNS server discovery, address prefix discovery, and neighbor unreachability detection.

In these roles, NDP replaces the Address Resolution Protocol (ARP) and the ICMPv4 Router Advertisement and Router Redirect messages.

Internet Group Management Protocol (IGMP)

TCP/IP systems can transmit packets to all of the systems on a subnet (as broadcasts), to specific individual systems on a network (as unicasts), to the nearest system on the network (as an anycast), or to groups of systems (as multicasts). Broadcasts, unicasts, and anycasts are relatively simple to implement, because the TCP/IP system simply sends its packets to an address with the appropriate format. Multicasting is more complicated, however.

Class D IP addresses ranging from `224.0.0.0` to `239.255.255.255` are reserved for multicasting purposes. A multicast transmission is simply a packet transmitted to one of those Class D addresses. The *Internet Group Management Protocol (IGMP)* is a specialized protocol for determining which systems are part of the multicast group that recognizes that address.

Unicasts are one-to-one transmissions, involving only a single source and a single destination. Broadcasts are one-to-all transmissions, with a single source and multiple destinations on a subnet. Anycasts are one-to-one-of-many transmissions that target a single router. A multicast is a form of one-to-many transmission that is designed to be more efficient than a broadcast, because it targets a specific group of systems.

For example, if an application wants to transmit a message to all of the routers on a network, it could conceivably use a broadcast, but this would cause two problems. First, the broadcast would have to be processed by all of the workstations on the network unnecessarily, and second, the broadcast would be limited to the local network.

Multicasts overcome both these problems, because only systems recognizing themselves as part of the host group represented by the multicast address process the message, and because routers can propagate multicast messages throughout an internetwork. However, for multicasting to function properly, the appropriate systems must be added to each host group, and the routers on the network must know which systems are in each host group. To become a member of a host group, a TCP/IP system uses the IGMP protocol to register itself with the routers on the local network.

Routers can also use IGMP to report their host group membership information to other routers. A router can therefore use IGMP for two purposes: to register its own group memberships and to exchange its group membership information with other routers. In addition to IGMP, there are also other protocols that routers can use to exchange group membership information, including Distance Vector Multicast Routing protocol (DVMRP), the Multicast Open Shortest Path First (MOSPF) protocol, and the Protocol Independent Multicast (PIM) protocol.

For a network to support multicasting, the following elements are required:

▶ All host group members and all of the routers providing internetwork access to the host group members must support IGMP.

▶ All of the routers providing internetwork access to the host group member must have a means of sharing their host group membership information, using IGMP or another protocol.

▶ All of the network interfaces in the routers must support multicast promiscuous mode. *Multicast promiscuous mode* is a special mode that causes the network interface adapter to process all incoming packets that have the multicast bit (that is, the last bit of the first byte of the destination hardware address) set to a value of 1.

TRANSPORT LAYER FUNCTIONS

Transmission Control Protocol (TCP) and User Datagram Protocol (UDP) are the two primary protocols used to transport data across a network. They both operate on the Transport layer of the OSI model, but they have distinctive differences.

The primary difference between these two is the delivery mechanism. Transmission Control Protocol (TCP) is a Transport layer protocol that provides guaranteed delivery with acknowledgments, sequence numbers, and flow control. User Datagram Protocol (UDP) is a Transport layer protocol that provides best-effort delivery without a guarantee.

TCP is a connection-oriented protocol. It starts with a three-way handshake process that ensures a connection is established before any data is transmitted.

UDP is a connectionless protocol that sends data using a best-effort method.

Application protocols use TCP, UDP, or both to transfer application data. These protocols identify the application generating the data using logical ports. For example, HTTP uses TCP port 80. When a system receives data using TCP port 80, it processes it as HTTP.

TCP

TCP provides guaranteed delivery by starting with a three-way handshake process. For example, say that Sally's computer wants to transfer information to Bob's computer. Before the data transfer starts, the computers establish a connection with each other using this three-way handshake. **See Figure 8-3.**

Sally's computer starts by sending a packet with the synchronize (SYN) flag set. When Bob's computer receives the packet, it responds with another packet with both the SYN and the acknowledge (ACK) flags set. Sally's computer then completes the three-way handshake by sending a third packet with the ACK flag set. A *flag* is a single binary bit set to a **1**. For example, the SYN flag is set by setting a specific bit in the packet to a **1**.

Figure 8-3 TCP

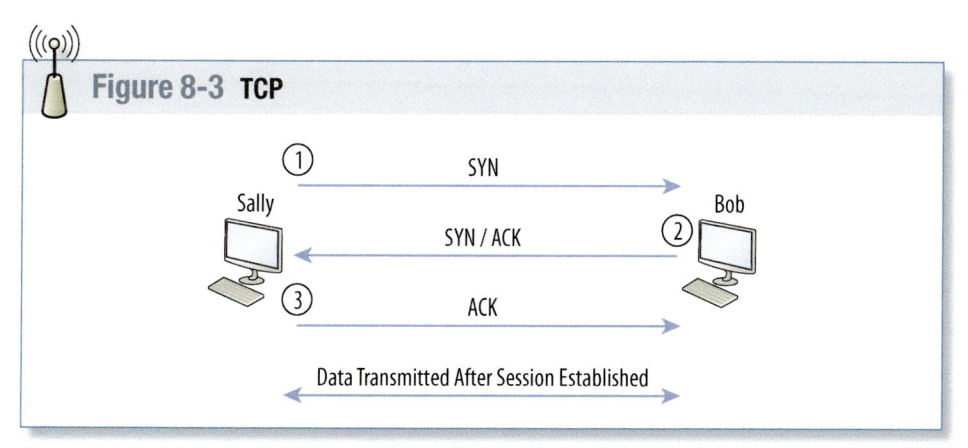

① SYN
Sally → Bob
SYN / ACK ②
③ ACK
Data Transmitted After Session Established

Figure 8-3. *TCP provides guaranteed delivery by starting with the three-way handshake process.*

At this point, both computers have an established session. They both have assurances that the other computer is operational and they are able to communicate. After the session is established, the two computers then proceed to transmit data.

Compare this handshake process to using different methods to get a message to a friend. One way is to make a phone call. This also uses a three-way handshake process:

1. Sally initiates the phone call to Bob.

2. Bob answers the call with "Hello, this is Bob." Sally recognizes Bob's voice and knows it is him.

3. Sally says "Hi. This is Sally." Bob recognizes Sally's voice and knows it is her.

The phone call may not be so formal between two friends. However, the conversation does not start until the phone connection is established.

TCP divides the data into smaller segments. For example, the data could be a 1 MB file. TCP could divide this file into 250 segments that are 4 KB. These 4 KB segments can travel over the network more efficiently than a single 1 MB file. TCP uses sequence numbers to track these segments.

When the data is segmented, each separate segment is assigned a different sequence number. The receiving computer then receives each of these segments and uses the sequence numbers to put the data back together in the correct order.

However, the sending computer does not just throw all of these segments onto the network and hope the other computer receives them. TCP coordinates this process between the two computers.

Imagine that Sally wants to download music from Bob's computer. The TCP handshake process starts the process. Next, the two computers decide on how big the individual segments can be, and how many segments they can send between acknowledgments. The *TCP sliding window* is the number of segments the computers can send at a time. The two computers have an established session. They have negotiated a segment size of 1,500 bytes and a sliding window of 3. When Sally's computer receives three segments, it verifies the data is intact and then sends an acknowledgment (ACK) message. Bob's computer then sends three more segments. **See Figure 8-4.**

If even a single bit is lost in the transmission, the data in the segment is no longer valid. A *cyclical redundancy check (CRC)* is an error-checking process used by TCP to verify that the data is intact in each segment. If the three segments are received without errors, Sally's computer sends an ACK packet saying, "Give me three more."

However, if any of the segments are missing or corrupt, Sally's computer sends a negative acknowledge (NACK) packet instead, requesting the missing or corrupt segment. If the sending computer receives a NACK, it retransmits the segments in the sliding window. Eventually, Sally's computer will receive all the segments and be able to reassemble them into the original file sent by Bob's computer.

When the computers are finished transmitting and acknowledging their data, they terminate the connection by exchanging another sequence of packets, these containing finish (FIN) flags.

Figure 8-4 TCP Sliding Window

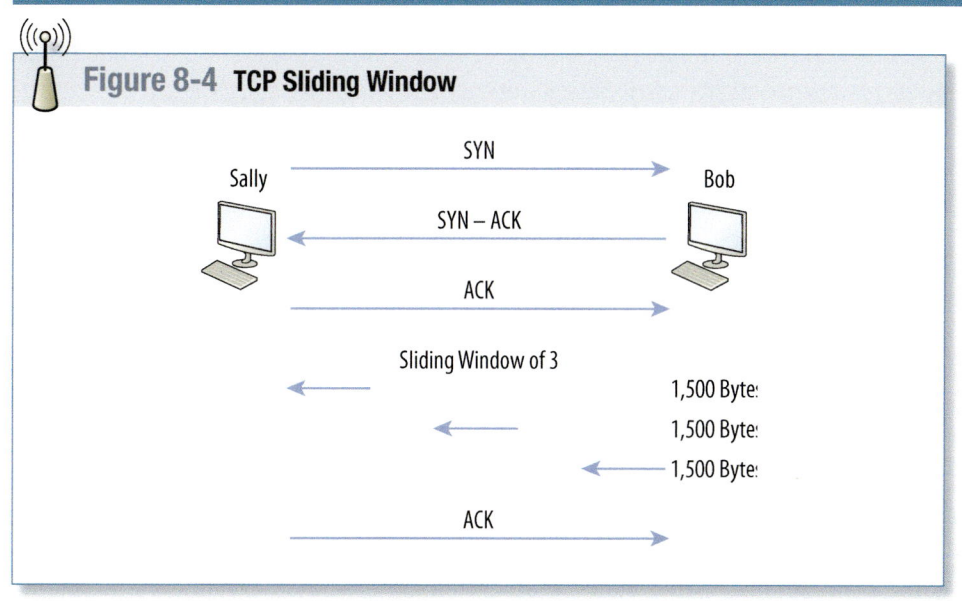

Figure 8-4. The TCP sliding window indicates the number of segments the computers can send at a time.

UDP

UDP is a best-effort protocol. Delivery is not guaranteed as it is with TCP, but UDP will do its best to get the data to its destination. UDP does not use a three-way handshake. It simply sends the data to the destination.

Imagine having a message to get to a friend as soon as possible. A call could be placed, but what if the friend does not answer? A message could be left. A text message could be sent. A letter could even be sent through the regular mail. However, none of these methods provides any assurance that the message was received. Still, a best effort is being made to pass on the message, and these methods normally work.

UDP does not have any of the overhead of TCP, and it does not use the TCP three-way handshake process to establish the session. It does not use periodic ACKs and NACKs to verify data was transmitted or request retransmissions of corrupt data.

UDP is often used for data that does not need guaranteed delivery, or when users want to avoid slow data transmissions because of the extra overhead required by TCP.

For example, streaming media such as streaming audio, streaming video, and Voice over IP (VoIP) all use UDP. These methods frequently lose packets here and there, but the overall message is still received. Many people have watched a video online that is occasionally jumpy or missing some of its audio. This is because it uses UDP and some of the packets are lost. Still, it is possible to get the overall message.

If TCP was used instead, the transmission would be a lot slower. To ensure that the full video was received, it might be possible to download the actual video file, instead of having it streamed. The file download would use TCP, and the entire file would be intact.

UDP is also used for short messages generated by protocols such as DNS.

> ✓ **FACT**
> Because UDP does not use the guaranteed delivery mechanisms of TCP, it is referred to as "unreliable." Also, it does not check for out-of-order messages.

✓ FACT
The IP protocol uses the IP address to get the packet from one computer to another over a network.

The Domain name system (DNS) is the means by which Internet domains are translated into IP addresses. DNS traffic consists primarily of name resolution requests and replies, which are simple, brief messages that fit in a single packet. Rather than go through the elaborate process of establishing and terminating a connection for each transaction, the systems simply send their messages. If a response is not forthcoming, they send them again. Even with a retransmission, the UDP transaction consists of less data than it would using TCP.

Even though UDP does not verify a connection before sending data or include a check for out-of-order messages, it does validate the data. UDP does use a checksum similar to how TCP uses a checksum. The checksum can indicate to the receiving computer that the data has been modified (perhaps by just dropping a single bit) and is not valid.

PORTS

Both TCP and UDP use logical port numbers to identify the contents of a packet. These port numbers help TCP/IP get the packet to the application, service, or protocol that will process the data once it arrives at the computer.

As an example, consider a home user who uses an ISP for access to the Internet, including e-mail. The user runs Microsoft Outlook to send and receive e-mail, and Microsoft Outlook has been configured with the names of both an SMTP server and a POP3 server. **See Figure 8-5**.

Sending and receiving e-mail are two separate processes. Sending SMTP e-mail consists of:

1. The client computer sends the e-mail to the SMTP server with a destination port number of 25.

2. The client computer assigns itself a random, unused source port number, such as 49152, and maps it to Microsoft Outlook for SMTP.

Figure 8-5 Ports

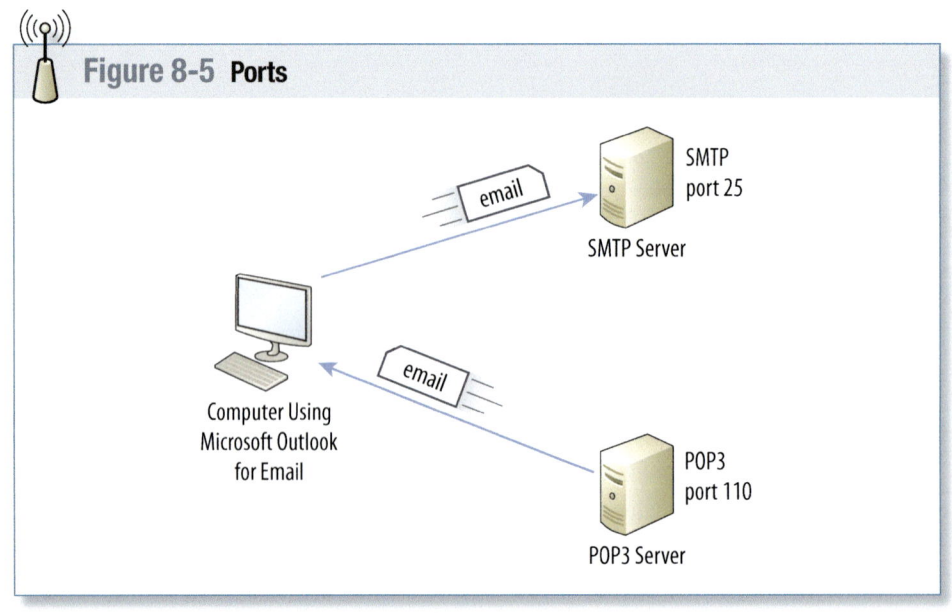

Figure 8-5. Ports are used to send and receive e-mail.

3. When the SMTP server receives the data from the client, it recognizes the destination port 25 as SMTP. It then forwards the data to the service handling SMTP.

4. After the e-mail is received, the server sends an acknowledgment to the computer using port 49152 to confirm the e-mail was received.

5. When the computer receives the packet with port 49152, it sends it to the Microsoft Outlook application. Outlook then moves the e-mail from the Outbox to the Sent folder.

The computer uses a similar process when it wants to download e-mail from the POP3 server:

1. The computer sends a request to the POP3 server with a destination port number of 110.

2. The computer assigns itself a random unused source port number, such as 49153, and maps it to Microsoft Outlook for POP3.

3. When the POP3 server receives the request, it recognizes the destination port 110 as POP3. It then forwards the request to the service handling POP3 requests.

4. The POP3 server then sends e-mail to the client using port 49153.

5. When the computer receives the data with port 49153, it sends it to the Microsoft Outlook application. Outlook then moves the e-mail into the Inbox folder.

Other applications use different, but similar, processes. The IP address gets the packet to the destination computer. The port then gets the packet to the correct application, service, or protocol on the target computer.

There are a total of 65,536 TCP ports and 65,536 UDP ports. The *Internet Assigned Numbers Authority (IANA)* assigns port numbers to protocols. It has divided the ports into different ranges. **See Figure 8-6.**

Port Traffic and Firewalls

In addition to using ports to get packets to the right protocol, application, or service, ports are also used to control traffic

For additional information, visit qr.njatcdb.org Item #1800

✓ FACT

IANA oversees the assignment of public IP addresses on the Internet.

Figure 8-6 Port Ranges for Well-Known, Registered, and Dynamic Ports

PORT NAME	PORT NUMBER	COMMENTS
Well-known ports	0 through 1023	These ports are associated with specific protocols or applications. Ports and protocols in the well-known port range are registered with IANA.
Registered ports	1024 through 49151	Some of these ports are registered with IANA for specific protocols, but this is not required. Computers can assign unused ports in this range for applications.
Dynamic ports	49152 through 65535	These ports are not registered with IANA and may be used for any purpose.

Figure 8-6. A full list of registered port numbers is available from IANA.

in a network. Firewalls can block traffic based on the TCP or UDP ports the packets are using.

For example, consider a network that wants to prevent any FTP traffic from being used on the network. FTP uses ports 20 and 21. A firewall can block FTP traffic. If the firewall receives any packet with either a source or destination port of 20 or 21, the firewall does not route the packet. **See Figure 8-7.**

Internally Used Ports and Protocols

Microsoft networks use several different ports and protocols on internal networks. It is important to understand what these ports and protocols are. Most protocols use either TCP or UDP, but some (such as DNS) use both. **See Figure 8-8.**

One of the primary reasons it is important to know the ports is for configuring firewalls. Firewall rules or exceptions can be created to allow or block the traffic based on the port. Firewall administrators have many of these ports memorized.

APPLICATION LAYER PROTOCOLS

TCP and UDP are the primary protocols used for data transmission at the Transport layer. However, there are several other protocols at the Application layer that are important to understand.

Hypertext Transfer Protocol

Hypertext Transfer Protocol (HTTP) defines how files on the World Wide Web (WWW) are requested by and transmitted to web browsers. For example, when Internet Explorer is used to access the site bing.com, the address bar shows the Uniform Resource Locator (URL) as http://www.bing.com. **See Figure 8-9.**

Some sites use encryption to protect the data transmission. For example, if a person purchases something over the Internet, that person provides information such as name, address, and maybe credit card data. This information needs to be protected as it goes over the Internet. *HTTP over Secure Sockets Layer (HTTPS)* is an Application layer protocol

> ✓ FACT
> HTTP is different from Hypertext Markup Language (HTML). HTML is the Internet standard for formatting and displaying documents on the Internet.

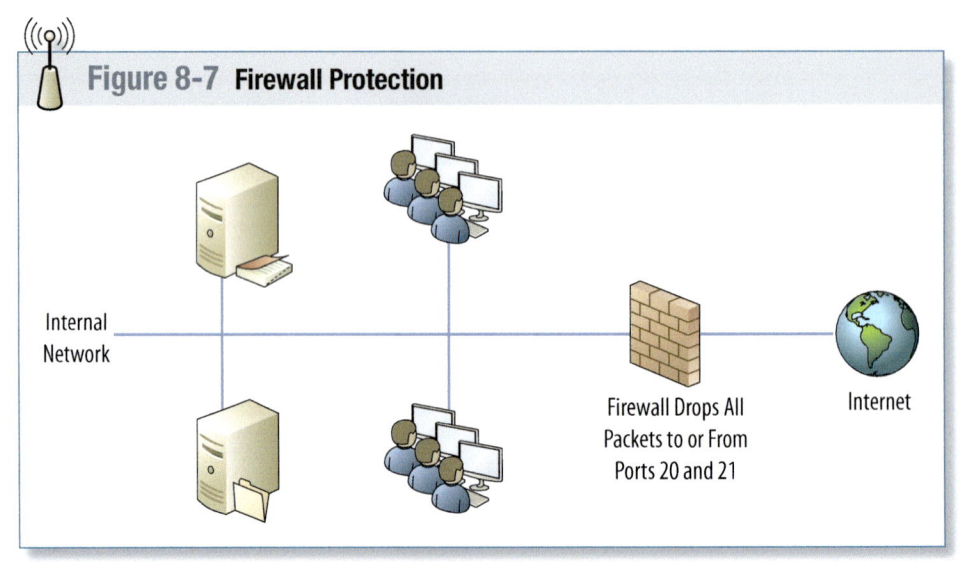

Figure 8-7 Firewall Protection

Firewall Drops All Packets to or From Ports 20 and 21

Figure 8-7. *A firewall is used to block traffic based on ports.*

Figure 8-8 Commonly Used Ports

PORT	TCP OR UDP	PROTOCOL	COMMENTS
20, 21	TCP	FTP	File Transfer Protocol.
22	TCP	SSH	Secure Shell.
23	TCP	Telnet	Can be secured with SSH.
25	TCP	SMTP	Simple Mail Transfer Protocol. Used to send e-mail.
110	TCP	POP3	Post Office Protocol. Used to receive e-mail.
143	TCP	IMAP4	Internet Message Access Protocol. Used when e-mail is stored on a server.
80	TCP	HTTP	Hypertext Transfer Protocol. Used for web pages.
443	TCP	HTTPS	Secure HTTPS. Commonly uses SSL for security.
53	TCP/UDP	DNS	Domain Name Service. Used to resolve names to IP addresses.
88	TCP	Kerberos	Primary authentication protocol used by Active Directory.
389	TCP	LDAP	Lightweight Directory Access Protocol (LDAP). Language used to communicate with Active Directory.
636	TCP	SLDAP	Secure LDAP. Uses SSL or TLS to encrypt LDAP communications.
161, 162	UDP	SNMP	Simple Network Management Protocol. Used to manage network devices such as routers and switches.
3389	TCP	Remote Desktop Services	Remote Desktop Services are used for Remote Assistance and remote desktops in a Microsoft network.
1723	TCP	PPTP	Point-to-Point Tunneling Protocol. Used in virtual private networks (VPNs).
1701	UDP	L2TP	Layer 2 Tunneling Protocol. Used in VPNs.

Figure 8-8. Administrators use port numbers when blocking services.

Figure 8-9 Address Bar

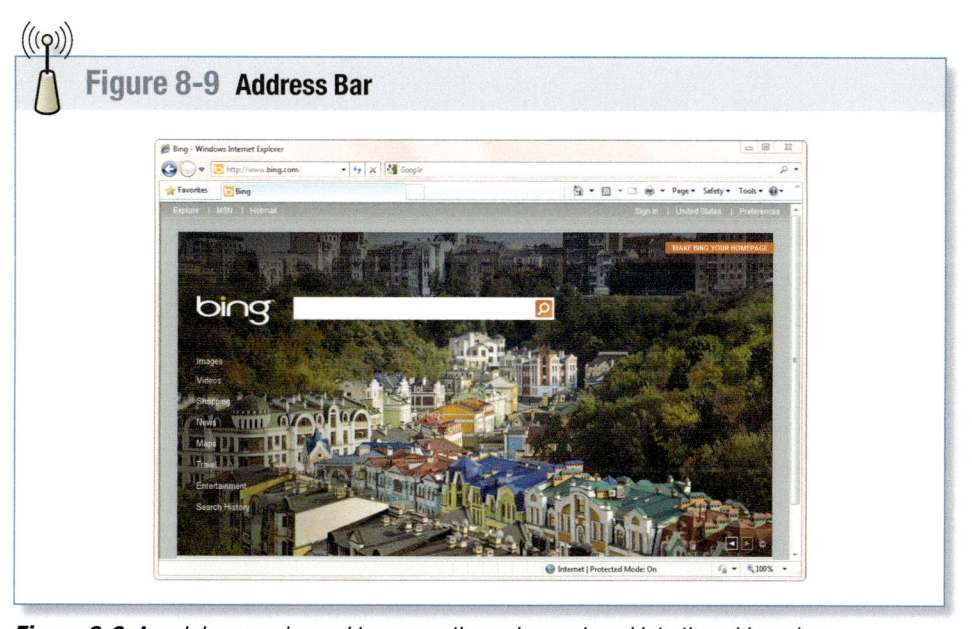

Figure 8-9. A web browser is used to access the webpage typed into the address bar.

✓ FACT
Encryption protocols scramble data from plain text into cipher text. Unauthorized users are not able to read the cipher text.

that handles encryption and decryption of secure data on the Internet.

It is possible to tell whether HTTPS is being used from the URL. Instead of HTTP, it will list it as HTTPS. Additionally, most web browsers include a lock icon somewhere on the page.

If HTTPS is not in the URL or the lock icon is not displayed, information that is entered and submitted to websites can be intercepted and read by eavesdroppers on the Internet.

HTTP uses TCP port 80 by default. HTTPS uses TCP port 443.

File Transfer Protocol

File Transfer Protocol (FTP) is used to upload and download files to and from computers on the Internet and within some internal networks. FTP uses TCP for guaranteed delivery of the files.

It is possible to access FTP from the command prompt from many operating systems, including Windows. `Get` commands download files, and `Put` commands upload files. Windows Explorer and File Explorer provide some basic FTP functionality, including drag-and-drop features. However, there are graphical applications that are shells for the command-line process, which make the process much simpler.

Graphical FTP clients enable the capability to browse the folders on the destination computer. A user can then pick which files to upload or download. Most FTP clients enable the capability to simply right-click a file and select to upload or download, depending on what the desired action is.

Most FTP servers require an account with a password before allowing files to be uploaded. This prevents malicious users from filling the FTP server with unwanted data. Additionally, FTP sites can limit permissions so that accounts can open and upload files only to certain folders.

Many FTP servers also allow users to download data anonymously. Users can use an account name of "anonymous" and then use an e-mail address as a password. The e-mail address is not verified to determine whether it is real, but it is often checked to ensure it is in the format of an e-mail address.

Trivial File Transfer Protocol

Trivial File Transfer Protocol (TFTP) is a scaled-down version of FTP that uses UDP as its transport protocol, which reduces overhead and keeps traffic to a minimum. In contrast, FTP uses TCP, providing guaranteed delivery of the files.

Network administrators often use TFTP when transferring configuration files to network devices such as routers and switches. TFTP is generally not used to communicate with FTP servers on the Internet because of its lack of data security features.

Telnet

Telnet is a command-line interface that provides bidirectional communication with network devices and other systems on the network. As a command-line interface, all commands are typed at a command prompt instead of using point-and-click methods within a Windows graphical user interface (GUI).

One of the benefits of Telnet is that it provides terminal emulation. In other words, it is possible to connect to a Telnet server remotely, and it acts as though the user is sitting right in front of the server accessing the local terminal. Telnet

✓ FACT
Many FTP clients are freely available. Search on the Internet for "download free ftp" to find them.

sessions include a Telnet server, a Telnet client, a Telnet window on the client (usually a command prompt) for issuing commands and viewing data on the server, and the Telnet protocol that transfers the commands between the two.

Remote Desktop Services

Remote Desktop Services (RDS) is an additional role included in Microsoft Windows servers to host applications or entire desktops that are accessible to users on the network.

For example, a client with limited processing power can connect to an RDS server and run a Windows desktop from the server. Even though the Windows desktop is running on the server, the end user has full access to all of the Windows capabilities on the older computer.

Similarly, a user might need to run a legacy application that is not compatible with his or her version of Windows. The RDS server can host the application, and the user can then run the application from the server without any compatibility problems.

The *Remote Desktop Protocol* is the same protocol used by Windows for Remote Assistance, which enables a help-desk professional to take control of an end user's desktop (with permission) and provide assistance. RDS uses TCP port 3389.

Secure Sockets Layer

Secure Sockets Layer (SSL) is an encryption protocol used for a wide assortment of purposes. SSL protects HTTP as HTTPS. SSL provides security in several key areas:

Confidentiality Secret data is protected from unauthorized disclosure through encryption. SSL encrypts data into cipher text to ensure that secret data remains secret.

Integrity Unauthorized users should not modify data. If they do, the data loses integrity and can no longer be trusted as valid. SSL helps ensure integrity by checking the data at different points to ensure it has not been modified.

Authentication Users and computers need to prove their identities. Based on their identities, a system grants or denies access based on permissions. However, the first step is authentication, in which the user or computer presents credentials that confirm its identity.

SSL uses digital certificates for confidentiality, integrity, and authentication. The *digital certificate* is a file that includes data used to encrypt the data for confidentiality. It also includes basic information to prove the identity of the certificate holder.

In recent years, many VPNs have emerged using SSL as a tunneling protocol. SSL-based VPNs have the advantage of being able to run using a web browser rather than requiring a separate VPN client program.

Transport Layer Security

Transport Layer Security (TLS) is a similar security protocol to SSL that can provide confidentiality, integrity, and authentication. RFC 2246 defined TLS in 1999, and TLS is designated as a replacement for SSL.

It is important to note that even though TLS came out more than ten years ago as a replacement to SSL, it still has not replaced it. Part of the reason for this is that SSL is a strong security protocol.

✓ **FACT**
Telnet is widely recognized as insecure since it transmits traffic in clear text. Secure Shell (SSH) has replaced Telnet in many applications.

✓ **FACT**
RDS was previously known as Terminal Services. Its name changed to RDS with Windows Server 2008 R2.

✓ **FACT**
The most recent version of SSL is 3.0, which was released in 1996.

✓ **FACT**
TLS has been upgraded. RFC 5246 defined TLS version 1.2 in August 2008. Version 1.3 is currently in a draft state.

Several protocols can use either SSL or TLS for security. For example, the Lightweight Directory Access Protocol (LDAP) can use either TLS or SSL for security.

Secure Shell

Secure Shell (SSH) is an encryption protocol that creates a secure encrypted session that other protocols can use. For example, SFTP is FTP encrypted with SSH. SSH has replaced Telnet in many applications. Telnet transfers data in clear text, while SSH encrypts the data. SSH is more secure than Telnet and more suitable for use on the Internet.

PuTTY (pronounced "putty") is an example application built on SSH. PuTTY is a free terminal emulator program that encrypts traffic with SSH. Many administrators use PuTTY to manage network devices such as routers and switches.

> ✓ FACT
> PuTTY is not an acronym. It is just a way of capitalizing the name that stuck.

Internet Protocol Security

Internet Protocol Security (IPsec) is an encryption protocol used to encrypt data transmitted over a network. IPsec provides two primary services:

Authentication The *Authentication Header (AH) protocol* is the protocol IPsec uses to prove the identity of the sender. This provides assurances to the computer receiving the traffic that a known computer sent it.

Encryption The *Encapsulating Security Protocol (ESP)* is the protocol IPsec uses to encrypt traffic. Only authorized users or computers are able to decrypt and read the traffic.

IPsec also provides data integrity. The receiving computer is assured that the data was not changed in transit.

Both IPv4 and IPv6 support IPsec, using one of two modes:

Tunnel Mode IPsec encrypts the entire IP packet (both data and headers). It encapsulates the original encrypted packet within another IP packet and then sends it across the network. VPNs use tunneling to protect the data. The L2TP/IPsec tunneling protocol is one of the popular VPN protocols.

Digital Certificates, PKI, and CAs

In the simplest terms, a digital certificate is just a file stored on a computer. However, this file has a lot of support behind it. Specifically, a Public Key Infrastructure (PKI) includes several elements to support digital certificates.

One of the core elements of a PKI is a certificate authority (CA). A CA is an organization or a service that issues, manages, and validates digital certificates. Many CAs operate on the Internet, and CAs can also operate on internal networks. If an entity (such as a user or computer) needs a certificate, the entity proves its identity to the CA and provides other information (and often money). The CA then issues a certificate.

This certificate helps verify the entity's identity and helps with other uses such as encryption and integrity. When the certificate is presented to a third party, the third party can then query the CA to verify the certificate is valid. As long as the third party trusts the CA, it can assume that any party with a certificate issued by that CA is who it claims to be.

For example, a website can purchase a certificate from a CA. When a user application (such as Internet Explorer) visits the website, the certificate is passed to the application. The application then queries the CA to verify that the certificate is still valid. If it is, a secure HTTPS session is created using data from the certificate.

Transport Mode Only the data is encrypted instead of the entire packet. The source and destination data (such as the IP addresses) within the packet are not encrypted. Transport mode is commonly used to encrypt data within internal networks.

Simple Mail Transfer Protocol

Simple Mail Transfer Protocol (SMTP) is the primary protocol used to deliver e-mail over the Internet and within internal networks. E-mail servers use SMTP to send e-mail to and receive it from other servers. Additionally, e-mail client applications use SMTP to send e-mail messages to their outgoing SMTP servers.

As an example of how systems commonly use SMTP, say that an e-mail application such as Microsoft Outlook is being used. This enables the user to connect with an e-mail server to send e-mail. The e-mail server receives the e-mail and then exchanges e-mail with other e-mail servers. **See Figure 8-10.**

Post Office Protocol v3

Post Office Protocol v3 (POP3) is a common protocol used to retrieve e-mail from an e-mail server. The current version is POP3.

As an example, say that Microsoft Outlook (or another e-mail client) is being used for e-mail at a home computer. The user connects to the Internet via an ISP, and the ISP provides e-mail access. When the user first configures an e-mail client, the user configures it with the names of the POP3 server and an SMTP server. The user's computer sends e-mail using the SMTP server, and it receives e-mail using the POP3 server.

The ISP's POP3 server receives e-mail addressed to the user and stores it there until the user connects to the Internet and contacts the server. When the user connects to the server, the server will then send all e-mail to the user's computer. Once the user's computer receives this e-mail, it is typically removed from the POP3 server.

Internet Message Access Protocol

Internet Message Access Protocol (IMAP) is a popular e-mail protocol that is more commonly used on internal networks rather than on the Internet. The current version is IMAP4.

The primary difference between POP3 and IMAP4 is that messages are not automatically downloaded to the client, and they can be retained on the

✓ FACT

The POP3 server can be the same server hosting SMTP. They do not have to be separate servers. This is common on smaller networks hosting both SMTP and POP3.

Figure 8-10 SMTP

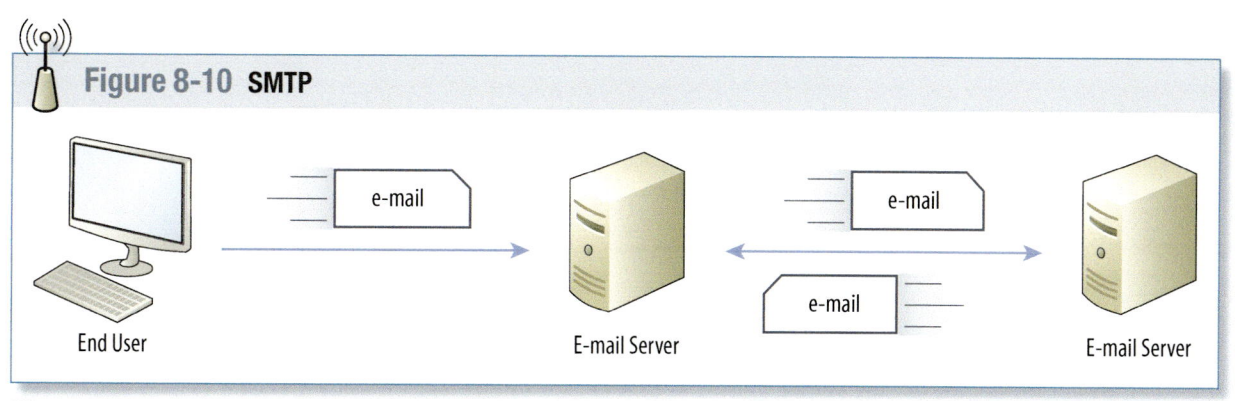

Figure 8-10. SMTP is used to send and receive e-mail.

server with IMAP4. An IMAP4 server enables users to view e-mail message headers individually. They can then choose which e-mail to open. For example, if a user is using a slow connection, he or she can choose to postpone opening an e-mail with a large attachment.

Since messages can be retained on the IMAP server, users can access the server from any computer in the network and still have access to the same e-mail. This is different from a POP3 server that downloads the messages to the user's computer when the user connects. With a POP3 server, if the user accesses the server with a different computer, the older messages are no longer on the server.

IMAP is useful for workers who roam the network and do not have a single computer they use all the time. The worker can connect to the IMAP server from any computer and access e-mail on the server.

IMAP also gives users the ability to manage their e-mail in folders. When the e-mail is retained on the IMAP server, users can move e-mail into different folders based on their preferences.

✓ FACT
RFC 4510 defines the latest version of LDAP, which is LDAP version 3.

✓ FACT
Messages do not have to be retained on the IMAP server. It is possible to configure the server so that e-mail is deleted after it is downloaded.

Lightweight Directory Access Protocol

Lightweight Directory Access Protocol (LDAP) is the protocol that systems use to query directories such as Microsoft's Active Directory Domain Services (AD DS). LDAP is derived from the Directory Access Protocol (DAP), which is part of a larger standard known as X.500.

It is easy to confuse the term directory since it has two meanings with computers. A directory is a domain directory in the context of LDAP. A directory can also be a folder on a disk drive, which has nothing to do with LDAP.

Domain Directory A *domain directory* is a database of objects such as users, computers, and groups. Administrators use the domain to manage users and computers. For example, Active Directory Users and Computers (ADUC) can be used for a domain named Wiley.com. The Servers organizational unit is selected, and several servers within the domain are indicated. Administrators use ADUC to manage the domain, and ADUC uses LDAP to query the AD DS database. **See Figure 8-11.**

Figure 8-11 Active Directory Users and Computers (ADUC)

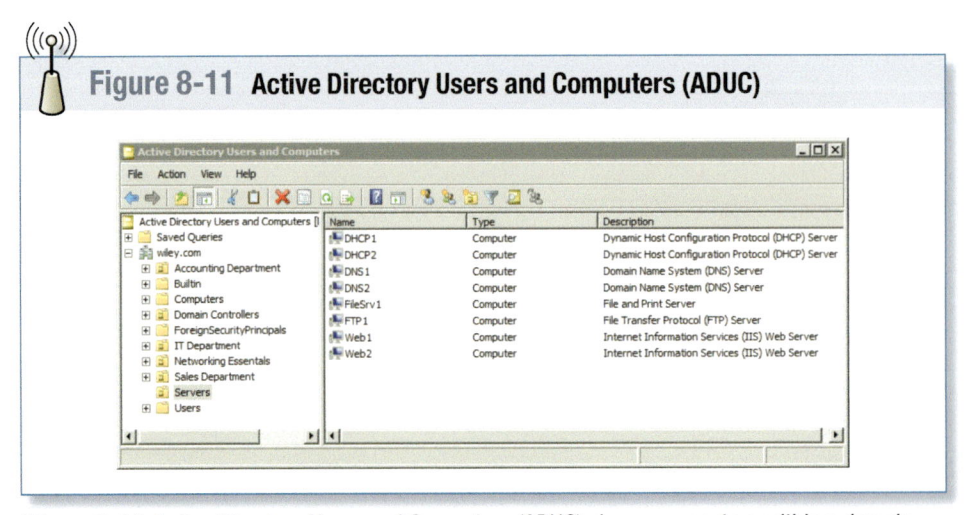

Figure 8-11. Active Directory Users and Computers (ADUC) shows computers within a domain.

Disk Drive Folders Windows disk drives can be examined by using Windows Explorer or File Explorer. Drives have folders that are also called directories. These folders have nothing to do with LDAP. Instead, these folders or directories are only on disk drives.

Although LDAP is integral to a Microsoft domain, it is also used in other non-Microsoft domains. Its purpose is the same, though. LDAP enables individuals to query the directory to locate and manage resources within the domain.

By default, LDAP transmits data across the network in clear text. *Eavesdropping* is the practice of using tools such as protocol analyzers or packet sniffers to capture this data and read it. This similar to a person listening in on another person's private conversation.

Secure LDAP (SLDAP) uses SSL or TLS to prevent attackers from using sniffers to capture data. Additionally, SLDAP uses digital certificates for authentication. This ensures that computers communicating with each other using SLDAP prove their identity prior to transferring data to each other.

LDAP uses TCP port 389 by default. SLDAP uses TCP port 636.

Kerberos

Kerberos is the primary authentication protocol used within a Microsoft domain and is managed as part of Active Directory. It was developed at the Massachusetts Institute of Technology (MIT) and is used in other non-Microsoft domains.

Kerberos uses a complex process of issuing time-stamped tickets to users after they log on. In simple terms, user accounts present these tickets when they try to access protected resources, such as a file or folder. If the tickets are valid, access to the resource is granted. This is similar to purchasing a ticket to watch a movie. The ticket enables the ticket holder to get in. Without a ticket, access is blocked.

These Kerberos tickets must be protected so that only specific user accounts can use tickets issued to them. Kerberos uses symmetric cryptography to encrypt the tickets.

Kerberos uses TCP port 88 by default.

Point-to-Point Tunneling Protocol

Point-to-Point Tunneling Protocol (PPTP) is a VPN protocol that provides a secure connection over a public network such as the Internet. PPTP is primarily used in Microsoft networks.

The *Point-to-Point Protocol (PPP)* is used for dial-up networking. PPTP extended PPP to make it useful for VPNs. The *Microsoft Point-to-Point Encryption (MPPE)* protocol encrypts the PPTP traffic.

PPTP uses port 1723.

Layer 2 Tunneling Protocol

Layer 2 Tunneling Protocol (L2TP) is a tunneling protocol used with VPNs. It is a combination of the Layer 2 Forwarding (L2F) protocol from Cisco and PPTP from Microsoft. However, L2TP is a standard used by more than just Cisco and Microsoft. IPsec is used with L2TP (as L2TP/IPsec) to provide security for the VPN connection.

L2TP uses UDP port 1701 by default.

Simple Network Management Protocol

Simple Network Management Protocol (SNMP) is a protocol used to manage network devices such as routers and

> ✓ **FACT**
> Authentication is used to prove identity. For example, a user could provide a username and password to authenticate within a domain.

switches. Many different applications are available that use SNMP.

One way to use SNMP is by installing SNMP agents on the network devices. The SNMP agents detect when specific events occur and generate a trap message to report back to a primary server that collects the information. Microsoft's System Center Operations Manager (SCOM) is an example of a server application used to monitor the health of devices on the network.

SNMP uses UDP port 161 by default. Since most of the traffic is diagnostic in nature, the guaranteed delivery of TCP is not required.

Summary

TCP and UDP are two primary protocols used to transport data across networks. TCP is connection-oriented and provides guaranteed delivery. UDP is connectionless and uses a best-effort delivery method. Many other application protocols are used within TCP/IP for a wide variety of purposes, including e-mail, web pages, encryption, interaction with Active Directory, and more. Application protocols use logical TCP and UDP ports for identification. IANA designates the port numbers for specific applications, and the first 1,024 are well-known ports.

Review Questions

1. Which of the following protocols is considered connection-oriented?
   a. ARP
   b. DHCP
   c. TCP
   d. UDP

2. DP traffic accepts the loss of some data.
   a. True
   b. False

3. What type of traffic commonly uses UDP?
   a. Streaming audio
   b. Streaming video
   c. Voice over IP
   d. All of the above

4. What is used to resolve an IP address to a MAC address?
   a. ARP
   b. DNS
   c. ICMP
   d. TCP

5. The three commonly used protocols for e-mail are SNMP, POP3, and IMAP.
   a. True
   b. False

6. L2TP is one of many tunneling protocols used for VPNs. What is used to encrypt L2TP traffic?
   a. IPsec
   b. L2F7
   c. PPTP
   d. TCP

7. The _____ protocol is used to manage multicast transmissions.
   a. ICMP
   b. IGMP
   c. LDAP
   d. SNMP

8. What port is used by RDS?
   a. 389
   b. 636
   c. 1701
   d. 3389

9. What port is used by LDAP?
   a. 25
   b. 389
   c. 1723
   d. 3389

10. What port is used by Kerberos?
    a. 25
    b. 80
    c. 88
    d. 443

Network Access with Switches

Switches are an important component in any wired network. Over time, switches have steadily replaced Ethernet hubs because of their extra capabilities. Additionally, advanced switches can perform router functions.

A technician must understand the details of how a switch works and its benefits over a hub. The technician also must understand the differences between managed and unmanaged switches, between layer 2 and layer 3 switches, and how to use a switch to create VLANS. Last, some basics about switch speeds and switch security are fundamentally necessary.

OBJECTIVES

- ▶ Explain how to connect multiple computers
- ▶ Describe various types of physical ports
- ▶ Identify various hubs and switches
- ▶ Compare unmanaged and managed switches
- ▶ Identify several basic switch transmission speeds
- ▶ List various common security options

CHAPTER 9

TABLE OF CONTENTS

Multiple-Computer Connections 172

Physical Ports 172
    Number and Type of Ports 174
    Ports in Drawings 174

Hubs and Switches 175
    Collision Domains 175
    Collision Domains with a Hubs 176
    Collision Domains with Switches 176
    Ports Mapped to MAC Addresses . . . 177

Unmanaged and Managed Switches . . . 179
    Unmanaged Switches 179
    Managed Switches 179
    Layer 2 and Layer 3 Switches 179
    Spanning Tree Protocol 180
    Managed Switches and VLANs 180

Switch Speeds 183
    Transmission Speeds 184
    The Uplink Port 184
    Backplane Speed 185

Security Options 185
    Port Security 185
    Hardware Redundancy 186

Summary 186

Review Questions 187

MULTIPLE-COMPUTER CONNECTIONS

Switches (or hubs) connect computers in a network. In larger organizations, routers connect multiple networks into an internetwork.

Consider an example network consisting of three separate subnetworks of 192.168.1.0/24, 192.168.5.0/24, and 192.168.7.0/24. Each of the subnetworks has a central switch connecting the devices. A router connects the three switches. **See Figure 9-1.**

Packets sent by computers on this network go through a switch first. The switch learns which computers are connected to which port. It uses this knowledge to determine the path for every packet it receives. In contrast, routers move packets between the subnetworks.

If the destination computer is on the same subnetwork, the switch forwards the packet directly to the destination computer. If the destination computer is on a different subnetwork, the switch forwards the packet to the default gateway router, which forwards it to the correct subnetwork.

Note that the switch is the central device for each subnetwork. It could be a hub, but for several reasons, switches have replaced hubs in many networks. Although most networks have switches connected, many network line drawings omit the icon of the switch. **See Figure 9-2.**

Though in a line drawing a switch may not be shown, the computers have to be connected to a central device. They cannot all be connected directly to the router.

PHYSICAL PORTS

Switches have physical ports into which physical cables are plugged. For example,

> ✓ **FACT**
> Switches track the location of the computers on networks. Routers track networks or subnetworks, not computers.

> ✓ **FACT**
> The /24 represents CIDR notation. It indicates the subnet mask has the first 24 bits set to a 1. In other words, the subnet mask is 255.255.255.0.

Figure 9-1 Multiple-Computer Connections

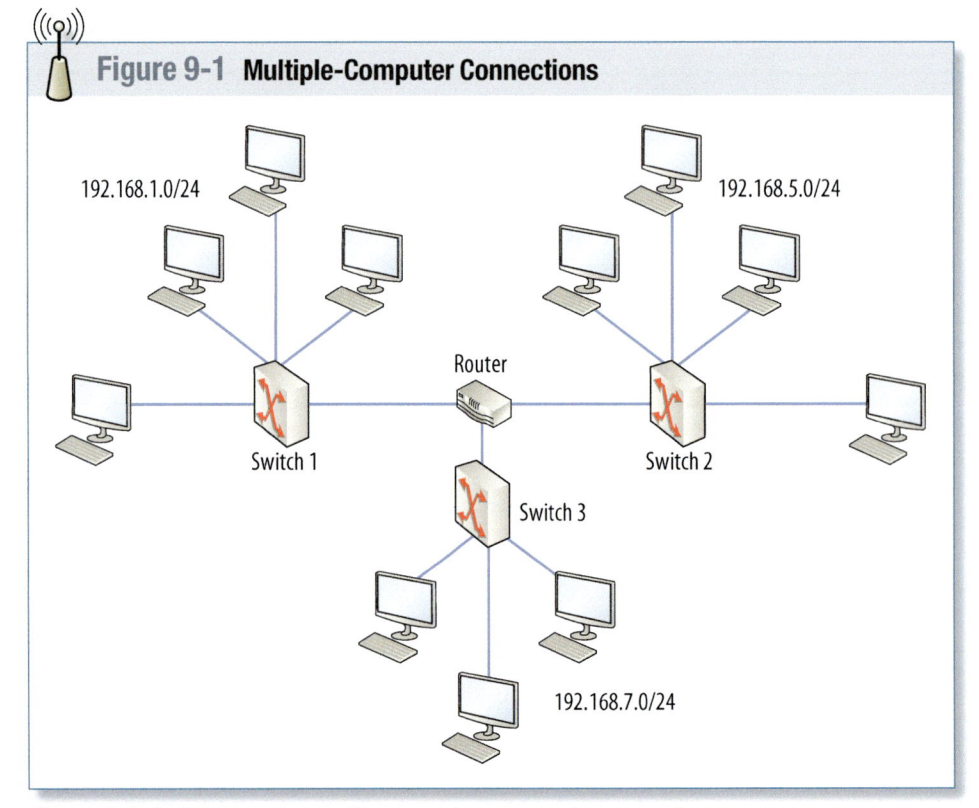

Figure 9-1. Switches can be used to connect computers in a network.

if the network is using twisted-pair cable, the switch will have 8P8C ports that accept cables with 8P8C connectors. If the cable is fiber optic, the switch has physical ports that accept the fiber-optic connectors.

Most switches that connect end-user computers have 8P8C ports since

---

Networks, Subnets, and Subnetworks

The terms network, subnetwork, and subnet are sometimes easily confused, and it is worth identifying the differences. In general, a network is two or more computers or other network devices connected together. When they are connected, they can share data and resources with each other. Both subnets and subnetworks are groups of computers with the same network ID. Additionally, routers separate subnetworks and subnets from each other. However, there is a subtle difference between these two.

A subnet is a network that started as a classful network and was divided into multiple subnets. For example, a single Class C network of 192.168.1.0/24 can be divided into four subnetworks of 192.168.1.0/26, 192.168.1.64/26, 192.168.1.128/26, and 192.168.1.192/26.

Subnetworks use different classful IP ranges without subnetting them. For example, one subnetwork can be created with an address in 192.168.1.0/24, and a different subnetwork could have an address range in 192.168.5.0/24. This creates two subnetworks without subnetting a classful IP address. This is actually very common on private networks. There are more than enough private, classful IP address ranges for even very large organizations to use them without subnettting.

Not everyone uses these terms in the same way. Technicians often call a subnetwork a "subnet" or simply a "network." Some insist that they are all called "networks." Some insist that a subnet is created only when a classful IP address range has been subnetted, and the rest are networks.

Although the terminology can be tricky, the primary message should still be clear. Switches connect and track computers within a subnetwork. Routers connect and track subnetworks.

Figure 9-2 Network Line Drawing

192.168.1.0/24

192.168.5.0/24

Router

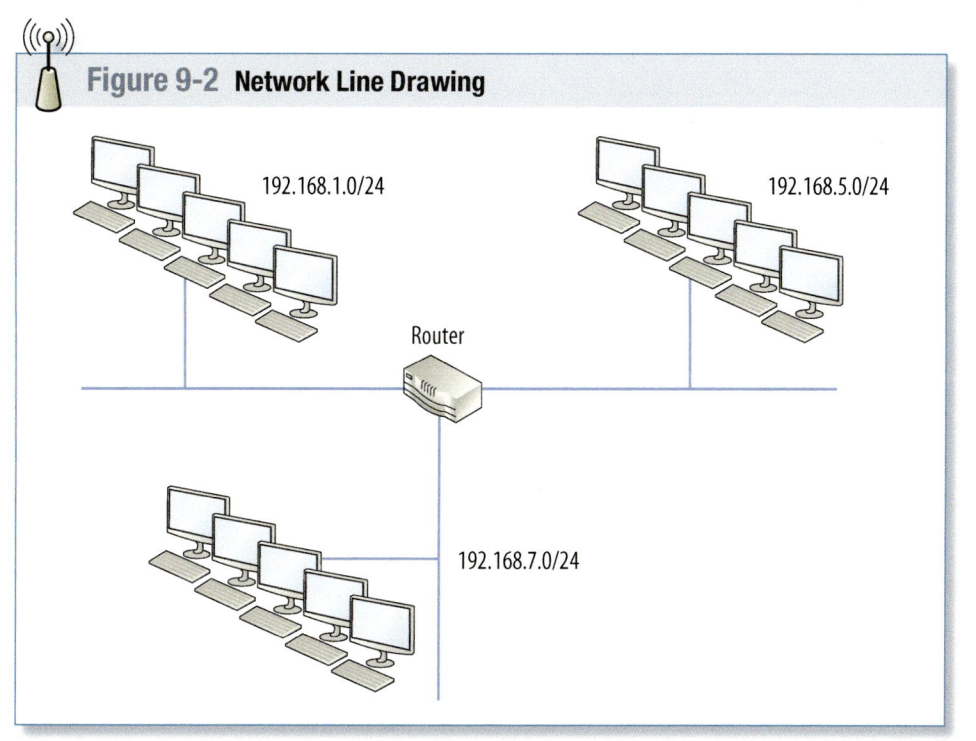

192.168.7.0/24

Figure 9-2. A network line drawing may omit the switches.

> **✓ FACT**
> Uplinks connect two switches together or connect the switch to a router.

> **✓ FACT**
> Different types of modules can be added to a modular switch. This includes modules for typical 8P8C ports, fiber-optic ports, wireless services, video services, and more.

twisted-pair is the most commonly used network medium today. Some switches include fiber-optic ports for uplinks.

Physical ports and logical ports are not the same thing. A physical port is something that can be touched and accepts a cable. A logical port is simply a number that is embedded in a Transmission Control Protocol (TCP) or User Datagram Protocol (UDP) packet. For example, HTTP is represented by a logical default port of 80. When the packet reaches the destination computer, the logical port identifies the service or application that will process the data.

Number and Type of Ports

The number of ports on a switch or hub varies according to the physical size of the device. This is usually enough for a small office/home office (SOHO) network. When selecting a device for a small business with 8 or 10 users, a 24-port device might be a reasonable investment, allowing for future growth.

Switches typically have from 8 to 64 ports. Switches can be purchased in two separate designs:

Form-Factor Switch A form-factor switch has a set number of ports built into the switch. The number of ports cannot be changed. Form-factor switches can have any number of ports, but 48 is the maximum for most form-factor switches. These switches are great when simplicity is required.

Modular Switch A modular switch starts with zero or perhaps a few ports and can expand to hundreds of ports. Plug-in modules are added to add ports. This is similar to a computer that can accept additional memory modules. For example, the computer may start with 1 GB RAM, but RAM can be added

when needs change. Similarly, a modular switch can be purchased with a module that includes eight ports, but then modules can be added to increase the number of available ports.

Selecting a switch design is a decision that can be based on several factors. For example, administrators must ensure that enough ports are available for the immediate needs while also considering future growth requirements. Most modular switches require programming, which adds administrative overhead, while many form-factor switches work right out of the box.

Ports in Drawings

When switches are included in network drawings or connection maps, the ports are usually labeled. This enables technicians to identify which port goes to which system. Switch ports are commonly labeled with E, F, or Gi, followed by a number.

E The first 10 Mbps port is labeled as E0. This indicates Ethernet port 0. Some manufacturers represent the first Ethernet port on a modular switch as E0/0, which represents the first port on the first module.

F The first 100 Mbps port is typically labeled as F0 or F0/0, a Fast Ethernet port. Compare this to the second port on the first module, which is labeled as F0/1. Fast Ethernet ports can also be labeled as Fa instead of just F.

Gi The first 1,000 Mbps port is labeled as Gi0/0, a gigabit port. Similarly, the first port on the second module is Gi1/0.

Based on how the ports are labeled in a network line drawing, additional information can be gleaned about the

> **✓ FACT**
> Many computer technologies use zero-based numbering, where the first item is a 0 instead of a 1. In a switch, the first port is a 0 instead of a 1.

switches. For example, the ports on switch 1 are labeled as E0 through E4, which indicates it is a 10 Mbps switch. Assume the ports on switch 2 are labeled as F0 through F4, indicating it is a 100 Mbps switch. If the ports on switch 1 are labeled as G0 through G3, it is a 1,000 Mbps switch. **See Figure 9-3.**

HUBS AND SWITCHES

A hub is a physical layer (layer 1) device that connects multiple network devices. When using a hub, bandwidth decreases as more devices are added to the network since all devices share bandwidth equally. Any data sent into one port of a hub goes out all other ports. With a switch, each port is separated from the others. The ports do not share bandwidth.

Hubs create a single collision domain and a single broadcast domain. Switches create multiple collision domains but share a common broadcast domain. Routers create separate collision and broadcast domains.

Collision Domains

A collision domain is a group of devices on the same segment that, when they transmit data, are subject to collisions. A hub creates a single collision domain, while a switch creates multiple collision domains.

The Carrier Sense Multiple Access with Collision Detection (CSMA/CD) media access control mechanism enables each device to determine when it can transmit data across the network. The devices listen for traffic, and if they do not hear any, they are free to send data. However, just as two people can start talking at the same time, two computers can start sending data at the same time.

Figure 9-3 Switch Port Identification

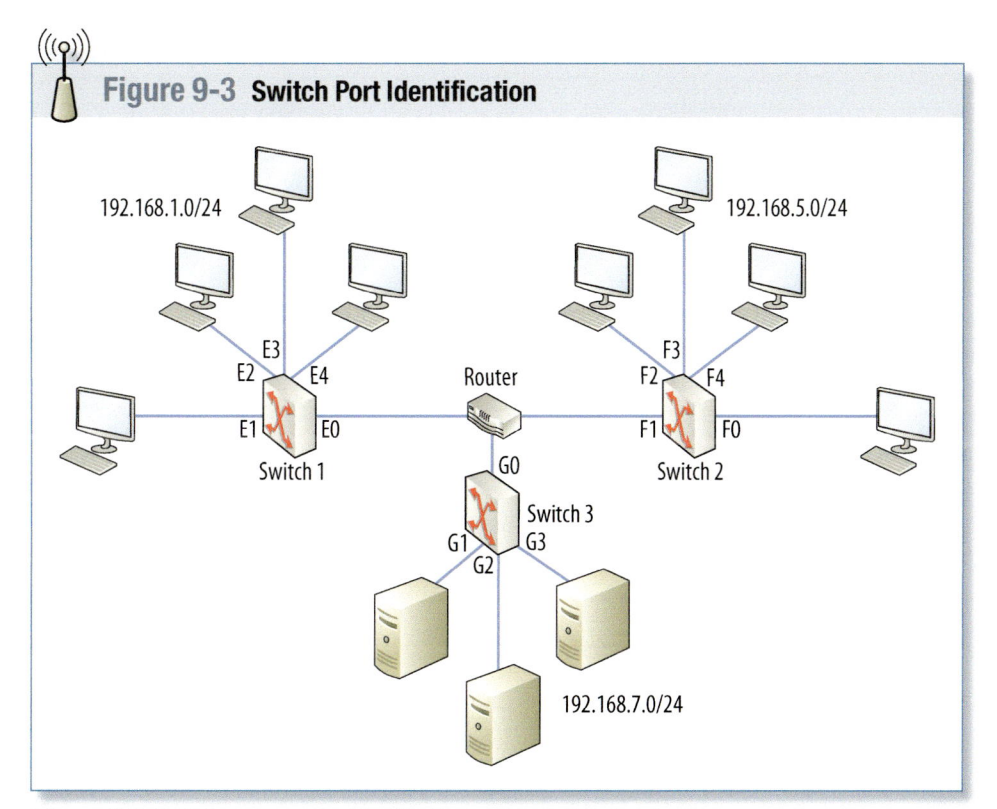

Figure 9-3. *Switch ports can be identified in a network line drawing.*

Consider an example network with several computers connected to the same segment. If PC-1 and PC-4 are both sending traffic at the same time, a collision occurs. **See Figure 9-4.**

CSMA/CD has a recovery mechanism for retransmission of lost data after a collision. First, the systems transmit a jamming signal on the segment, letting all devices know there has been a collision. Other devices postpone their data transmissions until the devices that had the collision resend their data. Once this process is complete, the network is reopened for business, and other devices may transmit their data.

Increasing the number of collision domains can help to reduce the number of collisions. Since a hub has a single collision domain and a switch creates a separate collision domain for each port, replacing hubs with switches helps to reduce collisions.

Collision Domains with a Hubs

A hub connects multiple computers into a single collision domain. In other words, all devices connected to a hub contend for equal access to the same segment.

Consider an example of four computers connected to a hub. Each computer uses CSMA/CD to listen before transmitting and can cause a collision if it sends data at the same time as another computer. **See Figure 9-5.**

The hub acts as the central point in the network, and any traffic destined for another device will go to the hub first. PC-1 sends data to PC-3. However, the hub forwards the packet to all of the computers connected to it. PC-3 will therefore process the packet while PC-2 and PC-4 will discard it. However, this packet can cause a collision if any other computer transmits data at the same time.

Collision Domains with Switches

A switch can create multiple collision domains. Unicast traffic sent from the source computer is passed only to the destination computer. The switch creates an internal connection between PC-1 and PC-3. PC-2 and PC-4 do not receive the data. **See Figure 9-6.**

If PC-2 sends data to PC-4 at the same time PC-1 is sending data to PC-3, it does not cause a collision. Instead, the

> ✓ **FACT**
> Collisions are a normal occurrence on an Ethernet network, but too many collisions on the network degrade the network's performance. Both computers must resend data each time a collision occurs.

> ✓ **FACT**
> A hub has no intelligence. All traffic received on one port is flooded to all other ports.

Figure 9-4 Collisions

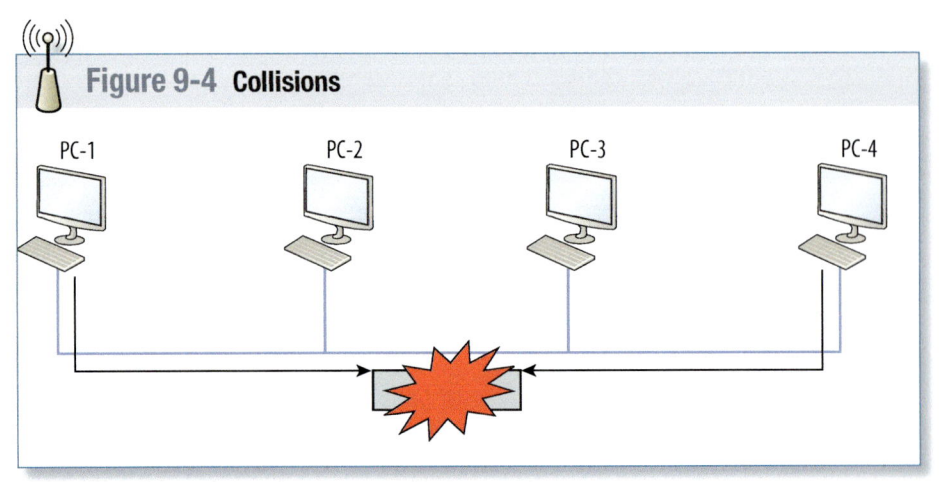

PC-1   PC-2   PC-3   PC-4

Figure 9-4. *If two computers are sending traffic at the same time, collisions occur on a collision domain.*

Figure 9-5 Collision Domain with Hubs

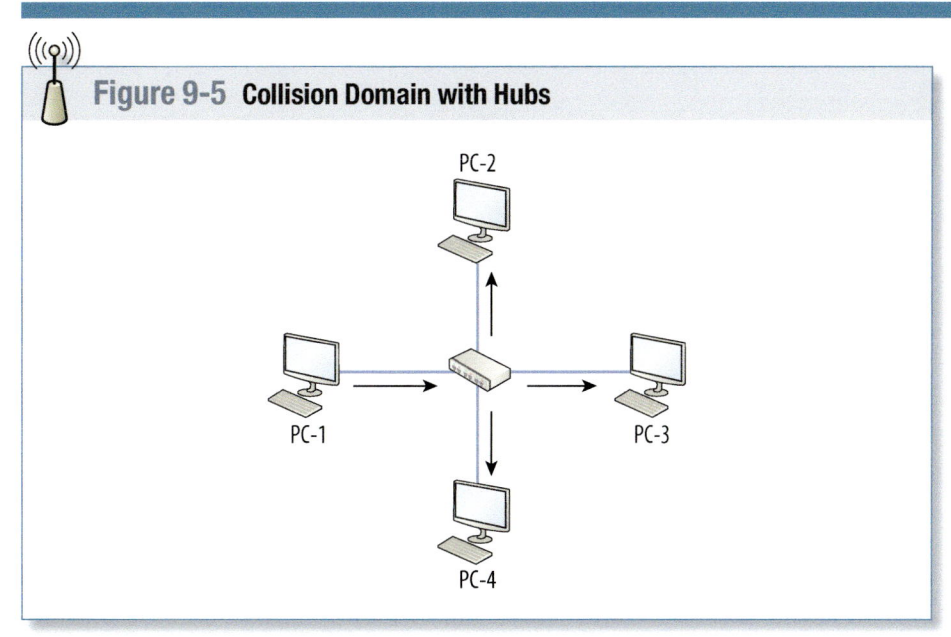

Figure 9-5. A single collision domain can be created with a hub.

Figure 9-6 Collision Domains with Switches

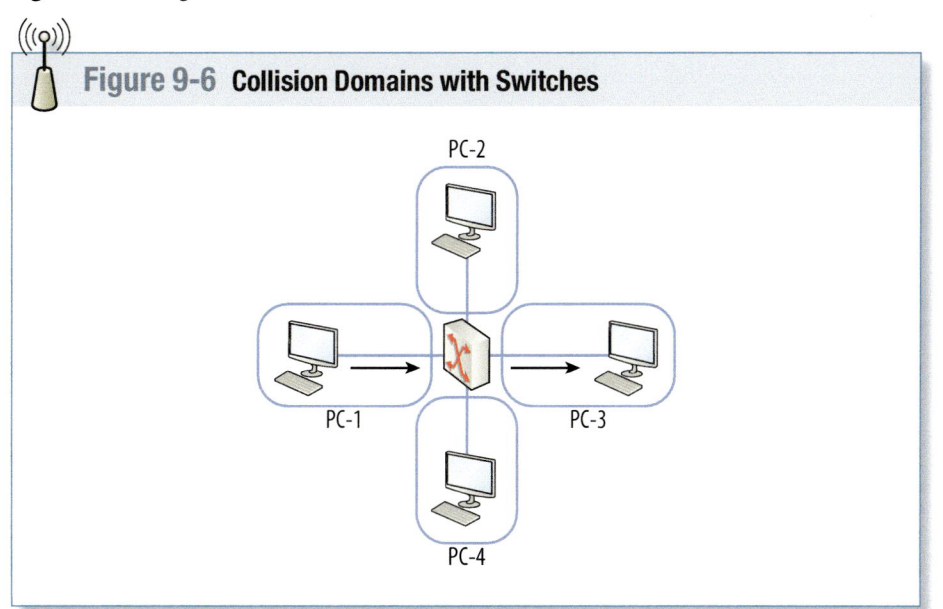

Figure 9-6. A switch can create multiple collision domains.

switch makes an internal connection between PC-2 and PC-4.

A switch maps physical ports to computers' media access control (MAC) addresses.

Ports Mapped to MAC Addresses

A MAC address is a 48-bit address expressed in a hexadecimal format. For example, a MAC address looks like 00-23-5A-33-C4-CA.

Every network interface adapter has a MAC address assigned to it. MAC addresses are typically burned into the card and are unchangeable, though some adapters enable modification of the MAC. Additionally, when a computer sends data to another computer, it always includes both its own IP address

✓ FACT

Four bits are used to represent each hexadecimal character.

178 Introduction to Network Technologies

> ✓ **FACT**
>
> Advanced switches (managed switches) are configurable. Configuration can be done with the MAC addresses of connected computers.

and its own MAC address as part of the source information.

A simple switch starts with very little knowledge when it is turned on. It knows what ports it has, but it does not know which computers are connected to the ports. However, as the switch processes incoming traffic from the computers, it learns and populates an internal MAC address table. A *MAC address table* is a list of the MAC address of each computer and maps them to the port to which they're connected.

Consider a line drawing showing MAC addresses for a four-port switch with a computer connected to each port. Right after the switch is turned on, when PC-1 sends data to PC-3, the switch does not know what port PC-3 is on, so it sends the data to all of the ports. However, the packet from PC-1 includes the MAC address of PC-1. The switch starts populating the MAC table by logging port number F0 with the MAC address of PC-1. See **Figure 9-7**.

When PC-3 answers, it includes the destination MAC address of PC-1 and the source address of PC-3. Again, the switch logs the MAC address of PC-3 with port F2 in the MAC address table. Since the switch knows that PC-1 is connected to port F0 (based on the MAC address), it internally switches the data from PC-3 to PC-1 on port F0. In a very short period, the switch will learn the MAC addresses of each computer, along with the associated ports.

When the switch maps the MAC addresses to ports as traffic passes, the table is updated dynamically. However, an administrator can configure a managed switch with specific MAC addresses as static entries. Switches can overwrite dynamic entries as time passes, but static entries remain.

Figure 9-7 Ports and MAC Addresses

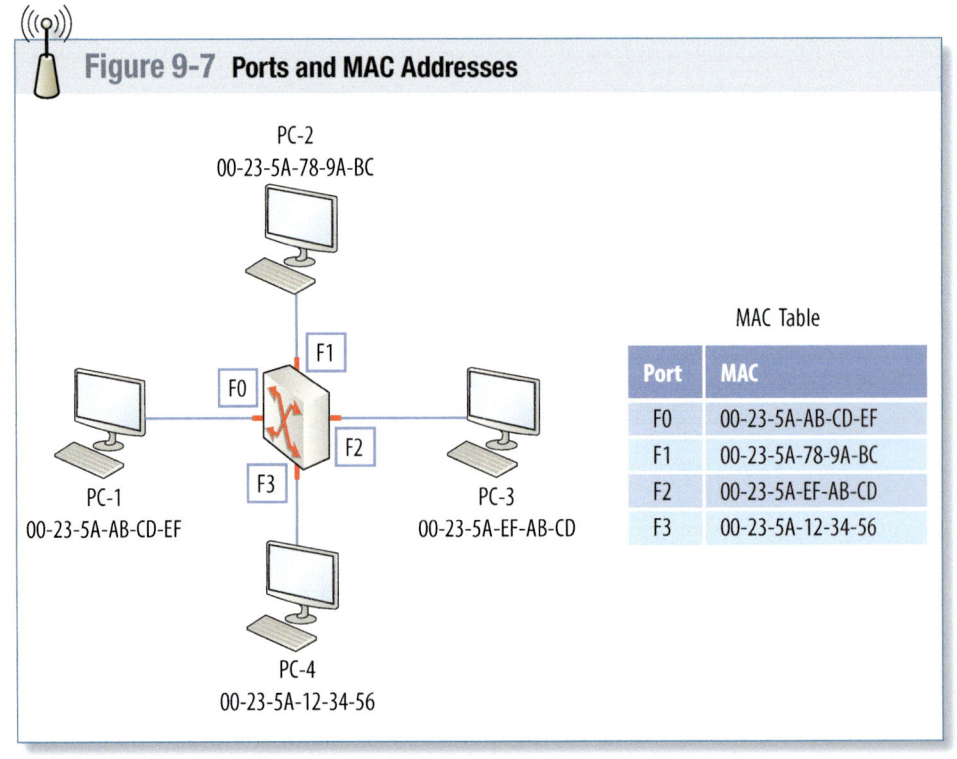

PC-2
00-23-5A-78-9A-BC

PC-1
00-23-5A-AB-CD-EF

PC-3
00-23-5A-EF-AB-CD

PC-4
00-23-5A-12-34-56

MAC Table

Port	MAC
F0	00-23-5A-AB-CD-EF
F1	00-23-5A-78-9A-BC
F2	00-23-5A-EF-AB-CD
F3	00-23-5A-12-34-56

Figure 9-7. A switch can map ports to MAC addresses.

UNMANAGED AND MANAGED SWITCHES

Switches can be configurable or nonconfigurable. If the switch must be customized in some way, then the administrator must be able to configure it. This requires additional knowledge on the part of the administrator and extra time. In other words, a configurable switch has more administrative overhead.

A *managed switch* is a configurable switch. An *unmanaged switch* is a nonconfigurable switch. Determining the necessary switch requires some basic knowledge of these two types of switches.

Unmanaged Switches

An unmanaged switch is just like a hub with respect to its administrative overhead. There is none. After taking the switch out of the box and plugging it in, it works.

The switch monitors the traffic from each of the ports and builds the MAC address table. The MAC address table maps the MAC addresses of the connected computers to their respective ports.

Even though an unmanaged switch does not require any administration, it does provide performance benefits over a simple hub. It still creates separate collision domains and increases performance on the network.

Unmanaged switches operate at the Data Link layer (layer 2) of the Open System Interconnection (OSI) model.

Managed Switches

In contrast to an unmanaged switch, a managed switch can be configured. Managed switches are commonly managed with protocols such as Telnet or Secure Shell (SSH), which administrators can use to monitor and configure the switch remotely. SSH encrypts the traffic so that it cannot be read if intercepted by a protocol analyzer or sniffer, while Telnet transmits in clear text.

A few of the management tasks that an administrator can perform include:

- Configure static entries in the MAC table
- Configure duplex settings (half-duplex or full-duplex) on ports
- Monitor performance of the switch using the Simple Network Management Protocol (SNMP)
- Configure the switch to send alerts called traps with SNMP when certain events occur
- Create a virtual local-area network (VLAN)
- Configure port mirroring

Although managed switches provide many more capabilities, they are also more expensive. Before spending the extra money for a managed switch, ensure that the administrative staff has the skills to support it.

Many managed switches can operate at the Data Link layer (layer 2) or the Network layer (layer 3) of the OSI model.

Layer 2 and Layer 3 Switches

The primary purpose of a layer 2 switch is to segment collision domains at layer 2 of the OSI model. Each port on the switch is segmented from the others. Layer 2 switches are hardware-based, which makes them extremely fast. They use the integrated circuitry on the device's main board to move data between ports at lightning speed.

> ✓ FACT
> Managed switches have configurable ports. Ports cannot be configured on an unmanaged switch.

> ✓ FACT
> Port mirroring sends a copy of all traffic on the switch to a single port. Administrators can capture traffic on this port for monitoring with a packet sniffer.

As expected, a layer 3 switch operates at layer 3 of the OSI model. It includes standard switching functionality, but also has the capability to route layer 3 traffic just as if it were a dedicated router. Although the router is a great layer 3 device, it can be slow, because additional processing of the packets by the integrated software is required. A hardware-based switch is quicker.

Only managed switches can route traffic on layer 3 like a router. Managed switches can also be configured to create VLANs.

A layer 2 switch creates separate collision domains. It does pass broadcasts, however, so broadcast traffic goes to all ports on the layer 2 switch. In other words, a layer 2 switch does not create separate broadcast domains.

In contrast, a router does not pass broadcasts. The router creates separate broadcast domains. However, a layer 3 switch acts like a router and creates separate broadcast domains. This VLAN capability can be used to ensure that traffic is routed only to certain ports on a switch without replacing the switch with a router.

Spanning Tree Protocol

Because layer 2 switches propagate broadcasts through all of their ports, it is possible for loops to occur on a network with multiple switches. This is sometimes called a broadcast storm. A *broadcast storm* occurs when broadcasts generated by one switch arrive at other switches, all of which forward the broadcasts in turn. Thus, the switches all start receiving the broadcasts that they themselves forwarded, resulting in an endless loop of broadcast transmissions that floods the network.

To prevent this from occurring, switches use an algorithm called the Spanning Tree Protocol (STP), which selectively blocks links between switches. The *Spanning Tree Protocol (STP)* enables a switch to discover a subset of the network topology that does not contain loops, eliminating the endless propagation of broadcasts. If a link in the subset becomes unavailable (because of hardware failure, for example), STP recalculates the topology and implements a new loopless pattern that makes the entire network accessible to the switches.

Managed Switches and VLANs

A virtual LAN (VLAN) is like a LAN inside of a LAN. However, just as the name implies, it only exists virtually; there is no dedicated physical hardware involved.

The benefits of creating a VLAN include:

▶ Improved LAN security, because broadcast traffic is limited to specific ports

▶ The capability to group workstations or servers based on needs, not on physical location

▶ Improved network performance for each separate broadcast domain

Assume that the administrator for an organization with several departments (including the Sales and Finance departments) recently noticed that financial data was leaked. The source of the leak is unknown, but security is being tightened everywhere that financial data flows. Traffic from the computers in the Finance department must be isolated with the least cost.

Consider a network configured with a switch before the creation of the

> ✓ FACT
> Routers route traffic on layer 3. A layer 3 switch can route traffic on layer 3 like a router.

VLAN. (Although the switch could actually be hosting many more computers, consider only four computers for simplicity's sake). The goal is to create two separate VLANs on this switch— one for Sales and one for Finance. **See Figure 9-8.**

VLANs need VLAN identifiers (VLAN IDs) and VLAN names. The Sales department will be on VLAN ID 2 using ports F0/0 through F0/11 and a network ID of 192.168.2.0/24. The Finance department will be on VLAN ID 4 using ports F0/12 through F0/24 with a network ID of 192.168.4.0/24. **See Figure 9-9.**

In this VLAN configuration, even though the switch is shown twice in a network line drawing, there is only one physical switch. However, it is using the

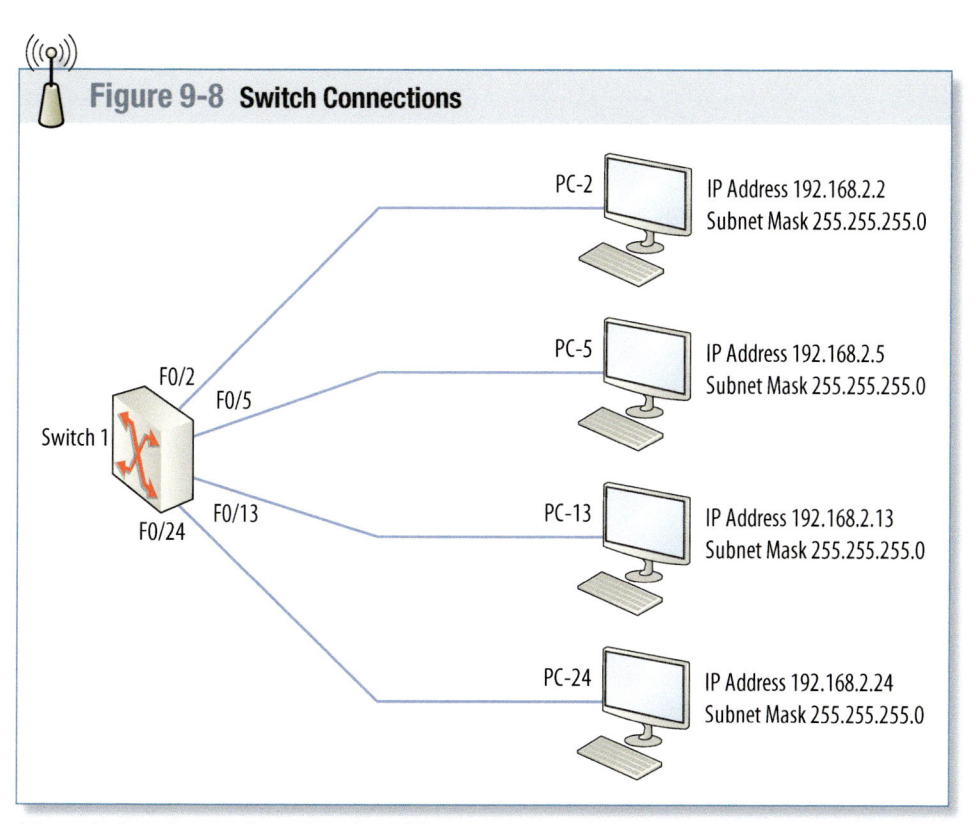

Figure 9-8 Switch Connections

Figure 9-8. Before configuration of a VLAN, computers can be connected with a switch.

Figure 9-9 VLAN Configuration

VLAN ID	VLAN NAME	PORT RANGE	SUBNET RANGE	SUBNET ID
2	Sales	F0/0-F0/11	192.168.2.0/24	192.168.2.0
4	Finance	F0/12-F0/24	192.168.4.0/24	192.168.4.0

Figure 9-9. Each individual VLAN must have a VLAN ID.

Introduction to Network Technologies

✓ FACT
Network IDs are derived from the IP address and subnet mask.

specific ports listed to create the two VLANs. **See Figure 9-10.**

To use the VLANs, ensure that the Finance department's computers are connected only to switch ports F0/12 through F0/24, and that its IP addresses are changed to use the network ID of 192.168.4.0/24. Additionally, ensure that all of the Sales department's computers are connected only to ports F0/0 through F0/12.

This scenario shows the basics of creating a VLAN, but VLANs can be much more complex. For example, VLANs can be created that span multiple switches.

Consider a scenario in which four departments are connected with five switches. Each switch is dedicated to a specific department, but as the company grows, more salespeople are added than the office space can support. Some salespeople are sitting in the office space where the HR switch is located. However, VLANs can be created so that these salespeople are virtually connected to the Sales department switch. **See Figure 9-11.**

Basic points to remember about VLANs:

- A VLAN must have at least two ports before traffic can flow, but it can have more.
  - Twenty-four two-port VLANs can be created on a 48-port switch.
  - Two 24-port VLANs can be created on a 48-port switch.
  - Any mixture can be created as long as each VLAN has at least two ports.
- VLANs can span multiple switches.

Figure 9-10 Switches and VLANs

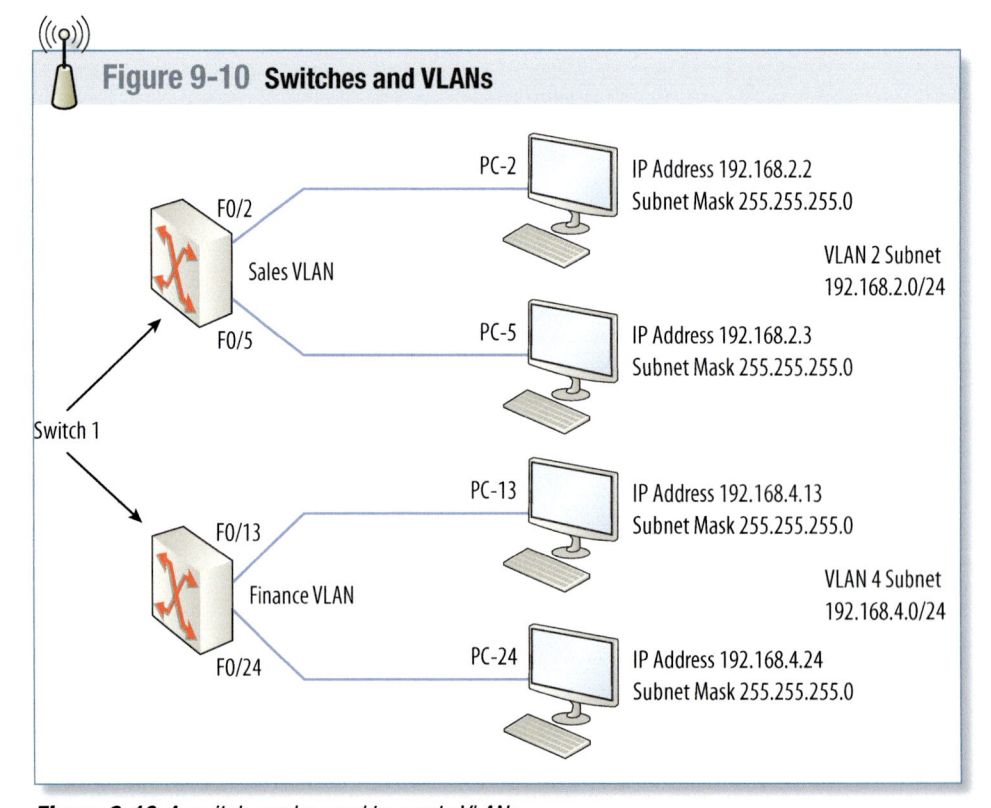

Switch 1
- F0/2 — Sales VLAN — PC-2 — IP Address 192.168.2.2, Subnet Mask 255.255.255.0
- F0/5 — PC-5 — IP Address 192.168.2.3, Subnet Mask 255.255.255.0
- VLAN 2 Subnet 192.168.2.0/24
- F0/13 — Finance VLAN — PC-13 — IP Address 192.168.4.13, Subnet Mask 255.255.255.0
- F0/24 — PC-24 — IP Address 192.168.4.24, Subnet Mask 255.255.255.0
- VLAN 4 Subnet 192.168.4.0/24

Figure 9-10. *A switch can be used to create VLANs.*

Figure 9-11 Departments Connected with Switches

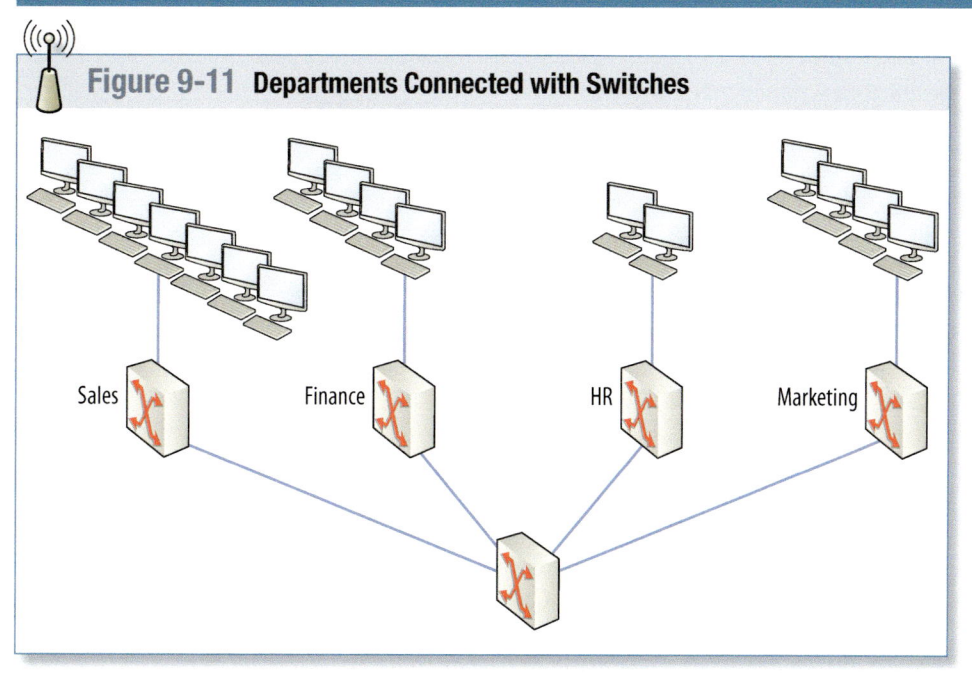

Figure 9-11. Multiple switches can connect different departments.

SWITCH SPEEDS

A major consideration when purchasing and managing a switch is the transmission speed. This indicates the bandwidth of the switch. *Bandwidth* is a measure of how much data a device such as a switch can process at a time. Speeds are rated in megabits per second (Mbps) or gigabits per second (Gbps), per physical port.

The terms bandwidth and speed are often used interchangeably. Compare this to traffic on a highway. A single-lane road may be able to handle 100 cars an hour. However, if traffic increases and the need arises to handle 1,000 cars an hour, the road can be widened and more lanes added. The first 100 cars may or may not reach their destinations any faster than before the road was widened, but as traffic increases, more cars will be able to get there faster because of the extra bandwidth of the road. Similarly, the amount of traffic a network can handle can be increased by increasing the bandwidth of the devices on the network.

The IEEE 802.3 standard identifies several basic speeds sometimes seen in networks today. Although more speeds are possible, there are some common speeds used with twisted-pair cables. **See Figure 9-12.**

Figure 9-12 Ethernet Speeds

PROTOCOL	SPEED	COMMENTS
IEEE 802.3	10 Mbps	10 million bits per second
IEEE 802.3u	100 Mbps	100 million bits per second
IEEE 802.3z	1,000 Mbps	1,000 million bits per second (1 gigabit)
IEEE 802.3an	10 Gbps	10 gigabits per second

Figure 9-12. The IEEE 802.3 standard identifies the speed capabilities for twisted pair cables.

There are three important speeds to pay attention to when looking at a switch:

▶ Transmission speed
▶ Uplink speed
▶ Backplane speed

Transmission Speeds

Switches are commonly represented with port speeds such as 10/100 Mbps or 100/1,000 Mbps. The 10/100 means that a port may operate at either 10 Mbps or 100 Mbps, and the 100/1,000 means that a port can operate at either 100 Mbps or 1,000 Mbps.

The limiting factors are the capabilities of the end devices and the cable grade that is used. In other words, CAT 5 twisted-pair cable cannot be used for 1,000 Mbps, though CAT 5E can.

Similarly, if a computer has a 10 Mbps network adapter, the switch can send data to the computer only at a maximum of 10 Mbps no matter how fast the switch is.

High-speed switches are available. Of course, they are more expensive. If a group of users needs to share large files, stream audio and video, or use Voice over IP (VoIP), it is worth getting the high-speed switches. If that alternative is chosen, ensure that the connecting cable and individual network adapters meet the same speed requirements.

If using a managed switch, it is possible to manually configure the speed of its individual ports. Some ports can be configured at 10 Mbps, some at 100 Mbps, and some at 1,000 Mbps, as long as the switch supports all of the speeds. Additionally, the ports can be configured individually for half-duplex or full-duplex communication.

A common option with many switches is autosense for speed between the PC and the port. In other words, the speed does not have to be set. The switch automatically determines the best settings for the optimal speed.

The Uplink Port

An *uplink port* is a special port on a switch used to connect the switch to another switch or to other devices. An *access link* is any other port on the switch other than the uplink port.

Uplink ports offer scalability by enabling the addition of switches to the network in a daisy chain. The uplink port can also be used to connect the switch to a router for access to other subnets.

The uplink port might be labeled as such. However, on some switches, the port is not labeled but instead shares the capability with another port. For example, on some smaller switches, one of the ports might include a push button labeled as MDI/MDI-X (for medium dependent interface/medium dependent interface crossover). When MDI is selected, it works as a regular port. When MDI-X is selected, it works as an uplink port. If the MDI/MDI-X button is not available, a crossover cable might be required.

> ✓ FACT
> Many switches use auto-sense to detect the speed of connected devices. For example, a switch rated at 100/1,000 can operate at either 100 Mbps or 1,000 Mbps on each port.

> ✓ FACT
> Most network hardware available today uses full-duplex communication. However, it is possible to downgrade a port to half-duplex for compatibility with legacy hardware.

> ### Uplink Ports and Crossover Cables
> An uplink port is wired so that a straight-through cable can be used to connect it to other switches or routers. However, if a regular port is used that is not wired to connect two switches, a crossover cable will be required.
>
> Many new switches can automatically sense whether a link needs a crossover or straight-through connection and configure the switch internally using Auto-MDIX. Other devices use an MDI/MDI-X button that can be toggled to change the port from straight-through to crossover.

Some switches offer the capability to bundle access links together to act as a single link between switches. If a switch has a bundling option, multiple ports can be configured together as a single link.

LACP (as defined in the IEEE 802.3ax standard) forms a single logical channel between devices with multiple physical links. For example, with LACP enabled, it is possible to bundle five 100 Mbps ports as one logical link. This gives an effective throughput of 500 Mbps.

Backplane Speed

Another speed to consider with switches is the backplane speed. *Backplane speed* is the internal speed of a switch. The faster this speed is, the better the overall performance of the switch.

Backplane speed applies only to modular switches, not form-factor switches. It measures how fast data is transferred between modules inside the switch.

Depending on the manufacturer, backplane speed might be measured at different points. *Application-specific integrated circuits (ASICs)* speed is the speed on the chassis where the modules plug in. This is similar in concept to the bus speed on a computer.

The second backplane speed measurement is port-to-port speed. This is slightly different from the ASICs speed. *Port-to-port speed* is the speed with which data moves between ports on the different blades of the same chassis.

SECURITY OPTIONS

Security is required on every network. Any time data must be kept secret, security is required. A company without any proprietary data quickly becomes a company without revenue.

The basic principle of in-depth security dictates that multiple layers of security are required. In addition, security must be reviewed and updated regularly.

Some common security steps that can be taken with switches include:

▶ Keep network hardware protected with physical security (in a locked room or a locked rack).

▶ Change default passwords on managed switches to a complex password.

▶ Never use blank passwords.

▶ Use a secure protocol (such as SSH) to remotely manage switches.

▶ Use SNMP version 3 (instead of just SNMP or SNMP version 2) for best security.

▶ Consider port security.

▶ Consider hardware redundancy for maximum availability of network resources.

Port Security

Port security helps restrict what devices can connect to ports on a switch. The danger is that someone might walk into an organization and simply plug a computer into an 8P8C jack in the wall and gain access to the network. That is disconcerting to both administrators and organization executives.

One method of port security is to configure each port with the MAC address of a specific computer. Only that computer can connect. If a device with a different MAC attempts a connection, the switch either refuses the connection or shuts down the port entirely. With five computers, this will not take much time. However, with 500 computers, it can be quite time-consuming and tedious.

An alternative is to configure the port to remember which MAC addresses it learned and to set a threshold for the

> ✓ FACT
> Most cables and network devices use full-duplex by default, but managed switches can be configured to match the existing hardware.

> ✓ FACT
> Security is never something that is done once and is over. It is an ongoing process.

> ✓ FACT
> Only managed switches can use port security. Unmanaged switches cannot be configured for specific MACs.

maximum number of addresses allowed. For example, the threshold may be set at one or two addresses, and any MAC learned after that would trip an alarm. The alarm may notify an administrator using an SNMP trap (or error message) or may shut down communication with the port.

Another element of port security is ensuring that unused ports are not enabled. Consider a 48-port switch that is cabled to 48 8P8C wall jacks in an organization. However, only 40 of the 8P8C jacks are currently used. The other jacks do not have computers attached. The switch ports where these jacks are connected should be disabled. This prevents someone from coming in, plugging in a computer to an empty jack, and accessing the network.

Hardware Redundancy

The switch in a typical network configuration presents itself as a single point of failure. If the switch fails, all of the computers connected through the switch lose connectivity to network resources. However, fault tolerance can be built in by adding hardware redundancy. *Fault tolerance* means that a failure can occur and a system can recover from it by itself. Fault tolerance can be implemented at the disk level, the server level, the site level, and more.

Hardware redundancy simply means that additional components are added to ensure that the failure of one component does not result in a complete failure. For example, many modular switches come with more than one power supply. If one power supply fails, the switch can continue to operate. In some switches, the power supplies are completely redundant, meaning that the switch will continue to operate with no loss of capability. In other switches, a failure of one power supply may affect only some of the ports on the switch, but not all of them.

Obviously, adding redundant capabilities costs more money. The majority of the time, the redundant component is not actively utilized but is instead there in case a failure occurs. When considering hardware redundancy for switches, the administrator must keep in mind how critical connectivity is for the devices.

It might be that a failure of a particular switch would result in immediate loss of critical resources and revenue. It would be worthwhile to add redundancy for this switch. On the other hand, the loss of a switch might not be critical. It might be possible to replace it within a day without any impact on the business's bottom line. In this case, the added cost of redundancy is not justifiable.

Summary

There are many capabilities and inner workings of switches. A switch uses the MAC addresses to create a MAC address table. It then creates multiple collision domains by sending data only to the destination device based on the MAC address instead of to all devices connected to the switch. Some switches are managed and can be configured, while other switches simply work by plugging them in. Some switches operate as layer 2 and layer 3. VLANs can be created within a managed switch. Port security can control what devices can connect to a switch.

Review Questions

1. A modular switch is expandable. Ports can be added by adding components.
   a. True
   b. False

2. It is necessary to identify what device is connected to the first port in a drawing of a 100 Mbps switch. How will the port be labeled?
   a. E0
   b. F0
   c. F1
   d. Gi1

3. Layer 2 switches create separate broadcast domains.
   a. True
   b. False

4. A network includes computers connected via hubs, and it is necessary to reduce the number of collisions to the least number possible. What should be done?
   a. Replace the hubs with bridges.
   b. Replace the hubs with firewalls.
   c. Replace the hubs with managed hubs.
   d. Replace the hubs with switches.

5. A managed switch requires less administrative overhead than an unmanaged switch.
   a. True
   b. False

6. A layer 3 switch functions just like a __?__.
   a. backplane
   b. hub
   c. router
   d. uplink port

7. What does a switch maintain to track the location of computers?
   a. A layer 3 table
   b. A MAC address table
   c. A managed table
   d. A routing table

8. A switch has 48 ports. How many VLANs can be created with it?
   a. 2
   b. 12
   c. 24
   d. 48

9. A 100 Mbps switch is configured to combine five ports using LACP. What is the effective throughput at the uplink?
   a. 100 Mbps
   b. 500 Mbps
   c. 600 Mbps
   d. 1,000 Mbps

10. To ensure that only known computers can connect to a switch, the administrator should implement port blocking.
    a. True
    b. False

Routing Networks

Routers expand networks by enabling administrators to create multiple subnets or subnetworks. Routers are the traffic cops on a network that direct all the traffic from one subnetwork to another. At the very least, a network needs a router to access the Internet, and most networks in organizations include multiple routers.

Although the administration of routers is minimal in small networks, administrators should know some basics. For example, a router needs to know the path to all subnets in the network and the path to the Internet. Routers include a routing table that lists all of these paths, and the routing table can be updated manually or automatically.

OBJECTIVES

- ▶ Describe how to connect multiple networks
- ▶ Explain various ways to manage traffic routes on a network
- ▶ What determines transmission speeds for sending data
- ▶ Describe basic router configuration tasks
- ▶ Compare routing protocols

CHAPTER 10

TABLE OF CONTENTS

Multiple Network Connections 190
    Hardware Routers and Software Routers. 191
    Default Routes 191
    Directly Connected Routes 192

Traffic Routing on a Network. 193
    Static Routes 193
    Dynamic Routing 195
    The Routing Table 197

Transmission Speeds 198

Router Configuration. 199
    Router Cabling 199
    The Console Session 201
    Password Configuration. 202
    Interface Configuration 204
    Saving Configuration Changes 206

Summary. 206

Review Questions 207

MULTIPLE NETWORK CONNECTIONS

Routers connect multiple networks so that computers on different networks can communicate with each other.

Some of the basic terms involved in router administration:

Hardware Router A *hardware router* is a dedicated hardware device that routes packets. For example, Cisco makes several different models of routers used on internal networks and the Internet.

Software Router A *software router* is a server that includes software used to route packets. For example, Windows Server 2012 R2 can be configured to function as a software router.

Routing Interface A *routing interface* is the interface where packets are received and transmitted. A router will have multiple interfaces but must have at least two.

Static Routing *Static routing* requires an administrator to manually add routes to different subnets. If there are any changes to the network effecting the routes, an administrator must manually modify the routes.

Dynamic Routing *Dynamic routing* allows the routers to automatically learn about changes in the network and alter their forwarding decisions based on those changes. Administrators must configure dynamic routing protocols, but after that, the routers take care of keeping routes up to date themselves.

Routing Protocols *Routing protocols* enable routers to communicate with each other and share routing information. Two of the routing protocols used on internal networks are Routing Information Protocol version 2 (RIPv2) and Open Shortest Path First (OSPF).

Routing Table A *routing table* is a table maintained within a router that identifies all known subnetworks and the paths to these subnetworks.

Consider a single router connecting four networks (labeled A, B, C, and D). In this context, each of the networks is joined into a single local area network (LAN). These joined networks are subnetworks in the overall LAN. **See Figure 10-1.**

Each of the subnetworks has its own network ID that is different from the

Figure 10-1 Router Connections

B 192.168.2.0/24

192.168.2.1/24

A 192.168.1.0/24

C 192.168.3.0/24

192.168.1.1/24

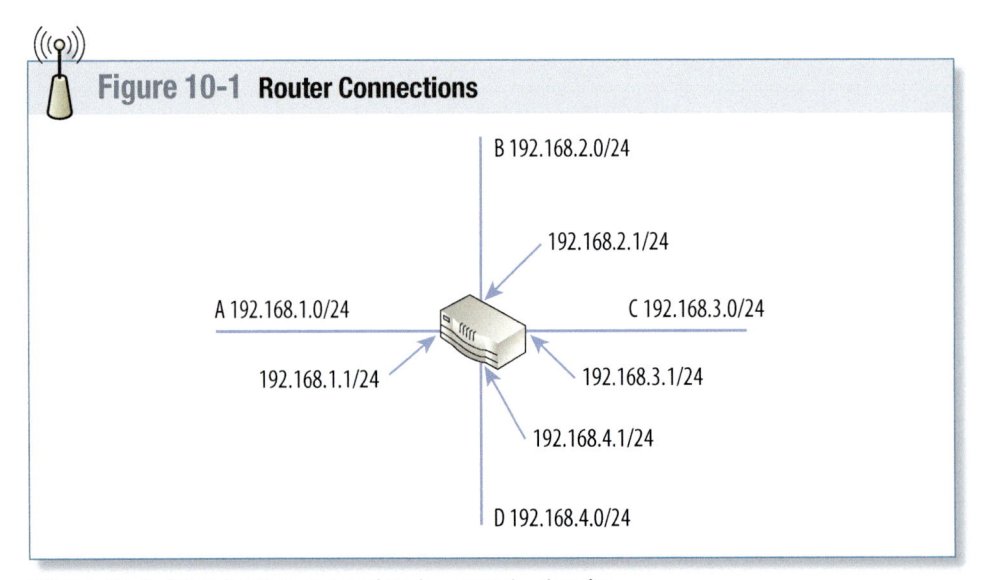

192.168.3.1/24

192.168.4.1/24

D 192.168.4.0/24

Figure 10-1. Administrators use routers to connect networks.

others. Additionally, the router has four separate interfaces with a single interface connected to each subnetwork.

Notice that the IP address assigned to the router interface in each of these subnetworks is the first IP address available. In other words, in the 192.168.1.0/24 subnetwork (A), the router interface is assigned the address of 192.168.1.1/24. It is common to assign the first address within a range to the router interface, but it is not required. No matter what address is assigned, though, it must have the same network ID as other devices in the subnetwork.

Hardware Routers and Software Routers

Routers can be dedicated hardware routers, or they can be software routers. Although most routers in production networks are hardware routers, it is possible to configure Windows Server 2012 R2 as a software router.

Windows Server 2012 R2 includes all the software components to configure it as a software router. The only item that is required is to have at least two network interface adapters installed. One adapter would connect to one network, and the other adapter would connect to another network.

Although it is much more common to use hardware routers, here are some examples of when administrators might want to use Windows Server 2012 R2 as a software router:

Development Environment Administrators can create a temporary subnetwork without purchasing additional hardware. For example, to isolate some computers in a separate subnet for testing or development purposes, a software router is an inexpensive alternative.

Replacement of Failed Router If a hardware router fails, in smaller environments administrators can temporarily replace it with a software router until a replacement is installed.

Default Routes

A *default route* is the path that IP traffic takes when another path is not identified. Most computers determine default routes based on their configured default gateway settings. **See Figure 10-2.**

Figure 10-2 Default Gateway

Sandy 192.168.1.57/24

Internet

Default Gateway for A 192.168.1.1/24

A 192.168.1.0/24

B 192.168.3.0/24

Router

Frank 192.168.1.94/24

Jim 192.168.3.25/24

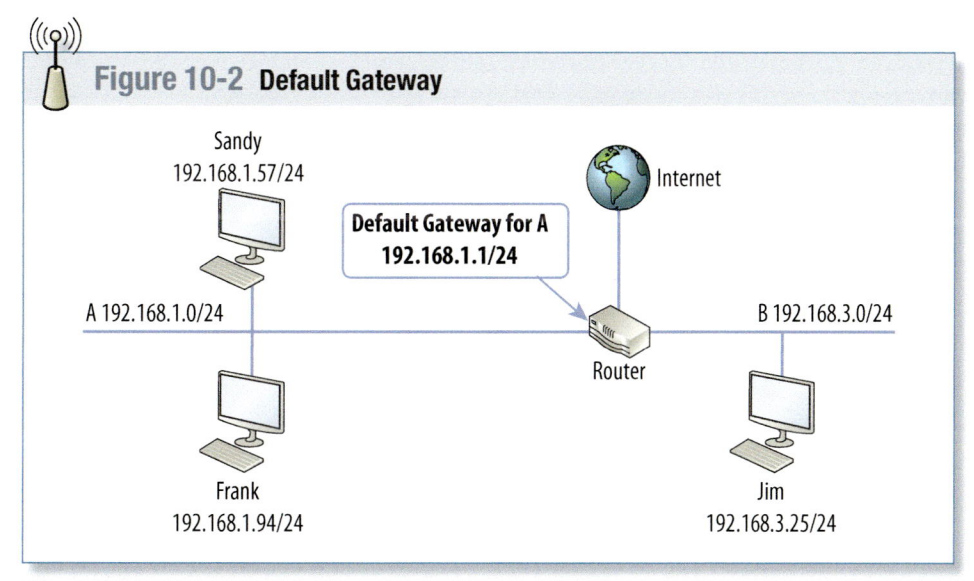

Figure 10-2. The default gateway moves traffic on a network.

Imagine that Sandy wants to send data to Frank. TCP/IP will determine Sandy and Frank's network ID. Since the network IDs of both computers are the same (192.168.1.0), TCP/IP sends the traffic directly to Frank's computer. **See Figure 10-3.**

Later, Sandy wants to send data to Jim, but Sandy and Jim's computers are on different subnetworks. TCP/IP determines this by calculating the network IDs. Since the network IDs are different, TCP/IP realizes it must send the data to another network.

More specifically, since the destination is on a different network, the sending system must forward the data to the addresses of the default gateway. The interface on router 2 with an IP address of 192.168.1.1 is the default gateway for subnetwork A. Note that the default gateway has the same network ID as other computers on subnetwork A.

Directly Connected Routes

A *directly connected route* is any subnetwork that is directly connected to a router, in other words the router has interfaces in those subnetworks. Each of the subnetworks is directly connected to a single router. Since the router is directly connected to these subnetworks, it inherently knows the path to them.

However, routers do not automatically know the paths to other subnetworks within a LAN. If a LAN includes multiple

Figure 10-3 Network ID of Local and Remote Computers.

COMPUTER	IP ADDRESS	SUBNET MASK	NETWORK ID
Sandy	192.168.1.57	255.255.255.0	192.168.1.0
Frank	192.168.1.94	255.255.255.0	192.168.1.0

COMPUTER	IP ADDRESS	SUBNET MASK	NETWORK ID
Sandy	192.168.1.57	255.255.255.0	192.168.1.0
Jim	192.168.3.25	255.255.255.0	192.168.3.0

Figure 10-3. When Network ID's are different then data must be transmitted to a different subnet.

Identify the Default Gateway and Routers

On Windows computers, the command-line tool ipconfig displays the default gateway.

At the command prompt, type ipconfig. The output will include information similar to the following:

```
Ethernet adapter Local Area Connection:

   Connection-specific DNS Suffix  . :
   IPv4 Address. . . . . . . . . . . : 192.168.1.57
   Subnet Mask . . . . . . . . . . . : 255.255.255.0
   Default Gateway . . . . . . . . . : 192.168.1.1
```

The default gateway identifies the IP address of the near side of the router. Note that this does not show the IP addresses of all the router's interfaces. It shows only the IP address of the interface on the same subnetwork.

routers with multiple subnetworks connected to these different routers, the routers must learn about the other subnetworks. This occurs either statically or dynamically.

TRAFFIC ROUTING ON A NETWORK

On a larger network with three routers and six subnetworks, the network includes both directly connected routes for each of the routers and indirectly connected routes. Is it possible to name the directly connected routes for each of the routers? **See Figure 10-4.**

- Router 1 knows about the directly connected subnetworks of B and C.
- Router 2 knows about the directly connected subnetworks of subnetworks A, C, D, and F.
- Router 3 knows about the directly connected subnetworks of subnetworks D and E.

For the network to be fully routed, each of the routers must know the paths to other networks. For example, if a computer in the F subnetwork needs to send data to a computer in the B subnetwork, router 2 must know how to get it there.

Static Routes

When a network includes multiple routers, routers can be statically configured or dynamically configured. Administrators may manually add routes to a router to create statically configured routes. In a dynamically configured router, the administrator configures routing protocols, and these routing protocols automatically discover subnets on the network.

If there are only a few routes to add and they are not likely to change, static routes can be the simplest method. It is also common to deploy some routers with some static routes, while others are configured dynamically.

The *Routing and Remote Access Service (RRAS)* is Windows Server software that allows a server to be configured as a router. For example, assume that one static route to the 192.168.5.0 destination has been created. The dialog box in the foreground of the Static Routes dialog box in RRAS shows the addition of a

Figure 10-4 Router Routes

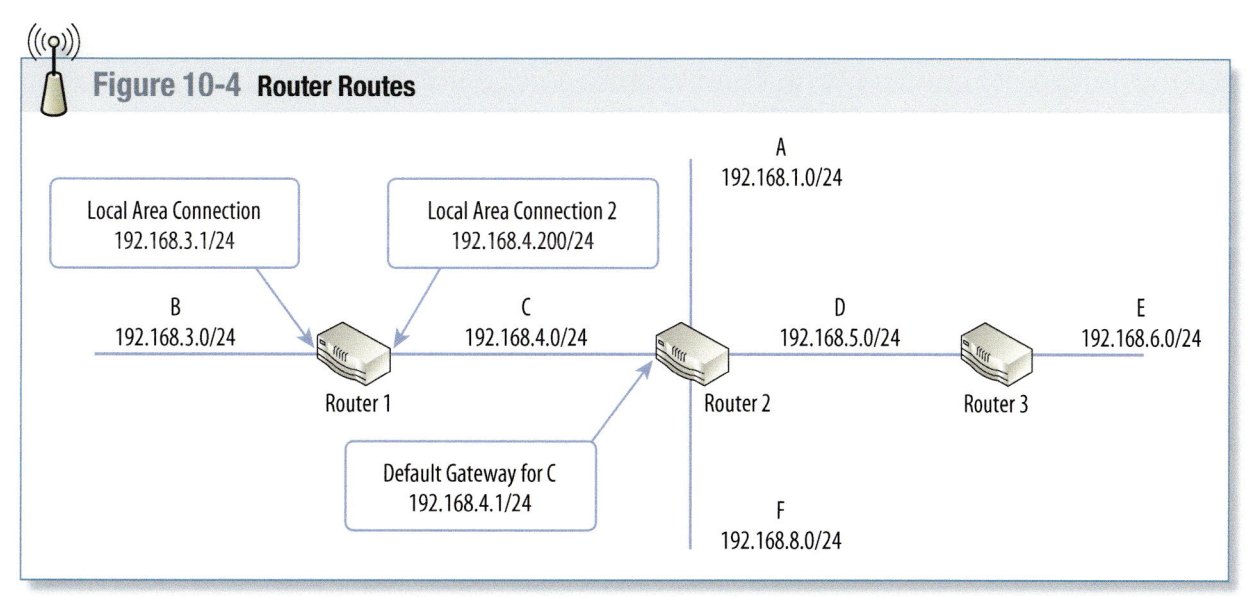

Figure 10-4. A multiple router network can have directly and indirectly connected routes.

194 Introduction to Network Technologies

> ✓ **FACT**
> It is also possible to create static routes to specific computers instead of to subnetworks. Routes to specific computers use a network mask of 255.255.255.255.

> ✓ **FACT**
> In large networks with multiple routers and multiple paths, the metric is important. It helps routers identify the best routes to different subnetworks.

second static route to the 192.168.8.0 destination subnet. **See Figure 10-5.**

There are several key pieces of information worth pointing out:

Destination The *destination* identifies the destination subnetwork. The router does not name these by letters, but instead uses the network ID.

Network Mask The *network mask* is the subnet mask of the network ID. Each of the subnetworks in the previous example has a CIDR notation of **/24**, indicating a subnet mask of 255.255.255.0.

Gateway The *gateway* is the IP address of the destination router's network interface. Notice in the example that the path to other subnetworks from router 1 is through router 2 (except for the directly connected routes of B and C). The IP address of the router's network adapter on subnetwork C is 192.168.4.1, so its IP address is entered here. It becomes the default gateway from router 1 and is also the default gateway for any computers on subnetwork C.

Interface The *interface* identifies the network adapter that should be utilized to connect to the specified destination by name. In this example, the adapter named `Local Area Connection` is on subnetwork B. The adapter named `Local Area Connection 2` is on subnetwork C.

Though there are only two static routes, many more static routes must be created to support the network in the example. **See Figure 10-6.**

- Router 1 needs four static routes (to A, D, E, and F) since it knows only about the two directly connected routes of B and C.
- Router 2 needs two static routes (to B and E) since it knows only about the directly connected routes of A, C, D, and F.
- Router 3 needs four static routes (to A, B, C, and F) since it knows only

Figure 10-5 Windows Server as a Router

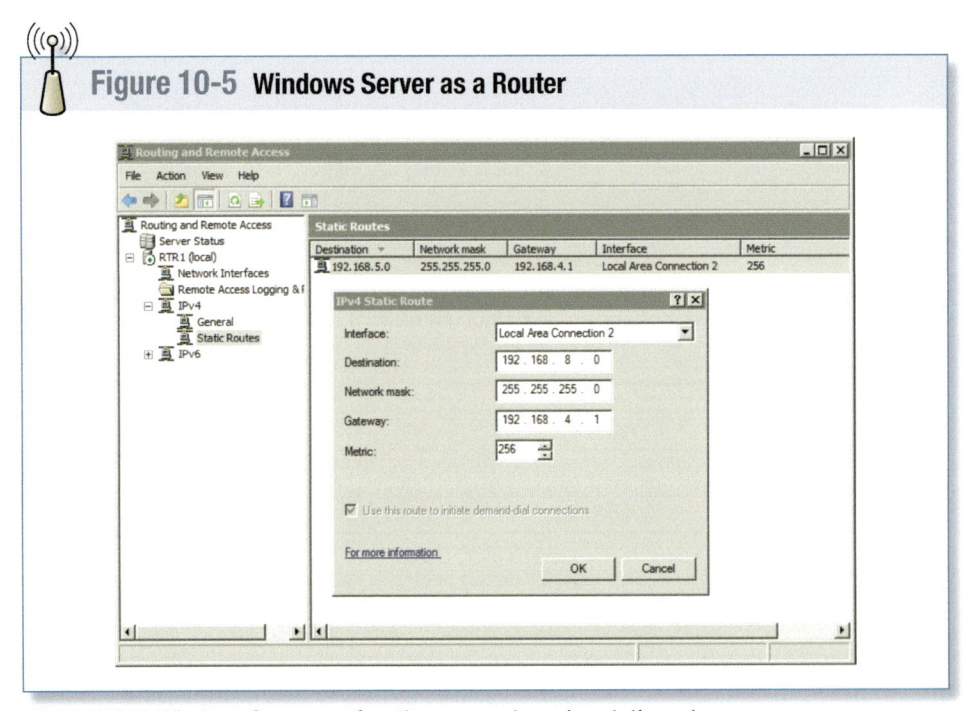

Figure 10-5. Windows Server can function as a router using static routes.

Figure 10-6 Router Routes

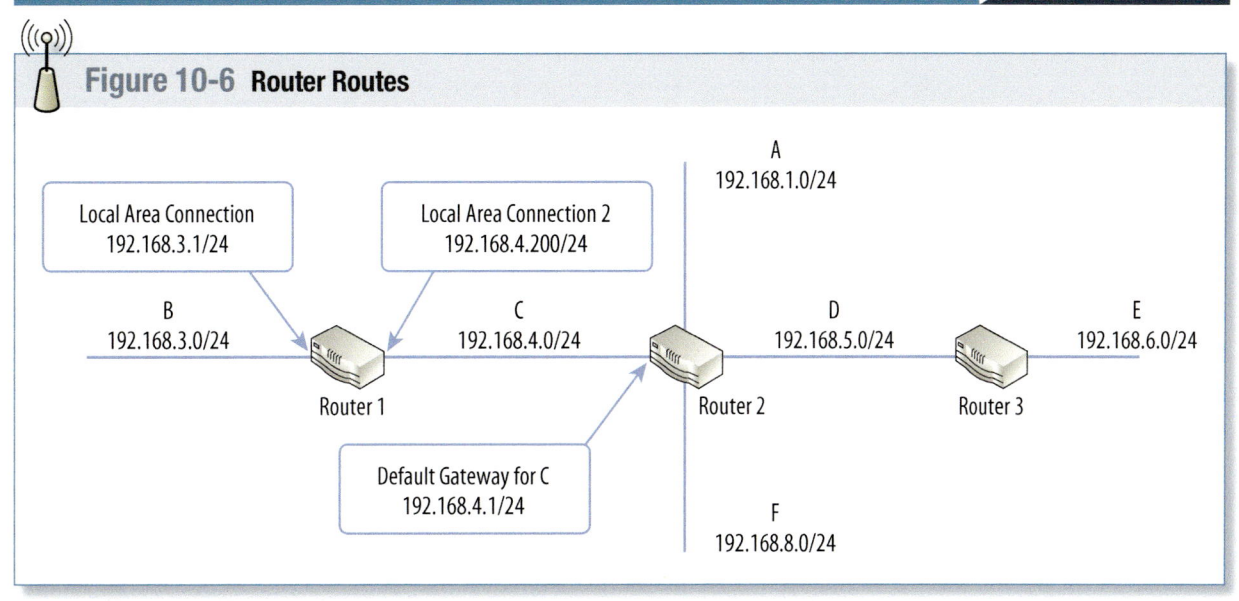

Figure 10-6. A multiple router network can have directly and indirectly connected routes.

- about the directly connected routes of D and E.
- This network has 10 static routes.

Instead of typing these static routes in manually, administrators can choose to configure dynamic routing.

Dynamic Routing

When dynamic routing protocols are enabled on routers, they learn the paths to other routers dynamically. After the protocols are configured, the routers talk to each other. Each router lets other routers know what it knows, and after a short time, each router knows the paths to all subnetworks on the network.

Static routing has very few features, while dynamic routing provides more benefits. **See Figure 10-7.**

Since dynamic routing provides more features than static routing, some administrators might wonder why they should ever use static routing. It depends in part on how many routers there are on the network. If there are only two routers, it is much easier to configure the static routes once rather than adding and configuring a routing protocol.

Figure 10-7 Compare Static and Dynamic Routing

FEATURES	STATIC ROUTING	DYNAMIC ROUTING
Discovery of remote networks (including new or changed networks)	Must be done manually	Done automatically
Information exchange with other routers	None	Information exchanged with routing protocol
Fault tolerance	None	Failure of routers can be detected and paths to networks modified (when multiple paths exist)

Figure 10-7. Dynamic routing has more benefits than static routing, specifically when multiple routers are used.

Routing Information Protocol version 2 (RIPv2)

Routing Information Protocol version 2 (RIPv2) is the primary routing protocol used on Windows Server. RIPv2 is very easy to add and configure on a Windows server running RRAS. The protocol is simply added and the network adapters it should operate on are identified. The routers then automatically share their routing information with each other.

RIPv2 has replaced RIPv1 on most networks because of its many benefits. **See Figure 10-8.**

Consider an example where the RRAS console on a Windows server has RIPv2 added. Both the `Local Area Connection` and `Local Area Connection 2` adapters have been added to the RIP node so that RIP operates on both. This router will now listen for RIP information from other routers and send RIP information to them through both of its interfaces. **See Figure 10-9.**

Although it is possible to use more advanced configurations of RIPv2, the default configuration will work for most networks as long as the settings are

> ✓ **FACT**
> RRAS will send RIPv2 data and listen for RIPv1 and RIPv2 data by default. However, it can be configured for any combination of RIPv1 and RIPv2.

Figure 10-8 Compare RIPv1 and RIPv2

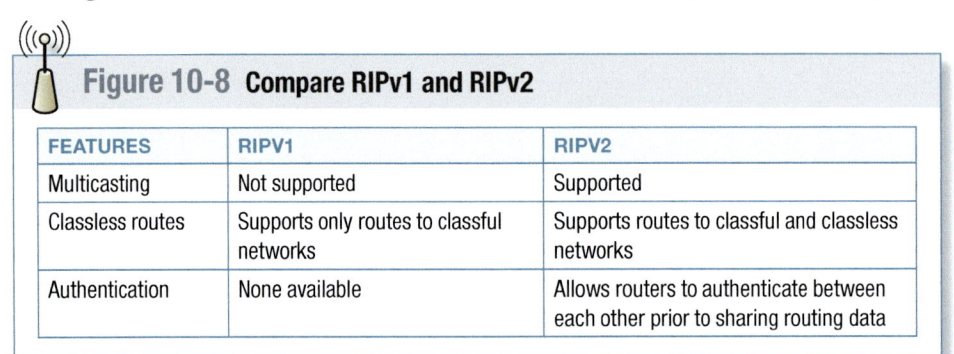

FEATURES	RIPV1	RIPV2
Multicasting	Not supported	Supported
Classless routes	Supports only routes to classful networks	Supports routes to classful and classless networks
Authentication	None available	Allows routers to authenticate between each other prior to sharing routing data

***Figure 10-8.** RIPv2 has replaced RIPv1 because of the additional benefits.*

Figure 10-9 RIPv2 Properties

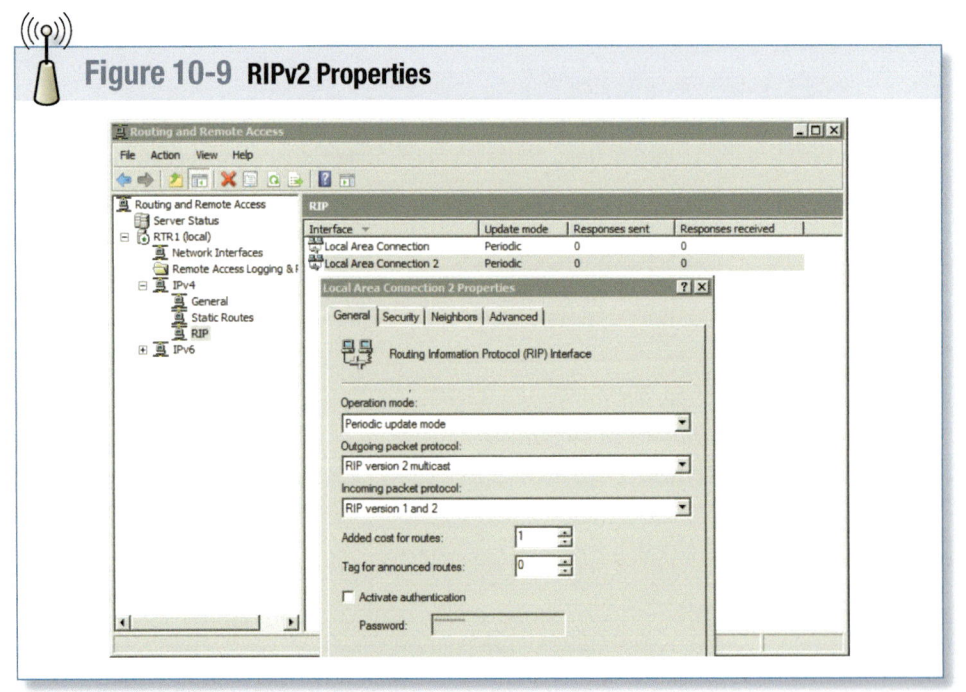

***Figure 10-9.** Configure RIPv2 properties on a network adapter.*

configured in the same way on other routers on the network. However, two settings are of primary importance:

- The "Outgoing Packet Protocol" is configured to use RIP version 2 broadcast by default. It is possible to change this to RIP version 2 multicast for better network performance.
- Authentication can be configured by clicking the "Activate Authentication" check box and adding a password. All routers will need the same password. This ensures that only routers that can authenticate can access the routing information.

Open Shortest Path First

Open Shortest Path First (OSPF) is a routing protocol used on internal networks with hardware routers. The method it uses to share and calculate the best routes between routers on the network is different from RIPv2. However, the result is the somewhat similar. With OSPF, routers on the network will learn all routes and will be dynamically updated when the network changes. OSPF takes many more factors into consideration when making routing decisions and is, therefore, a better routing protocol than RIPv2. But OSPF is more complicated to set up. RIPv2 and OSPF are not compatible with each other.

The Routing Table

Routers maintain routing tables that identify the paths to other networks.

For example, consider the routing table in the RRAS console of a Windows server with a few key items worth noting. The computer has two network adapters with IP addresses of 192.168.3.1 and 192.168.4.200. **See Figure 10-10.**

1. The network adapter connected to the 192.168.3.0/24 subnetwork has an IP address of 192.168.3.1. Note the

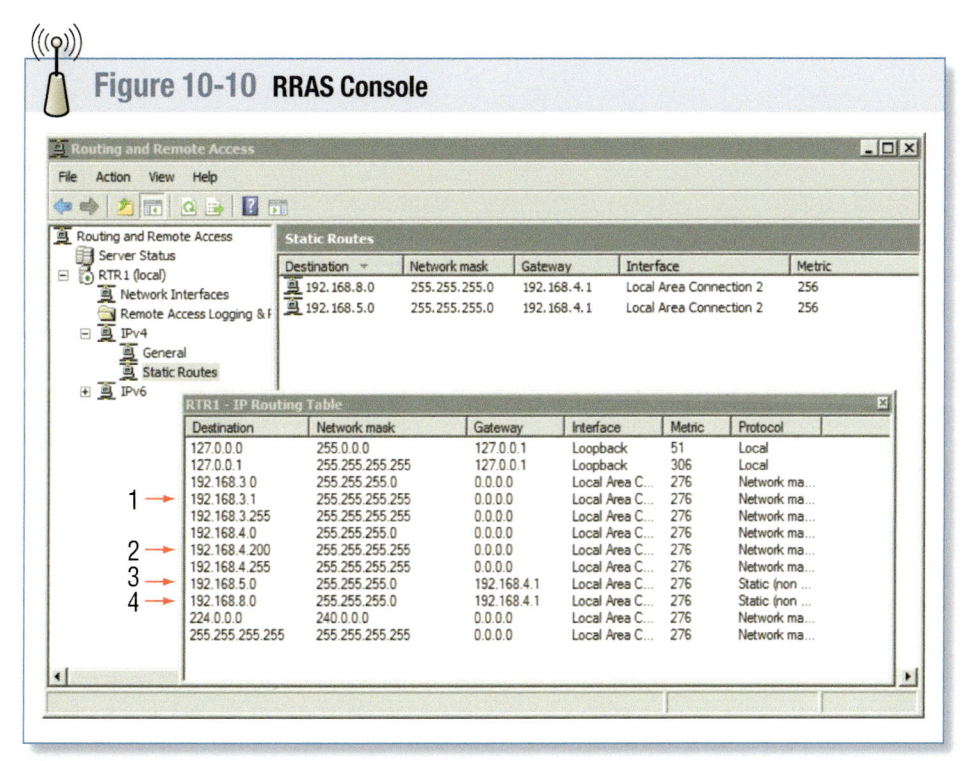

Figure 10-10. The RRAS console can display static routes in the routing table.

network mask is 255.255.255.255 and the gateway is 0.0.0.0. This helps identify it as a directly connected route.

2. The adapter connected to the 192.168.4.0/24 subnetwork has an IP address of 192.168.4.200. Note the network mask is 255.255.255.255 and the gateway is 0.0.0.0. This helps identify it as a directly connected route.

3. The path to the 192.168.5.0 subnetwork is via the 192.168.4.1 default gateway.

4. The path to the 192.168.8.0 subnetwork is via the 192.168.4.1 default gateway.

It is also possible to view the routing table for any individual system. For example, a partial output of the `route print` command executed at the command prompt on a Windows Server with two network adapters shows that the server is configured as a router connecting the 192.168.1.0/24 and 192.168.10-0/24 networks. **See Figure 10-11.**

TRANSMISSION SPEEDS

The maximum speed of a switch determines the maximum speed that computers can send data through the switch. Switch speeds are measured in bits per second (bps), such as 100 Mbps.

Figure 10-11 *route print* **Command**

```
===================================================================
Active Routes:
Network Destination        Netmask          Gateway       Interface  Metric
          127.0.0.0        255.0.0.0         On-link       127.0.0.1     306
          127.0.0.1  255.255.255.255         On-link       127.0.0.1     306
    127.255.255.255  255.255.255.255         On-link       127.0.0.1     306
        192.168.3.0    255.255.255.0         On-link     192.168.3.1     276
        192.168.3.1  255.255.255.255         On-link     192.168.3.1     276
      192.168.3.255  255.255.255.255         On-link     192.168.3.1     276
        192.168.4.0    255.255.255.0         On-link   192.168.4.200     276
      192.168.4.200  255.255.255.255         On-link   192.168.4.200     276
      192.168.4.255  255.255.255.255         On-link   192.168.4.200     276
        192.168.5.0    255.255.255.0     192.168.4.1   192.168.4.200     276
        192.168.8.0    255.255.255.0     192.168.4.1   192.168.4.200     276
          224.0.0.0        240.0.0.0         On-link       127.0.0.1     306
          224.0.0.0        240.0.0.0         On-link   192.168.4.200     276
          224.0.0.0        240.0.0.0         On-link     192.168.3.1     276
    255.255.255.255  255.255.255.255         On-link       127.0.0.1     306
    255.255.255.255  255.255.255.255         On-link   192.168.4.200     276
    255.255.255.255  255.255.255.255         On-link     192.168.3.1     276
===================================================================
```

Figure 10-11. The *route print* command displays the routing table on a Windows computer.

Routing Tables and Memory

The information for routing tables is stored in the router's memory. Routing tables on Internet routers can contain tens of thousands of entries. These tables consume a significant amount of memory.

Similarly, routers are also measured in bits per second, and the maximum speed of the router is the maximum speed that switches can send data through the network. One important consideration is the speed of routers that accept traffic from multiple switches.

On a network with multiple routers and switches, different routers must accommodate different amounts of data. For example, the router to the Internet might have high usage if all users regularly access the Internet. Even the router right below it (which is central to the network) will likely be substantially busier than other routers. The activity depends on where resources are placed in the network and the amount of usage for each of the subnetworks. **See Figure 10-12.**

Common speeds for wired routers are 100 Mbps and 1,000 Mbps. However, just as switches have increased in their speed capabilities, higher-performance routers can be purchased in the 10 gigabits per second (Gbps) range. Of course, higher speeds cost more money.

ROUTER CONFIGURATION

Routers and multilayer switches are complex devices that must be configured before they can operate effectively.

Router Cabling

To connect a router to a network, administrators can use cables of various types, depending on the connections being made. Hardware routers are available with many combinations of connectors.

Figure 10-12 Router Traffic

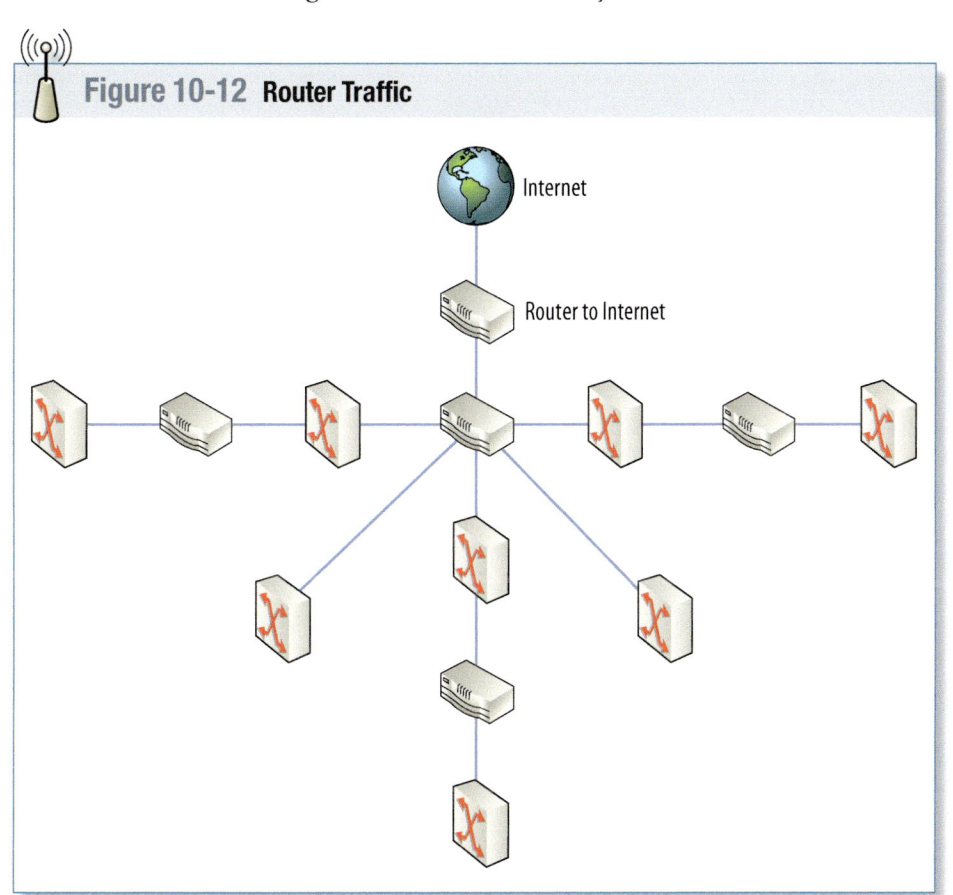

Figure 10-12. Routers on a multiple-router network with access to the Internet must support different amounts of traffic.

Modular routers can support virtually any combination of cable connections, while non-modular routers can be more limited.

When integrating a router into an enterprise LAN, Ethernet cables are the most common solution for connections to other routers, switches, and host computers. However, there are alternatives, especially when connecting routers to other devices, such as CSU/DSUs.

Router-to-Router Cabling

Connections between two routers typically use one of two cable types. The simplest is an Ethernet connection, in which both routers have female 8P8C sockets and use a patch cable to join the two. However, unlike a connection to a switch, a crossover cable is required to connect one router to another.

Ethernet connections between routers are easy to install and require relatively little configuration. The other alternative for connecting one router to another is a serial cable connection.

Many routers have serial connectors, but they are not all of the same type. Some have a large D-connector called a DB-60, while many newer models have a smart serial connector with 26 pins. The serial cables used for router connections are available with different interfaces at the other end, supporting different types of devices.

All serial connections have a data terminal equipment (DTE) end and a data communications equipment (DCE) end. *The data terminal equipment (DTE)* end plugs into the serial port on a terminal or server. The *data communications equipment (DCE)* end plugs into network components like a router or a modem, and provides clocking for the interface, which is critical to the proper function of the interface. To connect two routers together using a serial interface, an administrator must configure one router (and only one) to provide the clocking signal.

Connecting two routers together using their serial connectors requires either a cable designed for that purpose or two cables with connectors that join in the center. The single-cable solution has appropriate connectors at both ends, with one end marked DTE and the other marked DCE.

The alternative is to purchase two cables with proper connectors for the routers at the DTE end, and compatible connectors at the other (DCE) end, which can connect together. For example, cables with V.35 DCE connectors are available in both male and female form factors. By purchasing one of each, the administrator can plug the two V.35 connectors together in the middle and then plug the other ends into the two routers, forming what is known as a *back-to-back connection.*

Router-to-CSU/DSU Cabling

A connection between a router and a separate CSU/DSU always uses a serial cable, with the router functioning as the DTE and the CSU/DSU as the DCE. Because the CSU/DSU is typically supplied by the service provider responsible for the telecommunication link, administrators often do not know what type of connector the unit uses. They must purchase an appropriate cable with the correct connector for the router on the DTE end and the CSU/DSU on the DCE end.

The 34-pin V.35 connector is commonly found on many CSU/DSU units, but they might use other connectors as well (or instead). Modular routers can often accept CSU/DSU units in the form of plug-in modules. In this case, administrators do not have to concern themselves with cable connections.

Router-to-Switch Cabling

Connections between routers and switches usually use Ethernet connections with twisted pair cables and 8P8C connectors. If the devices have Ethernet ports running at different speeds, the connections between routers and switches should always use the fastest speed available, because they are likely to carry the most traffic.

Connections between routers and switches use straight-through Ethernet cables, not crossovers, as when connecting two routers together. Switches are often connected to one to another by using trunk links. A *trunk link* carries traffic from multiple VLANs and, therefore, needs to be as high-capacity as possible.

PC-to-Router Cabling

Except for hardware routers that have switched ports built into them, the only reason to connect a router directly to a PC is to manage it using a Telnet connection. To do this, use a crossover Ethernet cable. It is also necessary for the router interface and network adapter in the PC to have IP addresses on the same subnet. It is then possible to establish a Telnet connection that administrators can use to manage the router.

This type of management connection is usually only used for remote management, because routers typically have a connector for a console cable that is used instead.

The Console Session

Hardware routers do not have monitors or keyboards, so to configure and manage them, administrators must connect a PC to the router and run a program on the PC that provides an interface to the router's functions. Administrators can do this by connecting the PC's serial port to the router's console port, or by connecting the router to the network with an Ethernet cable.

Once the appropriate cables are connected and the interfaces configured, the administrator must run a communications program on the PC that establishes a connection to the router and provides the management interface.

Console Cable

To connect to a router from a PC using a serial console cable, administrators must install and run a terminal emulation program. Older Windows versions up to Windows XP included a free program of this type called HyperTerminal, but this program was omitted from more recent versions of Windows. HyperTerminal is now available as a retail product, or it is possible to purchase or download an equivalent terminal emulator.

After installing and running the terminal emulator, the administrator must configure it with appropriate settings for the connection. The first step is typically to select the serial port on the PC to which the console cable is connected.

Most computers on the market today have only a single serial port, if they have any at all, but at one time PCs could have as many as four. The standard designations for serial ports are COM1, COM2, COM3, and COM4, and the terminal emulation software will have some means of selecting one of these.

The next step is to configure the port settings on the PC so that they match those of the router at the other end. For Cisco routers, the standard terminal settings are as follows:

▶ Bits per second: 9600
▶ Data bits: 8
▶ Parity: None

- Stop bits: 1
- Flow control: None

Once the terminal settings are correctly configured, it should be possible to connect to the router and receive a router prompt or a setup dialog box.

Telnet

To manage a router through a network connection, administrators must use a network communications program, such as Telnet. All current Windows versions include the Telnet client program, but it is not installed by default, so administrators must install it from the Windows Control Panel.

To establish a connection using Telnet, the TCP/IP interfaces on both devices must be configured with compatible IP addresses. If the IP addresses of the PC and the router are not located on the same subnet, there must be another router on the network that forwards traffic between the two.

To connect to the router from the PC using Telnet, administrators use a command such as:

`telnet 192.168.5.5`

If the interfaces are properly configured, the response received from the router will likely be similar to:

`Connecting to 192.168.5.5`

Routers typically require a password to allow a Telnet connection. If an administrator has not configured a virtual terminal (vty) or line password on the router, a message such as:

`Password not set, connection refused`

If the router is configured with a password, a prompt to enter it appears. When the administrator supplies the correct password, a router prompt or a setup dialog box appears.

Initial Setup

When a console or a network connection to a router is successful, the first thing that appears is a router prompt (`Router>`) or, if the router is new or has no startup configuration file, a different prompt appears:

`--- System Configuration Dialog ---`
`Would you like to enter the initial configuration dialog? [yes/no]:`

Password Configuration

Configuring passwords is one of the first things an administrator should do on a new router. Cisco routers have several different passwords, which administrators must configure individually.

Prompt Interpretation

To access the various passwords on the router, an administrator must know how to navigate through its prompt hierarchy. The initial connection displays the user mode prompt:

`Router>`

To configure passwords, the administrator must navigate to the privileged mode prompt, and then to the configuration mode prompt. On a new router with no passwords set, it is possible to access these modes by typing the proper commands, without the need for authentication.

To switch to privileged mode, type enable at the router prompt. The prompt then changes to:

`Router#`

Then, to switch to configuration mode, type configure terminal (or just config t) to change the prompt to:

`Router(config)#`

In this mode, administrators can begin to set passwords.

Privileged (Enable) Passwords

A *privileged mode password* (sometimes referred to as an enable password) prevents anyone not possessing the password from switching the router into privileged mode with the `enable` command.

To set a privileged mode password, use a command: (where *XXXXX* is replaced by the password to be assigned):

`enable password XXXXX`

Enable Secret Passwords

When an administrator creates a privileged mode password using the `password` command, the password itself is stored in plain text, leaving it vulnerable to compromise. In addition, if the router's startup configuration file is stored on a server and accessed using the Trivia File Transfer Protocol (TFTP), the password is transmitted over the network in clear text as well, making it more vulnerable.

To create a password that is automatically stored and transmitted in encrypted form, use the `secret` command, where *XXXXX* is replaced by the password to be assigned:

`enable secret XXXXX`

Line Passwords

While the `password` and `secret` commands protect the router's privileged mode, administrators can also assign what are called line passwords. A *line password* is a password that requires administrators to authenticate when accessing the router through a console or network connection. Administrators can assign different ones for each connection type.

To set line passwords, switch from the global configuration mode (entered with the `enable` and `config t` commands) to line configuration mode, using one of three commands:

```
line console
line vty
line aux
```

These commands select the mode for console connections, Telnet connections, and modem connections, respectively.

In addition to specifying the type of connection, an administrator must also specify the number of the logical connection to configure. Routers have multiple logical connections for each physical connection, the number of which varies. Add the number of the line to configure on the same command line as the connection type, as follows:

`line vty 1`

To enter a line configuration mode for all of the available lines, use a command like the following:

`line vty 0 15`

This command switches a router with 16 `vty` lines to the configuration mode for all of the lines at once.

When entering a line configuration mode, the router prompt changes to the following:

`Router(config-line)#`

Once in the correct line configuration mode, specify a password using the password *XXXX* command, where *XXXXX* is the password to be assigned:

`password XXXXX`

Finally, to configure the router to prompt for a password when an administrator connects using that line, use the login command:

```
login
```

Interface Configuration

By definition, routers are connected to two or more networks, and administrators must configure the interfaces to those networks with appropriate IP addresses and other settings. It is also necessary to enable each interface, so that the router is accessible.

Interface Mode Access

To configure a router interface, an administrator must switch to interface mode, using a sequence of commands:

```
enable
config t
interface XXX
```

Replace the *XXX* with an interface identifier. An *interface identifier* can be a media type, a slot number, a port number, or a combination of these, depending on the router.

To identify the interfaces on a router, use the `show interfaces` command. For example, the interface fa0/0 command switches the router to interface mode for the first Fast Ethernet interface in a router's slot 0:

```
interface fa0/0
```

After switching to interface mode, the router prompt appears:

```
Router(config-if)#
```

Once in this mode, the administrator can issue the commands to configure the interface.

IP Address Assignment

Once in the appropriate interface mode, an administrator can assign an IP address to the interface using the command:

```
ip address 192.168.3.1 255.255.255.0
```

The `ip address` command must be followed by a valid IP address for the subnet to which the interface is connected, as well as the correct subnet mask for that subnet.

DHCP

If a router interface is to receive an IP address from a Dynamic Host Configuration Protocol (DHCP) server on the network, then an administrator must enable the DHCP client for the interface.

```
ip address dhcp
```

Enabling the Interface

Issuing the `ip address` command for an interface does not turn it on. To do this, the administrator must issue an additional command after configuring the IP address:

```
no shutdown
```

`no shutdown` is the command to enable the interface. As a result of this command, a message appears indicating that the state of the interface has changed to *up*.

Interface Verification

To display the interface configuration settings, issue the `show interface` (or `show int`) command from the privileged mode prompt, or issue the `do show interface` do (or `do show int`) command from the configuration mode prompt. This command displays the configuration for all of the router's interfaces. An interface identifier can also be specified on the command line to display one interface's configuration.

The display generated by the `show int` command for the `fa0/0` interface includes:

```
FastEthernet0/0 is up, line protocol is down (disabled)
  Hardware is Lance, address is 0001.9638.3073 (bia 0001.9638.3073)
  Internet address is 192.168.5.5/24
  MTU 1500 bytes, BW 100000 Kbit, DLY 100 usec,
     reliability 255/255, txload 1/255, rxload 1/255
  Encapsulation ARPA, loopback not set
  ARP type: ARPA, ARP Timeout 04:00:00,
  Last input 00:00:08, output 00:00:05, output hang never
  Last clearing of "show interface" counters never
  Input queue: 0/75/0 (size/max/drops); Total output drops: 0
  Queueing strategy: fifo
  Output queue :0/40 (size/max)
  5 minute input rate 0 bits/sec, 0 packets/sec
  5 minute output rate 0 bits/sec, 0 packets/sec
     0 packets input, 0 bytes, 0 no buffer
     Received 0 broadcasts, 0 runts, 0 giants, 0 throttles
     0 input errors, 0 CRC, 0 frame, 0 overrun, 0 ignored, 0 abort
     0 input packets with dribble condition detected
     0 packets output, 0 bytes, 0 underruns
     0 output errors, 0 collisions, 1 interface resets
     0 babbles, 0 late collision, 0 deferred
     0 lost carrier, 0 no carrier
     0 output buffer failures, 0 output buffers swapped out
```

WAN Interfaces

Serial-based wide area network (WAN) interfaces use the same TCP/IP protocols as LAN interfaces, but at the Network Access layer, they are different. As a result, administrators have additional settings to configure, such as those that identify the Network Access layer protocol the devices use and configuring clocking.

Both ends of a serial connection must use the same protocols for the devices to communicate successfully. The protocols at the Network Access layer specify how the data generated by the upper-layer protocols should be encapsulated for transmission over the connection. Administrators must therefore ensure that the serial interface on the router is using the same encapsulation type as the device at the other end of the connection.

To configure the encapsulation type for a particular interface on a router, switch to the interface mode prompt and issue a command such as:

```
encapsulation frame-relay
encapsulation hdlc
encapsulation ppp
```

The High-level Data Link Control (HDLC) protocol is the default for connections between two Cisco routers.

Once the two connected devices are configured to use the same encapsulation type, there is at least one other remaining configuration task which concerns signal clocking.

If the serial connection is linking a router to a CSU/DSU, then the router always functions as the DTE and the CSU/DSU as the DCE. Because the clocking signal must always come from the DCE, the CSU/DSU is responsible for it. However, if connection is between two routers, then the router at the DCE end of the cable must provide the clocking signal.

To configure a router interface to generate clocking signals, issue the clock rate *XXXXX* command from the interface mode prompt:

```
clock rate XXXXX
```

The *XXXXX* in this case is the clocking rate in bits per second. The value to use depends on the type of interface, but the most common setting is 64,000.

Saving Configuration Changes

After making the configuration changes, an administrator must save them or they will not persist when the router is restarted. To save the changes to the startup configuration file, the administrator must return to the privileged mode prompt and issue another command:

`copy running-config startup-config`

If this is a new router, and there is no startup configuration file, a prompt appears asking for a destination filename. The default name for the file is `startup-config`.

Summary

Networks with Internet access include at least one router that provides a path to the Internet. Most networks include more than one router that provides paths to multiple other subnetworks. Each router includes a routing table that includes the paths to these other subnetworks. The routing table can be updated manually by an administrator (static routes), or routing protocols can update the routing table automatically (dynamic routing). Two routing protocols that can be used on internal networks are RIPv2 and OSPF. An administrator can configure a Microsoft server as a software router by adding and configuring the Routing and Remote Access Service (RRAS). Hardware routers require configuration before they are fully operational. Administrators must make the appropriate cable connections, access the router's management interface, specify IP addresses and passwords, configure the router's network interfaces, and add routes at a minimum.

Review Questions

1. A router is configured in a network that includes multiple other routers. What routes does a router know by default?
   a. Directly connected routes
   b. Dynamic routes
   c. Routes added to the routing table
   d. Static routes

2. An administrator has added a second router to a network that includes three subnets. What is the easiest way to ensure that both routers know the routes to all subnets?
   a. Add dynamic routes.
   b. Add OSPF to each router.
   c. Add RIPv2 to each router.
   d. Add static routes.

3. A router determines the best path to another subnet based on the highest cost metric.
   a. True
   b. False

4. A network includes more than 50 hardware routers. What can an administrator configure on these routers so that they will share routing information with each other?
   a. ARP
   b. DHCP
   c. OSPF
   d. Telnet

5. Windows Server 2012 R2 server supports the RIPv2 and OSPF routing protocols.
   a. True
   b. False

6. Which of the following settings must an administrator configure on the DCE end of a serial cable that does not need to be configured on the DTE end?
   a. Clock rate
   b. Encapsulation type
   c. IP address
   d. Subnet mask

7. At which of the following prompts will issuing the `enable` command generate a password prompt?
   a. `Router>`
   b. `Router#`
   c. `Router(config)#`
   d. `Router(config-line)#`

8. Which of the following is not one of the connection types that an administrator can specify when issuing the `line` command?
   a. `aux`
   b. `console`
   c. `tel`
   d. `vty`

Resolving Names to IP Addresses

Computers are named with hostnames and NetBIOS names. Each of these name types has special features. For example, computers on the Internet use hostnames, but internal networks can use either hostnames or NetBIOS names. Additionally, TCP/IP uses several different methods to resolve these names to IP addresses. The primary method used to resolve hostnames to IP addresses is with Domain Name System (DNS) servers. The primary method used to resolve NetBIOS names to IP addresses is with Windows Internet Naming System (WINS) servers. However, additional methods exist. It is important to understand the different types of names, the types of name resolution, and the steps TCP/IP uses to resolve names to IP addresses.

OBJECTIVES

- ▶ Identify the elements used to resolve names to different types of addresses on a network
- ▶ Describe how to resolve names to IP addresses
- ▶ List the steps in the name resolution process

CHAPTER 11

TABLE OF CONTENTS

Types of Names Used in Networks....210
- Hostnames..................... 210
- Hostnames, URLs, and FQDNs..... 211
- NetBIOS Names................. 212
- NetBIOS Names Versus Hostnames . 212
- Computer Names................ 214

Types of Name Resolution 215
- Domain Naming System 216
- Host Cache 219
- Hosts File.................... 220
- WINS 221
- NetBIOS Cache 222
- lmhosts File 222
- Broadcast Name Resolution 223
- Link-Local Multicast Name Resolution 223

Steps in Name Resolution.......... 223
- The Steps in Hostname Resolution 223
- The Steps in NetBIOS Name Resolution 224

Summary........................ 226

Review Questions 227

TYPES OF NAMES USED IN NETWORKS

Computers work with numbers. At the lowest level, the computers use ones and zeros assigned to individual bits. Every single piece of data that flows through a computer is reduced to simply ones and zeros.

However, humans just do not think that way. Instead, they think in words. It may be challenging to memorize the MAC addresses or IP addresses of favorite websites. However, it would be far easier to name some favorite websites.

Thankfully, computers can also use names. However, there are many different elements built into networking to convert these names into the numbers used by the computers.

The progression of how names are resolved to different types of addresses are:

Names Computers are assigned names, and a computer can usually be reached in a network by using the name. These names can be hostnames or NetBIOS names, or both, depending on where they are located. Only hostnames are utilized on the Internet, but both hostnames and NetBIOS names can be used on internal networks.

IP Addresses IP addresses are assigned to the network interface adapters of computers. The IP address is used at the Network layer of the OSI model to route traffic between subnetworks. Name resolution methods resolve the computer name to an IP address.

MAC Addresses The *media access control* (*MAC*) address or physical address uniquely identifies the network adapter. Each device on a network has an interface with a different MAC address. The MAC address is used at the lower levels of the OSI model.

Bits Bits are the lowest level of data. Data streams to and from computers using bits of ones and zeros.

The types of names given to computers and other network devices are either hostnames or NetBIOS names. There are several characteristics that highlight the differences between hostnames and NetBIOS names. **See Figure 11-1.**

Hostnames

A *hostname* is a user-friendly string of characters (or label) assigned to a computer or other network device. Hostnames are the primary name type used

Figure 11-1 Comparing Hostnames and NetBIOS Names

CHARACTERISTICS	HOSTNAMES	NETBIOS NAMES
Length	Up to 255 characters	Fifteen readable characters; sixteenth character identifies a service
Location	On Internet and internal networks	Only on internal networks
Primary name resolution method	Domain Name System (DNS)	Windows Internet Naming Service (WINS)
Namespace	Hierarchical (part of fully qualified domain name)	Flat namespace (single-level names only)

Figure 11-1. *There are many differences between hostnames and NetBIOS names.*

today. They are the only types of names used on the Internet and the primary name type used on many internal networks.

Hostnames can be as long as 255 characters. They can contain letters, numbers, periods, and hyphens.

A *fully qualified domain name (FQDN)* is the full computer name when it is a host and part of a domain. Consider the hostname and FQDN of a computer running Windows Server 2012 R2. The FQDN in a Windows system can be up to 255 characters, as long as no more than 63 characters are used between each period. **See Figure 11-2.**

Note that the computer name is Server01. This is the hostname. The computer is a member of a domain named adatum.com. The full computer name (or FQDN) is server01.adatum.com.

Hostnames, URLs, and FQDNs

A *uniform resource locator (URL)* is the address used to access Internet resources such as websites. It includes the protocol and the fully qualified domain name (FQDN). For example, to reach the website www.bing.com, the URL http://www.bing.com could be used. The protocol is Hypertext Transfer Protocol (HTTP), and the FQDN is www.bing.com.

www is the hostname, and it represents a web farm of computers that respond to that name. It is not necessary to use www, though. Instead, the address may be entered as **http://bing.com**, and it will work. Of course, the protocol can also be skipped in a web browser. For example, **bing.com** can simply be entered and the browser assumes that HTTP is being used and fills that in.

✓ **FACT**
Windows limits the length of hostnames in Windows systems to 63 characters. However, it is recommended to limit the length to 15 characters for compatibility with NetBIOS names.

✓ **FACT**
The hostname of a Windows computer can also be viewed from the command prompt by typing **hostname** and pressing Enter.

Figure 11-2 Computer Name

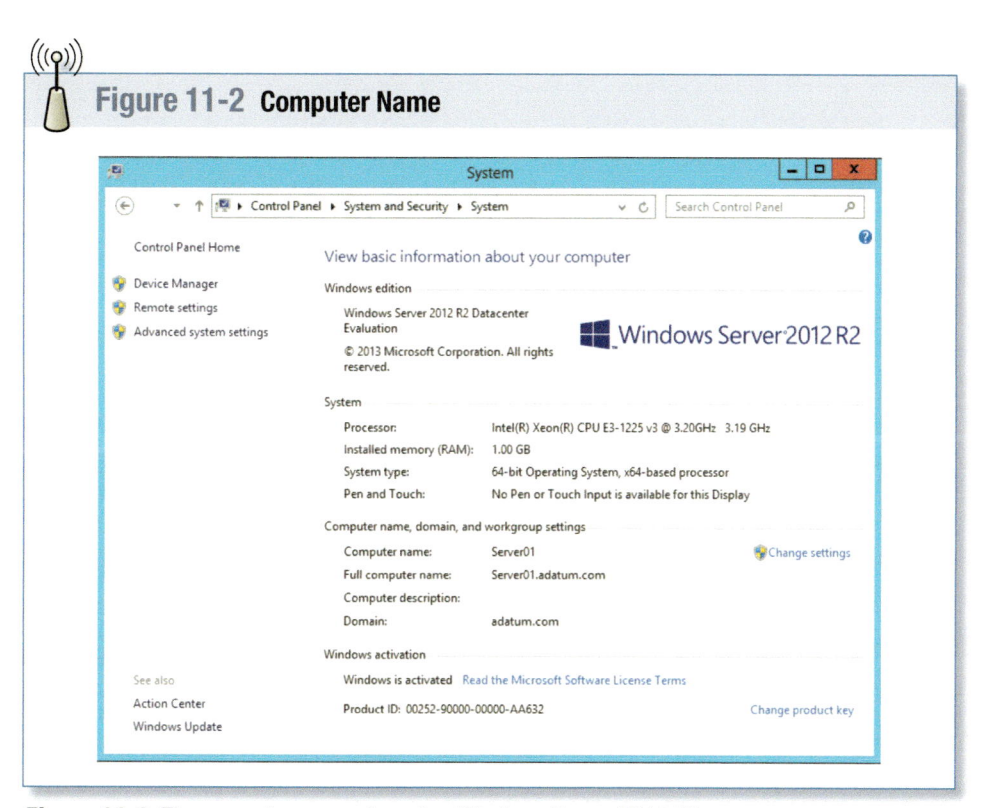

Figure 11-2. The computer name viewed on Windows Server 2012 R2.

✓ FACT

The value <00> means something different when it is a UNIQUE type than when it is a GROUP type. UNIQUE <00> indicates the workstation service, and GROUP <00> indicates the domain name.

✓ FACT

Duplicate NetBIOS names result in errors and communication problems between these computers. All computers on the same network need unique names.

DNS supports multiple computers with the same name and can resolve name requests to different servers in a round-robin fashion. DNS also supports alias names to enable computers to respond to different names. A single computer can be registered in DNS with multiple different names, and each name will resolve to the same IP address.

NetBIOS Names

A *Network Basic Input/Output System (NetBIOS) name* is a 15 character-long unique identifier that NetBIOS services use to point to network resources. Even when the actual name is shorter (such as PC1), the NetBIOS name is padded with trailing spaces to make the name 15 characters long.

The NetBIOS name includes a hidden sixteenth byte. This sixteenth byte is a hexadecimal number that identifies services running on the system. Other systems and applications on the network use this information to determine how they can communicate with a system.

Values for the sixteenth byte of a computer's NetBIOS name identify services running on desktop and server operating system computers or provide other information about the computer. **See Figure 11-3.**

In addition to tracking the name of the computer, NetBIOS tracks the name of the workgroup or domain that a computer has joined.

NetBIOS names registered by a Windows system can be viewed by entering the **nbtstat -n** command at a command prompt.

Consider the output of the nbtstat -n command on a Windows Server 2012 R2 server named SUCCESS1. This computer is a domain controller within the networking.mta domain. **See Figure 11-4.**

Note the output for the computer name (SUCCESS1) has specific hex values listed. Similarly, the output for the domain name (NETWORKING from networking.mta) has hex values listed. These values can also have meaning for the domain. **See Figure 11-5.**

Many more NetBIOS services can be assigned to any computer. The important point to grasp is that each computer can have multiple NetBIOS names. Different NetBIOS names have different hex values to provide information about the computer.

NetBIOS Names Versus Hostnames

A Windows server's name is SUCCESS1. Is this a hostname or a NetBIOS name? The answer is that it is both. When a computer is named, the name is used for both the hostname and the NetBIOS name.

If the name is 15 characters long or fewer, the computer will have the same hostname and NetBIOS name. However, if the hostname is more than 15 characters, Windows truncates the name to the first 15 characters for use as the NetBIOS name. This is important because it is possible to inadvertently give different computers duplicate NetBIOS names.

Consider the hostnames and the resulting NetBIOS name derived from the hostname. Some of the hostnames are more than 15 characters, resulting

Figure 11-3 NetBIOS sixteenth byte

HEXADECIMAL VALUE	MEANING	COMMENTS
00	Workstation service	Creates and maintains client network connections to other computers on the network
20	File server service	Indicates the computer can share files and printers over the network
23/24	Microsoft Exchange	Identifies a server hosting Microsoft Exchange

Figure 11-3. Examples of the value and meaning of the NetBIOS sixteenth byte for computer names

Figure 11-4 Output of nbtstat - n command

```
C:\>nbtstat -n

Local Area Connection:
Node IpAddress: [192.168.3.1] Scope Id: []

            NetBIOS Local Name Table

     Name              Type         Status
   ---------------------------------------
     SUCCESS1       <00>  UNIQUE    Registered
     NETWORKING     <00>  GROUP     Registered
     NETWORKING     <1C>  GROUP     Registered
     SUCCESS1       <20>  UNIQUE    Registered
     NETWORKING     <1B>  UNIQUE    Registered
```

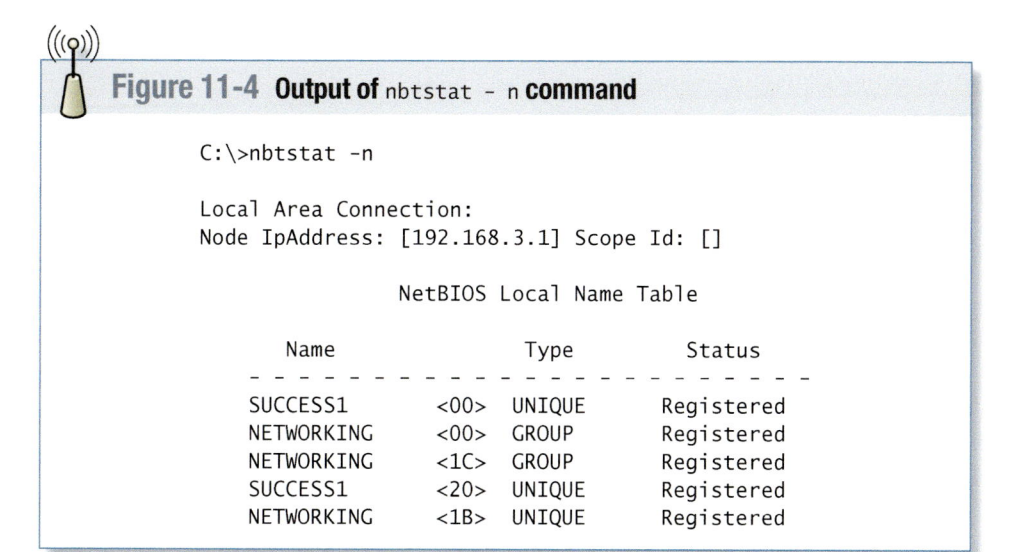

Figure 11-4. Entering the nbtstat -n command will display the NetBIOS names registered by Windows.

Figure 11-5 Examples of the Value and Meaning of the NetBIOS Sixteenth Byte for Domain Names

HEXADECIMAL VALUE	MEANING	COMMENTS
00	Domain name	Indicates the name of the domain
1C	Domain controller	Indicates that the server is a domain controller in the domain
1B	Domain master browser	Indicates the computer is hosting the Domain Master Browser role, which is used by NetBIOS services in the network

Figure 11-5. The hexidecimal value of the sixteenth bit has meaning to administrators.

214   Introduction to Network Technologies

in duplicate NetBIOS names. **See Figure 11-6.**

Note that the shorter computer names are identical as both hostnames and NetBIOS names. However, since NetBIOS truncates the longer computer names to only the first 15 characters, some of the computers have duplicate NetBIOS names.

Computer Names

To view a Windows computer's hostname, view its NetBIOS name, or modify the computer's name:

1. Open the System control panel applet and click Advanced System Settings. The System Properties sheet appears.
2. Select the Computer Name tab and click Change. The Computer Name/Domain Changes dialog box appears. **See Figure 11-7.**
3. Click More. The DNS Suffix and NetBIOS Computer Name dialog box appears. **See Figure 11-8.**

> ✓ FACT
>
> It is possible to modify the primary DNS suffix. However, it is not possible to modify the NetBIOS name.

Figure 11-6 NetBIOS Names Derived from Hostnames

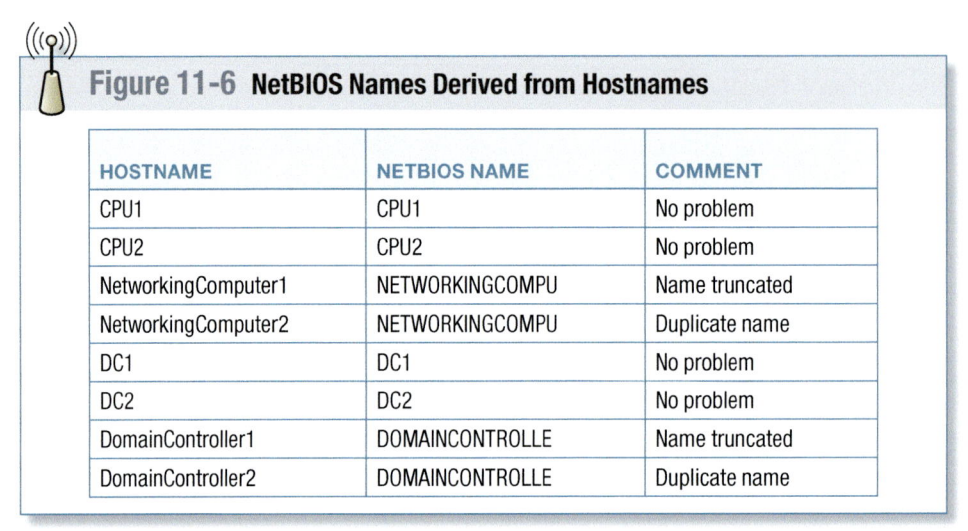

HOSTNAME	NETBIOS NAME	COMMENT
CPU1	CPU1	No problem
CPU2	CPU2	No problem
NetworkingComputer1	NETWORKINGCOMPU	Name truncated
NetworkingComputer2	NETWORKINGCOMPU	Duplicate name
DC1	DC1	No problem
DC2	DC2	No problem
DomainController1	DOMAINCONTROLLE	Name truncated
DomainController2	DOMAINCONTROLLE	Duplicate name

Figure 11-6. When a hostname is longer than 15 characters, Windows will truncate the name to match the NetBIOS name.

Figure 11-7 Computer Name/Changes Dialog

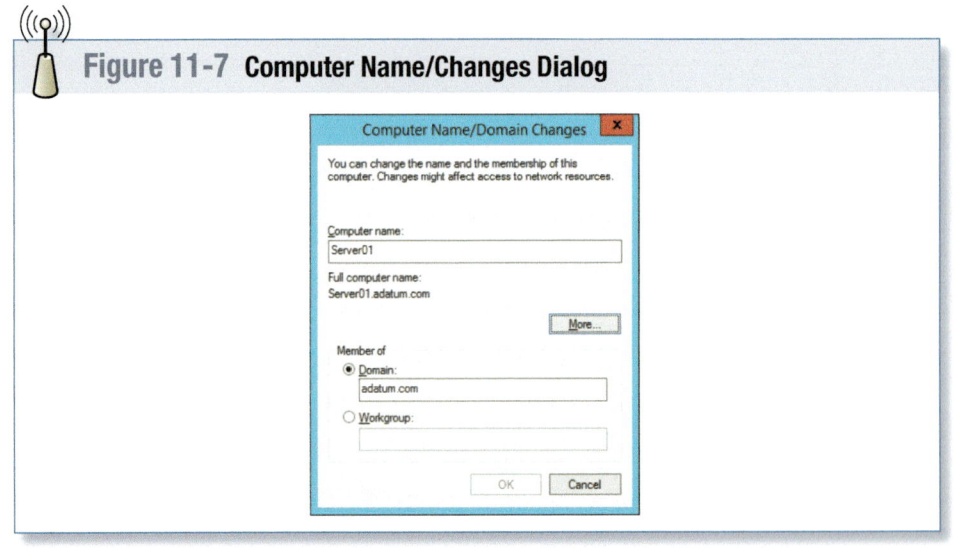

Figure 11-7. Computer names can be viewed from the Computer Name/Domain Changes dialog box.

Note that it is possible to view the DNS suffix of the computer and the NetBIOS computer name on this page. The suffix is automatically added when a computer joins a domain, and the NetBIOS name is automatically created from the computer name.

4. Click Cancel. If desired, the computer name can be modified by changing it on the Computer Name/Domain Changes page.

5. Close all windows.

TYPES OF NAME RESOLUTION

Before communication between TCP/IP systems can take place, names must be resolved to IP addresses, both on the Internet and within internal networks. TCP/IP systems can use eight types of name resolution. **See Figure 11-9**.

Figure 11-8 DNS Suffix and NetBIOS Computer Name Dialog

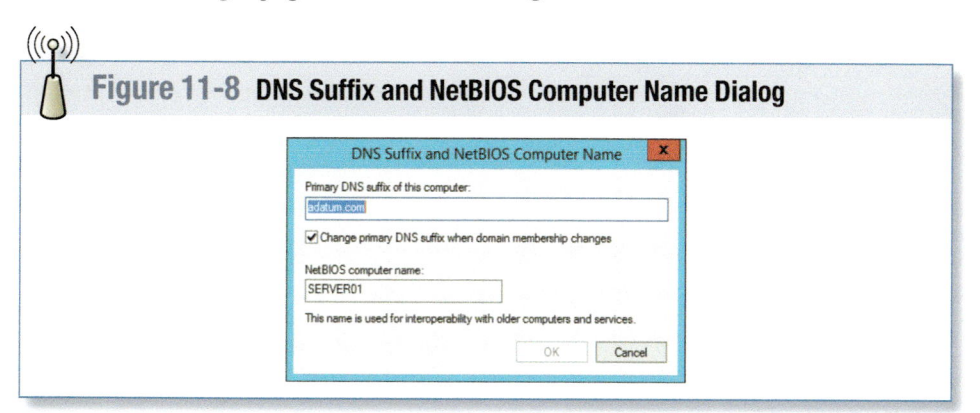

Figure 11-8. NetBIOS names can be viewed from the DNS Suffix and NetBIOS Computer Name dialog box.

Figure 11-9 Name Resolution Methods

NAME RESOLUTION METHOD	RESOLVES (HOSTNAMES, NETBIOS NAMES, OR BOTH)	COMMENTS
Domain Name System (DNS) server	Hostnames (Windows Server 2008 DNS can be configured to resolve NetBIOS names using GlobalNames zones)	DNS servers are on the Internet and internal networks. Microsoft domains require DNS.
Host cache	Hostnames	Can be viewed with the `ipconfig /displaydns` command.
Hosts file	Hostnames	Located in the `c:\windows\system32\drivers\etc\` folder by default.
Windows Internet Name Service (WINS)	NetBIOS names	WINS servers are located only on internal networks.
NetBIOS cache	NetBIOS names	The NetBIOS cache can be viewed with the `nbtstat -c` command.
`lmhosts` file	NetBIOS names	Located in the `c:\windows\system32\drivers\etc\` folder when used.
Broadcasts	Both	The system simply sends a broadcast transmission with the name asking the owner to reply with its IP address.
Link-local multicast name resolution (LLMNR)	Hostnames	This is a newer method similar to broadcast transmissions that works on internal networks.

Figure 11-9. There are eight types of name resolutions available with TCP/IP.

Domain Naming System

The Domain Naming System (DNS) is a service that resolves hostnames to IP addresses. These can be names of computers within an internal network or names of computers on the Internet. Clients send name resolution requests to a DNS server, and the DNS server responds with the IP address. The client computer then uses the returned IP address as the destination IP address for data traffic.

Consider the DNS console on a server running Windows Server 2012 R2. It shows several host (A) records with their names and IP addresses. When a system queries a DNS server with the name of a computer, the DNS server checks to see whether it has a matching host (A) record for the requested name in its database, and if it does, it returns the IP address to the requesting client. **See Figure 11-10.**

Note that the server also has a reverse lookup zone. A *reverse lookup zone* uses pointer (PTR) records to do reverse lookups. In other words, it is possible to pass the IP address to the DNS server and retrieve the name of the computer with that IP address.

DNS servers host multiple types of records beyond the A records. For example, an Active Directory domain must have service (SRV) records to locate domain controllers in the network. **See Figure 11-11.**

A DNS server that holds records for a specific namespace (such as the **networking.mta** namespace) is authoritative for that namespace. In other words, it knows all the computers and the IP addresses in that namespace. If it does not have a record for one of these computers, no one else will either.

If a DNS server is not authoritative for a namespace, it can still resolve names by forwarding the name request to other DNS servers.

DNS is hierarchical. No single server knows the names and IP addresses for all the computers on the Internet. Instead, DNS servers are authoritative for different namespaces. **See Figure 11-12.**

> ✓ **FACT**
> Host records are sometimes listed as host (A) and other times as A (host). However, they are the same.

> ✓ **FACT**
> Reverse lookup zones are optional. Some DNS servers do not host reverse lookup zones or support reverse lookups.

Figure 11-10 DNS Console

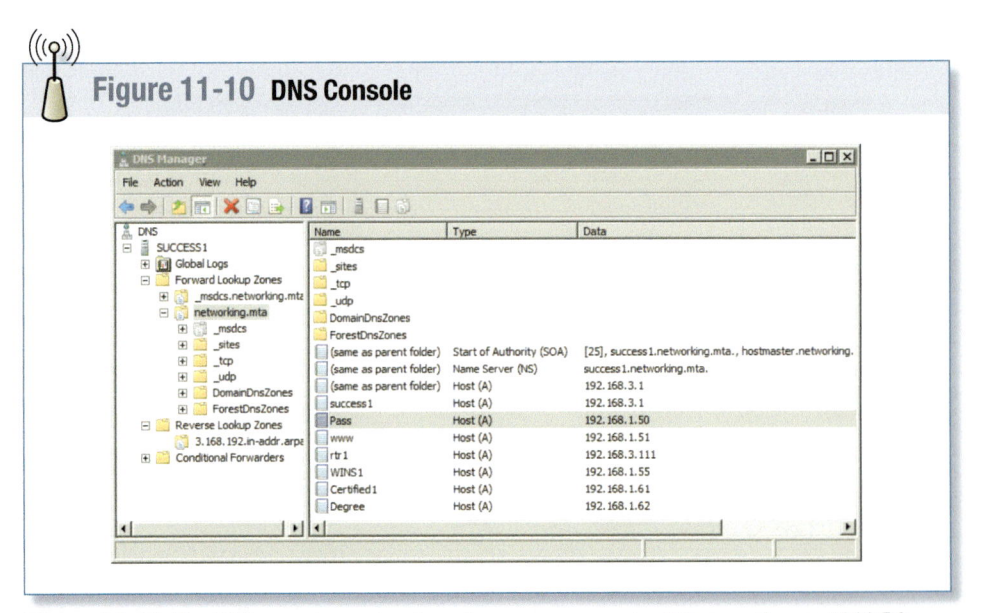

Figure 11-10. *The DNS console can be viewed on a server running Windows Server 2012 R2.*

Figure 11-11 Common DNS Records

RECORD TYPE	USE
A (host)	Resolves hostnames to IPv4 addresses.
AAAA (host)	Resolves hostnames to IPv6 addresses.
PTR (pointer)	Resolves IP addresses to hostnames.
CNAME (alias)	Resolves one hostname to another hostname, which enables multiple computer names to be resolved to the same IP address.
MX	Used for mail exchange servers (e-mail servers).
SRV	Required by Active Directory to locate servers running specific services (such as domain controllers).
NS	Identifies DNS name servers.

Figure 11-11. The DNS record type gives information about how that record is being used.

✓ FACT
SRV records for hostnames perform a function similar to that of the sixteenth byte of NetBIOS names. Both identify specific services running on computers.

Figure 11-12 DNS Server on the Internet

- DNS Root Servers (13 in the World)
- Top-Level Domain DNS Servers (.com, .net, .org, .biz, and so on)
  - .com
- Second-Level Domain DNS Servers
  - microsoft.com
- Third- and Lower-Level Domain DNS Servers
  - training.microsoft.com

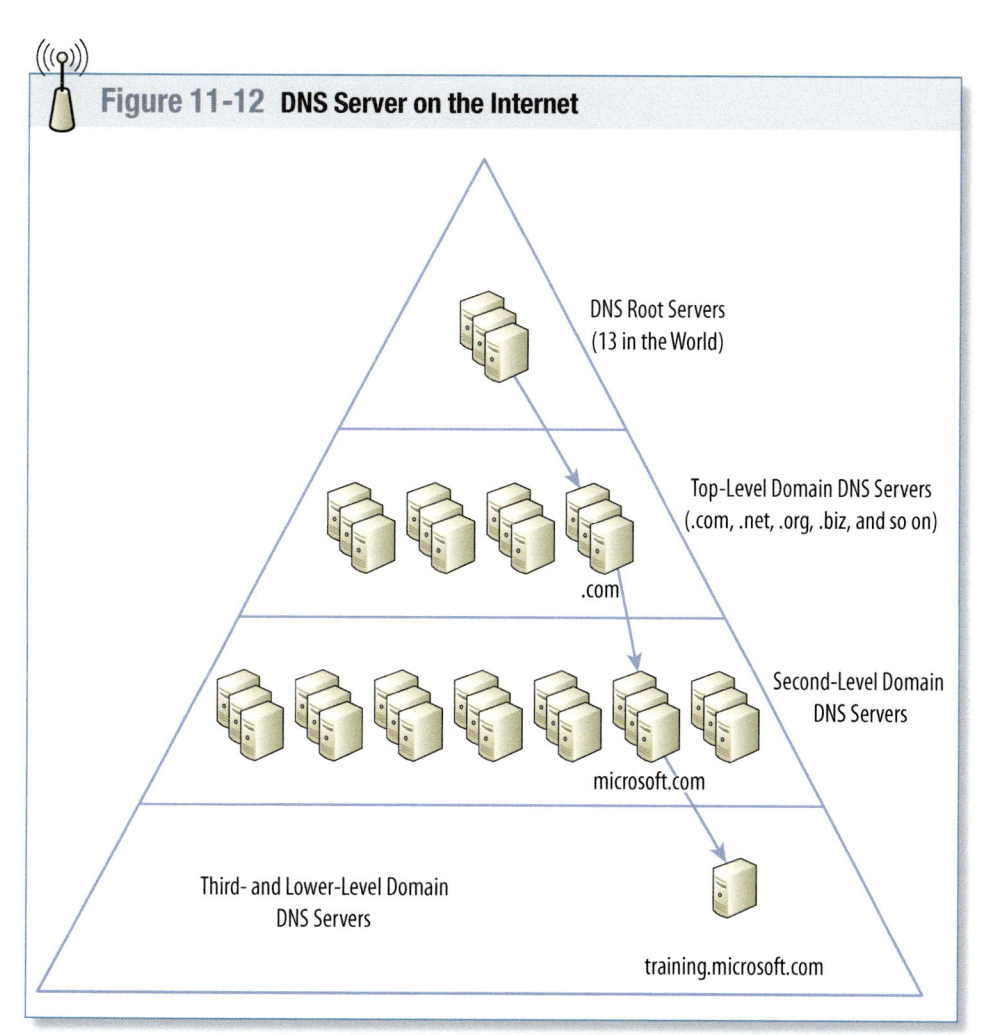

Figure 11-12. DNS servers on the Internet are authoritative for different namespaces.

The levels of hierarchy are:

DNS Root Servers At the top of the hierarchy are DNS root servers. There are only 13 DNS root servers in the world. These servers know only the addresses of DNS servers that are authoritative for top-level domains (such as .com, .net, and .org). They will not know the address of `training.microsoft.com`. However, they will know the address of the DNS servers that are authoritative for the .com namespace.

Top-Level Domain DNS Servers Top-level domain DNS servers know the addresses of second-level domain DNS servers in their namespace. For example, a .com DNS server knows the addresses of servers that are authoritative for the `Microsoft.com` namespace. However, a .com DNS server does not know anything about .net, .org, or any other top-level domain namespace.

Second-Level Domain DNS Servers Below the top-level domain DNS servers are the second-level domain DNS name servers. These servers are authoritative in the second-level DNS namespace. For example, Microsoft has several servers that are authoritative in the `Microsoft.com` domain. The DNS servers in the `Microsoft.com` namespace know only about `Microsoft.com`. They do not know anything about other namespaces such as `sybex.com`.

Third- and Lower-Level Domain DNS Servers Third-level and lower-level domain DNS servers are possible. However, these are needed only when the FQDN includes these lower levels. For example, Microsoft may have a DNS server dedicated to the `training.microsoft.com` namespace. This DNS server can resolve all the hostnames in the `training.microsoft.com` namespace. Other companies may not have third-level DNS servers. Instead, the second-level server resolves the names for all the company's resources on the Internet.

DNS queries to the Internet start with a query to one of the DNS root servers. For example, if a client is trying to reach a web server named www.sybex.com from an internal network named `networking.mta`. The record for the www.sybex.com web server will not be on the internal DNS server. However, the internal DNS server can make queries to the Internet to retrieve the name.

This is accomplished by: **See Figure 11-13.**

1. The client passes the request to the DNS server to resolve www.sybex.com. Assume that the DNS server has just turned on and does not have any information except for the address of the DNS root servers.

2. Since the top-level domain is .com and the DNS server does not have the IP address of a DNS server in the .com domain, it queries a DNS root server. The DNS root server responds with the IP address of a DNS server that is authoritative for the .com namespace.

3. Next, the internal DNS server queries the .com DNS server for the address of a DNS server that is authoritative in the sybex.com domain. The top-level domain DNS server responds with an IP address.

4. The internal DNS server then queries the sybex.com DNS server for the IP address of the web server named www.

> ✓ **FACT**
> Multiple DNS servers are available for each of the top-level domains. If one DNS server in the .com namespace fails, others in the .com namespace can still answer queries.

> ✓ **FACT**
> If the DNS server has been on for a while, it will have cached information, and it may be able to skip some of the steps.

Figure 11-13 DNS Query on the Internet

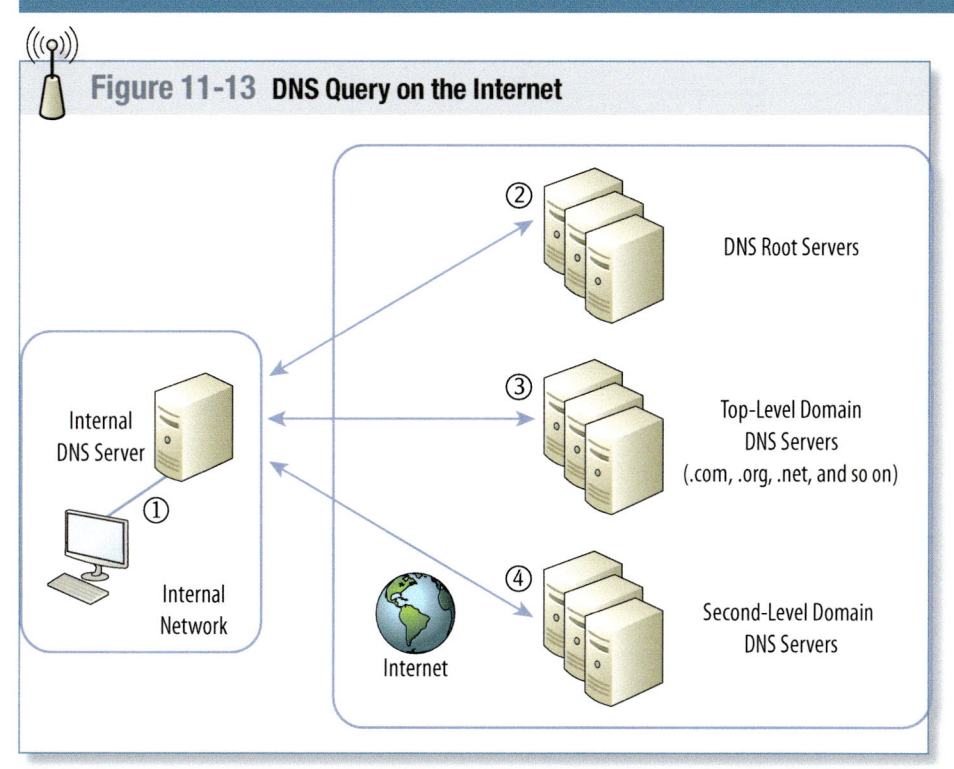

Figure 11-13. Several steps are required to resolve a DNS query on the Internet.

5. Finally, the internal DNS server returns the IP address it obtained from the sybex.com DNS server to the client computer.

DNS servers cache responses in their internal memory. In other words, after a DNS server queries a root DNS server for an address of a .com DNS server, it keeps this information. The next time it needs to query the .com server for an address, it just looks in the cache for this information.

Host Cache

Every time a computer receives a name resolution response from a DNS server, it places the result in the localhost cache. The *host cache* is an area of memory on any computer that is dynamically updated with hostnames and their corresponding IP addresses.

DNS resolver cache is another name for the host cache, since many of the entries are created when DNS is queried to resolve a hostname. However, the host cache also includes data from the hosts file.

The host cache on any computer can be viewed from the command prompt by entering the **ipconfig /displaydns**

Resolving NetBIOS Names with DNS

Windows Server 2012 R2 supports a type of zone called a GlobalNames zone. In networks where there are very few NetBIOS applications, GlobalNames zones can be used for single label names, just as if they were NetBIOS names.

When a GlobalNames zone is created, DNS can resolve both hostnames and NetBIOS names. GlobalNames zones apply only to internal Microsoft networks. On the Internet, DNS can only resolve hostnames.

✓FACT

The host cache on an end user's computer is different from the DNS cache on a DNS server. However, they work in the same way. Cached data does not need to be queried again.

220 Introduction to Network Technologies

command on a Windows Server 2012 R2 server. **See Figure 11-14.**

Note that the first record is a PTR reverse lookup record for the local computer using the loopback address of 127.0.0.1. The second record is an A (host) record for localhost record that is mapped to the loopback address of 127.0.0.1. These two entries will usually be displayed for any Windows computer.

The third record for bing.com is a record returned from a DNS server and placed in the cache. Notice that it has a Time To Live section. Every record returned from a DNS server includes this section, and it indicates how long the data will remain in the cache. The value of 580 indicates that it will remain in the cache for another 580 seconds. Any queries to bing.com will use this IP address as long as it remains in the cache. After the timeout period, it is removed from the cache and requires another query to DNS to resolve it.

Hosts File

The *hosts file* is a simple text file located in the c:\windows\system32\drivers\etc folder by default that maps the

> ✓ FACT
>
> Cached items can be removed from the host cache with the ipconfig /flushdns command. However, this does not remove items in the cache from the hosts file.

Figure 11-14 ipconfig /displaydns **command**

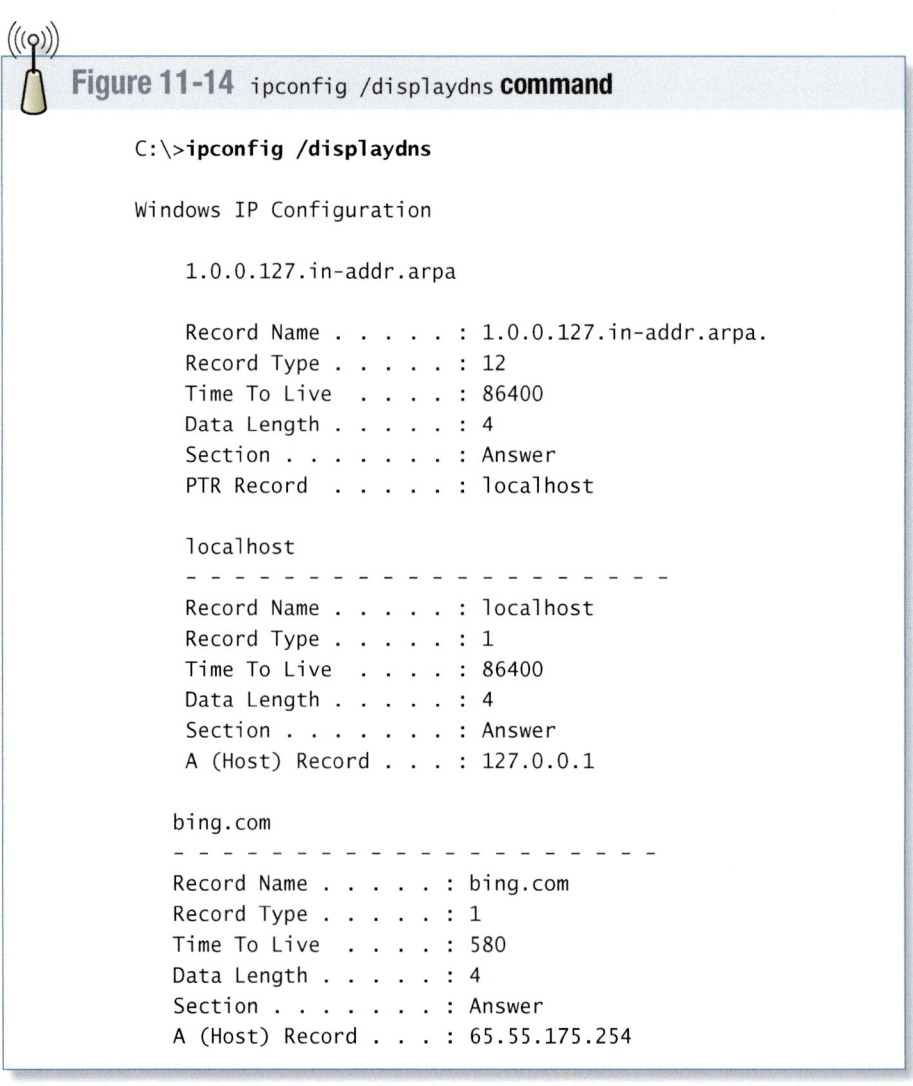

```
C:\>ipconfig /displaydns

Windows IP Configuration

    1.0.0.127.in-addr.arpa

        Record Name . . . . . : 1.0.0.127.in-addr.arpa.
        Record Type . . . . . : 12
        Time To Live  . . . . : 86400
        Data Length . . . . . : 4
        Section . . . . . . . : Answer
        PTR Record  . . . . . : localhost

    localhost
    ----------------------
        Record Name . . . . . : localhost
        Record Type . . . . . : 1
        Time To Live  . . . . : 86400
        Data Length . . . . . : 4
        Section . . . . . . . : Answer
        A (Host) Record . . . : 127.0.0.1

    bing.com
    ----------------------
        Record Name . . . . . : bing.com
        Record Type . . . . . : 1
        Time To Live  . . . . : 580
        Data Length . . . . . : 4
        Section . . . . . . . : Answer
        A (Host) Record . . . : 65.55.175.254
```

Figure 11-14. *Output of the* ipconfig /displaydns *command provides valuable information.*

names of computers to IP addresses. The benefit is that mapped records in this file are automatically placed in the host cache. **See Figure 11-15.**

Note that the beginning of the file consists of comments preceded by hash marks (#). The only two entries are the `127.0.0.1` and `::1` lines. These lines map the localhost name to the IPv4 loopback address of `127.0.01` and to the IPv6 loopback address of `::1`.

All entries in the hosts file are immediately placed in the cache, and they stay there permanently. Hosts file entries do not time out and fall out of the cache.

WINS

Windows Internet Name Service (WINS) is a service that can be added to a server to resolve NetBIOS names to IPv4 addresses. A WINS server can resolve only NetBIOS names, not hostnames.

WINS servers can be found on internal Microsoft networks. Non-Microsoft networks might include NetBIOS servers to resolve NetBIOS names, but they can be other types of NetBIOS servers. WINS is Microsoft's implementation of a NetBIOS server.

Recall that DNS is hierarchical. It uses multilevel names such as root level, top level, and so on. Because of this, DNS is highly scalable. DNS on the Internet efficiently resolves the IP addresses of billions of computers. It works as efficiently with these billions of computers as it will on an internal network with just a few dozen computers.

> ✓ **FACT**
> Entries can be placed in a hosts file to bypass DNS queries for specific hosts. If the entry is in the hosts file or in the cache, DNS is not queried.

> ✓ **FACT**
> When `ping localhost` is entered at the command prompt, the localhost name is resolved to either 127.0.0.1 or ::1, depending on whether `ping` is using IPv4 or IPv6.

Figure 11-15 The hosts file

```
# Copyright (c) 1993-2006 Microsoft Corp.
#
# This is a sample HOSTS file used by Microsoft TCP/IP for Windows.
#
# This file contains the mappings of IP addresses to host names. Each
# entry should be kept on an individual line. The IP address should
# be placed in the first column followed by the corresponding host name.
# The IP address and the host name should be separated by at least one
# space.
#
# Additionally, comments (such as these) may be inserted on individual
# lines or following the machine name denoted by a '#' symbol.
#
# For example:
#
#      102.54.94.97     rhino.acme.com        # source server
#       38.25.63.10     x.acme.com            # x client host

127.0.0.1        localhost
::1              localhost
```

Figure 11-15. *The hosts file is not commonly used, but it is still present in Windows machines.*

In contrast, WINS is not hierarchical. Instead, it is a flat database that supports only single-level names. WINS does not scale well and could not possibly work with billions of computers. As more computers are added to a WINS server, it can get bogged down.

Since DNS performs so much better than WINS, and because WINS does not support IPv6, WINS is being phased out. However, it is still being used on many networks today since many applications still use NetBIOS names.

The TCP/IP properties can be viewed for a network interface adapter on a Windows system. Note that the name of a DNS server can be configured on the same page as the one used to configure the IP address of the adapter. However, it is necessary to click the Advanced button and select the WINS tab to add the IP address of a WINS server. **See Figure 11-16.**

If a network includes multiple WINS servers, the IP addresses of each one can be added.

NetBIOS Cache

Just as any hostname that DNS resolves is placed in the cache, NetBIOS names that WINS resolves are also placed in the cache. DNS names are placed in the host cache, and WINS names are placed in the NetBIOS cache.

Use the `nbtstat -c` command to view the cache of a Windows server that recently resolved the name of a file server (named `FS1`) to the address of 192.168.1.117. **See Figure 11-17.**

The `Life [sec]` column lists how long (in seconds) the entry will remain in the cache. It is similar to the `Time To Live` entry for the hosts cache. After the time expires, the entry will fall out of the cache, and another NetBIOS query will be needed to resolve the IP address. The NetBIOS cache can be flushed with the `nbtstat -R` command.

lmhosts File

The `lmhosts` file is similar to the hosts file except that it maps NetBIOS names to IP addresses. Although the `lmhosts`

> ✓ **FACT**
>
> Although the IP address of DNS and WINS servers can be configured manually, most networks use DHCP to configure these addresses automatically.

Figure 11-16 WINS Configuration

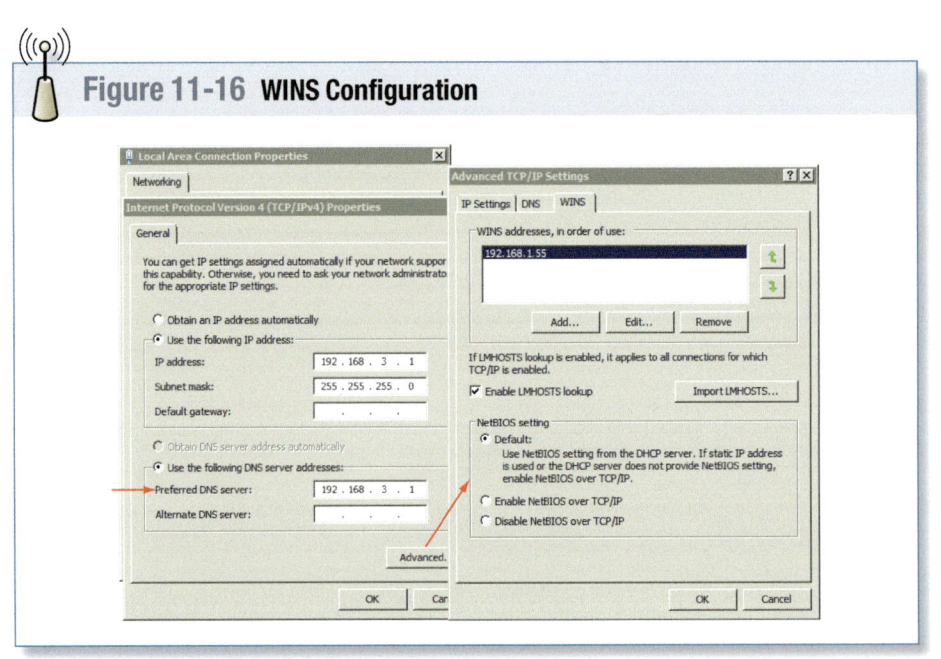

Figure 11-16. A computer can be configured to use WINS.

Figure 11-17 Output of `nbtstat -c`

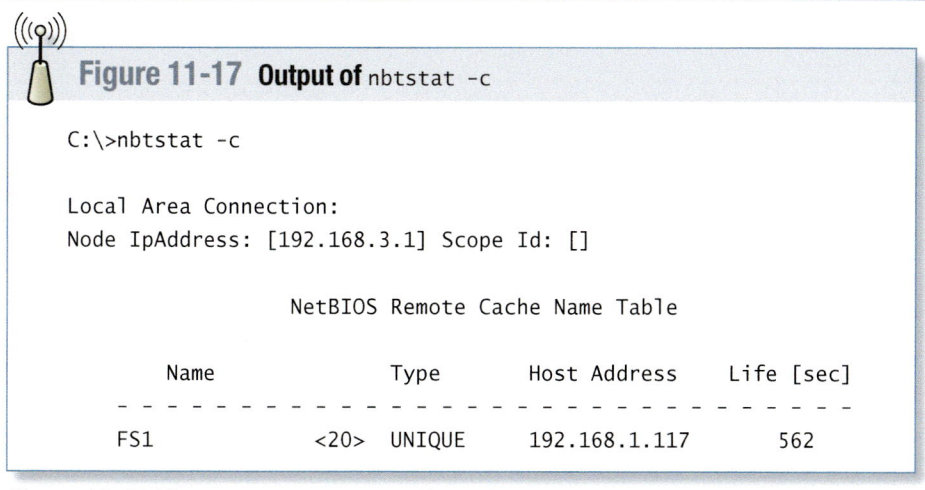

```
C:\>nbtstat -c

Local Area Connection:
Node IpAddress: [192.168.3.1] Scope Id: []

              NetBIOS Remote Cache Name Table

        Name            Type        Host Address    Life [sec]
        ---------------------------------------------------
        FS1             <20> UNIQUE 192.168.1.117      562
```

Figure 11-17. The `nbtstat -c` command is used to view the cache of a Windows server.

file was used quite often in the early days of Microsoft networking, it is rarely used today. Current Windows products do not even include a working `lmhosts` file in operating systems.

The `lmhosts.sam` file (a sample `lmhosts` file) can be viewed in the same location as the hosts file: `c:\windows\system32\drivers\etc`. An `lmhosts` file must be created to use it. The name of the file must be `lmhosts` without any extension.

Broadcast Name Resolution

Another method of resolving names is with the use of broadcast transmissions. A *broadcast transmission* is a transmission of data to every device on a network. A system can simply broadcast a request message containing a name on the segment. Any host that has that name replies with its IPv4 address.

Recall, though, that broadcasts do not pass routers, so the use of broadcasts for name resolution works only when the computers are on the same segment.

Link-Local Multicast Name Resolution

Link-local multicast name resolution (LLMNR) is similar to broadcast, but it can resolve both IPv4 and IPv6 addresses. It works for hosts on the same local link.

Automatic Private IP Addresses (APIPA) are assigned to DHCP clients when they cannot reach a DHCP server. APIPA do not include DNS server addresses, and the primary method of name resolution for APIPA clients is via broadcasts. If a system is using a link-local IPv6 address, LLMNR can be used in place of DNS for name resolution. It will work for other hosts that have the same link-local address prefix.

STEPS IN NAME RESOLUTION

Applications and services on networks need the resolution of computer names to IP addresses. Some of the applications and services are host-based, and some are NetBIOS-based. Some expect that the computers have hostnames, and some expect that the computers have NetBIOS names. This is important because it affects the steps of the name resolution process.

The Steps in Hostname Resolution

When an application or service assumes that a name is a hostname, it uses this procedure to resolve it:

> ✓ **FACT**
> The application or service determines the steps used in name resolution, based on whether it expects a hostname or a NetBIOS name.

1. Windows first checks to see whether the queried name is the same as its hostname. If so, it uses its own IP address.
2. Next, Windows checks the host cache. If the name is in the cache, it does not check any further.
3. If the name is not in the cache, Windows queries DNS. If a system is configured with both a preferred and an alternate DNS server, it queries the preferred DNS server. An alternate DNS server is queried only if the preferred DNS server does not respond.
4. If the DNS servers fail to resolve the name, Windows checks the NetBIOS name cache.
5. If the name is not in the NetBIOS name cache, Windows queries a WINS server. If the system is configured with multiple WINS server addresses, Windows queries each WINS server until it either resolves the name or runs out of WINS servers to query.
6. If WINS does not resolve the name, Windows attempts to resolve the name using broadcast transmissions. This succeeds only if the computer is on the local subnet.
7. Last, Windows checks the lmhosts file, if it exists.

This is the procedure if the application assumes that the name is a hostname. However, if the application assumes that the name is a NetBIOS name, then it performs the steps in a different order.

The Steps in NetBIOS Name Resolution

If the application or server assumes that the name is a NetBIOS name, it uses this procedure to resolve it.

1. Windows checks the NetBIOS name cache.
2. If the name is not in the NetBIOS name cache, Windows queries the DNS servers for a name in a Global-Names zone (GNZ).
3. If the DNS server is not using GNZ or it cannot resolve the name, then Windows queries a WINS server. If the system is configured with multiple WINS server addresses, Windows queries each WINS server until it either resolves the name or runs out of WINS servers to query.
4. If WINS does not resolve the name, Windows attempts to resolve the name using broadcast transmissions. This succeeds only if the computer is on the local subnet.
5. If the broadcast transmissions cannot resolve the name, Windows then checks to see whether the queried name is the same as the computer's NetBIOS name.
6. Next, Windows checks the host cache.
7. Last, Windows generates standard DNS queries.

Although the preceding steps are the default, different steps and orders are possible. Windows systems use NetBIOS over TCP/IP (NetBT). The NetBT node type can be modified to use different combinations. **See Figure 11-18.**

To view which NetBT node type the system is configured to use, enter the **ipconfig /all** command. Note in a portion of the resulting output that the node type is listed as Hybrid. This shows that it will use WINS by default and then use broadcast transmissions. See **Figure 11-19**.

Figure 11-18 NetBIOS over TCP/IP (NetBT) Node Types

TYPE	COMMENTS
B-node (broadcast)	Sends only a broadcast
P-node (peer-to-peer)	Queries only a WINS server
M-node (mixed)	Combines B-node and P-node; uses broadcast by default
H-node (hybrid)	Combines B-node and P-node; uses WINS by default
Microsoft enhanced B-node	Uses broadcast and then the lmhosts file

Figure 11-18. RFC 1001 and RFC 1002 define NetBIOS node types.

Figure 11-19 Partial output of ipconfig /all

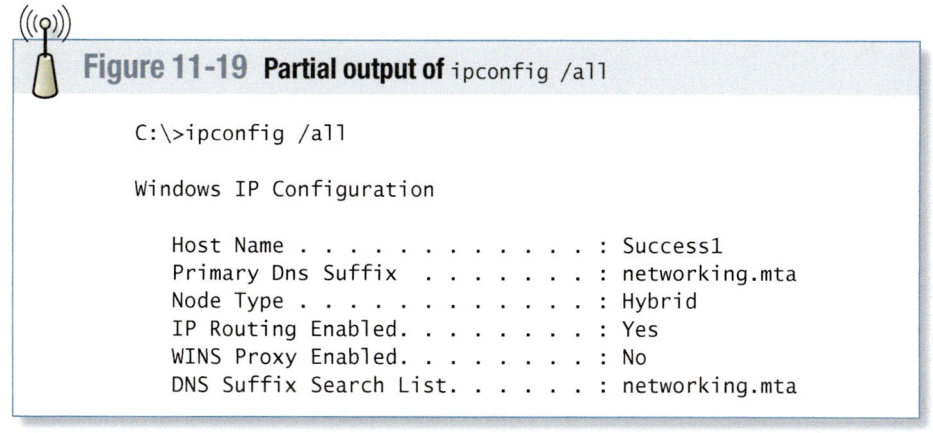

```
C:\>ipconfig /all

Windows IP Configuration

    Host Name . . . . . . . . . . . . : Success1
    Primary Dns Suffix  . . . . . . . : networking.mta
    Node Type . . . . . . . . . . . . : Hybrid
    IP Routing Enabled. . . . . . . . : Yes
    WINS Proxy Enabled. . . . . . . . : No
    DNS Suffix Search List. . . . . . : networking.mta
```

Figure 11-19. The node type hybrid indicates that the computer will use WINS by default.

Summary

The two types of computer names are hostnames and NetBIOS names. Computers on the Internet use hostnames. Internal networks use either hostnames or NetBIOS names. The primary name resolution method for hostnames is DNS. The primary name resolution method for NetBIOS names in Microsoft networks is WINS. Other name resolution methods include the host cache, the hosts file, the NetBIOS cache, the `lmhosts` cache, the broadcast, and the LLMNR.

Review Questions

1. What is the type of name used for computers on the Internet?
   a. DNS name
   b. Hostname
   c. NetBIOS name
   d. WINS name

2. The primary name resolution method for NetBIOS names is DNS.
   a. True
   b. False

3. Any entries in the Windows hosts file automatically appear in the host cache.
   a. True
   b. False

4. How can the host cache (or DNS resolver cache) be viewed?
   a. Enter `ipconfig /displaydns` at the command prompt
   b. Enter `ipconfig /flushdns` at the command prompt.
   c. Enter `nbtstat -c` at the command prompt.
   d. Enter `nbtstat -n` at the command prompt.

5. The Windows Internet Naming Service (WINS) operates on the Internet.
   a. True
   b. False

6. What command could be entered at the command prompt to remove DNS resolved entries from the host cache?
   a. Enter `ipconfig /displaydns` at the command prompt.
   b. Enter `ipconfig /flushdns` at the command prompt.
   c. Enter `nbtstat -c` at the command prompt.
   d. Enter `nbtstat -n` at the command prompt.

7. A system has an IPv6 address with a prefix of `fe80`. It does not have an IPv4 address. To resolve the computer name to an IP address, the system uses LLMNR.
   a. True
   b. False

Network Security

Security is an important consideration with any network. Some areas of a network are more vulnerable to attacks than other areas. This increased risk requires increased security. Different areas of a network are categorized in zones, with varying levels of security required in different zones.

The Internet is the riskiest zone. Internal networks (or intranets) are the safest. Between these two, it is possible to create perimeter networks as a buffer zone. One of the primary methods of separating the zones is by using firewalls. It is important to understand these different zones, as well firewalls in general, and Microsoft firewalls in particular.

OBJECTIVES

- ▶ Identify several risks on the Internet
- ▶ Describe how to implement security on an intranet
- ▶ List various methods that a firewall uses to protect a network
- ▶ Explain the security protection a perimeter network provides
- ▶ Describe the purpose of an extranet

CHAPTER 12

TABLE OF CONTENTS

Risks on the Internet 230
    Malware 230
    Botnets. 230

Intranets 231
    Network Address Translation (NAT) . . 231
    Proxy Servers. 233

Firewalls. 235
    Windows Server Firewall 236
    Early Windows Firewalls. 236
    Current Windows Firewalls 236

Perimeter Networks. 238
    Reverse Proxy Server. 241
    Guest Networks 241

Extranets 242

Summary 244

Review Questions 245

RISKS ON THE INTERNET

The Internet is the largest network in the world and continues to grow by leaps and bounds with no end in sight.

Since the internet is so large, network administrators must consider the security risks from all sources. To do this it is much easier to address network security zones.

The Internet Is the Riskiest Security Zone Attackers from anywhere in the world can attack computers on the Internet, and they do. A recent study found that in a ten month period malware (malicious software) authors created 20 million new strains of malicious software (an average of 63,000 a day). Infected systems can join massive botnets and participate in attacks on other computers.

All Internet Addresses Are Public Internet Protocol (IP) addresses used on the Internet are public IP addresses. In other words, they are accessible from any other computer with access to the Internet. In comparison, IP addresses on internal networks are private.

The Internet Is TCP/IP Based The TCP/IP protocol suite is the standard used on the Internet. Most internal networks use the same TCP/IP protocol suite for easy interaction on the Internet.

The World Wide Web (WWW) Travels Over the Internet The primary protocol used to transfer web pages is the Hypertext Transfer Protocol (HTTP). Note that the WWW itself is not the Internet. Rather, the WWW is like a truck delivering goods, and the Internet is the highway that the truck travels on. Other protocols traveling over the Internet include the File Transfer Protocol (FTP) and Simple Mail Transfer Protocol (SMTP).

Malware

Malicious software (malware) includes viruses, worms, Trojan horses, and other software designed with malicious intent. In the early days of computers, malware would often cause harm to a user's computer (such as destroying data or destroying a user's hard drive). Some were less malignant and simply popped up a message like "Legalize Marijuana" on a certain day.

However, malware has changed. Today, the primary purpose of most malware is to have a computer join a botnet.

Botnets

Botnet is short for "robot network," implying an automated network. *Zombie* is the term for an infected computer that becomes a member of a botnet. The terms clone and zombie are interchangeable. Botnets are networks of these clones or zombies that can be secretly controlled at will by the attackers. Attackers manage computers on the Internet with command-and-control software that can issue orders to them. These zombies check in periodically and do the bidding of the attacker. It is not unusual for the attackers to have almost as much control of the user's computer as the user does.

Zombies may send spam on behalf of the attackers, steal identities, or steal financial data. Zombies also participate in massive distributed denial of service (DDoS) attacks on the Internet. A *distributed denial of service (DDoS)* is a simultaneous attack on a single system or server by multiple attackers.

Any computer with access to the Internet (even computers within private networks) can become a zombie. Users are often unaware that their computers are infected as zombies. Indeed, this is one of the strengths of botnets. They do not harm the user's computer, but instead enlist it in their army. Today, it is not unusual for a botnet to have tens of thousands or even millions of zombies at its beck and call.

The best defense is antivirus software that is always on and regularly updated.

INTRANETS

An *intranet* is a private LAN that uses TCP/IP protocols, the same protocols found on the Internet, to share resources within the network.

From a network security perspective, the intranet is the safest network security zone. It includes clients on the internal network and has substantially fewer risks than computers placed directly on the Internet. Administrators control these computers and can implement many layers of security on them.

However, do not think that computers within an intranet are risk free. They are not. The only way to keep a computer free of risks is to leave it powered off.

Intranets have private IP addresses. The usable private IP address ranges are:

▶ 10.0.0.1 through 10.255.255.254
▶ 172.16.0.1 through 172.31.255.254
▶ 192.168.0.1 through 192.168.255.254

Recall that private IP addresses can only be used on internal networks and they are never used on the Internet. However, most users within intranets need to access the Internet. Since private IP addresses are used on intranets and public IP addresses are used on the Internet, networks need some method of connecting the two, such as NAT.

Network Address Translation (NAT)

Network Address Translation (NAT) is a service that translates private IP addresses to public IP addresses and translates public addresses back to private ones.

On a private intranet with connectivity to the Internet via a router that is running NAT, the router does basic routing, and the NAT service translates the private and public IP addresses. **See Figure 12-1.**

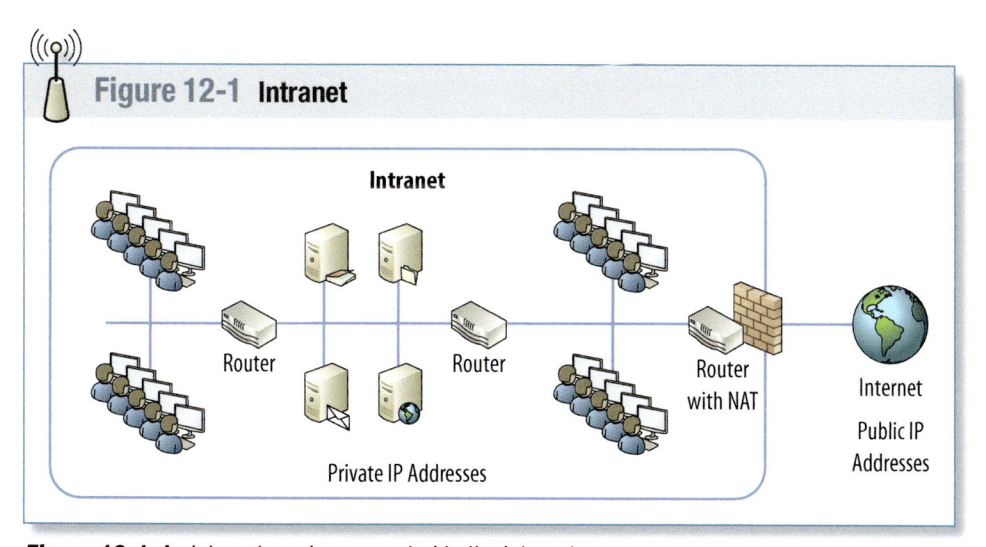

Figure 12-1. *An intranet can be connected to the Internet.*

All the computers on the intranet have private IP addresses, and the Internet has public IP addresses. The router with NAT has a private IP address assigned to the interface connected to the intranet, and a public IP address assigned to the interface connected to the Internet.

Port Address Translation (PAT) is a popular way that NAT is implemented. PAT is sometimes called Network Address Port Translation, or NAT Overload, but more often than not, it is simply called NAT.

Imagine that a user named Dawn on the intranet is trying to access `Bing.com` via the router. NAT takes the following actions:

1. The NAT router receives the request and logs the source IP address and port (from Dawn's computer) and logs the destination IP address and port (for `Bing.com`) in an internal table.

2. NAT then creates a new packet to forward the request to `Bing.com`. It keeps the destination IP and port but changes the source IP address to a public IP address. It also changes the source port to an unused port. At this point, the NAT table has only one entry. **See Figure 12-2.**

3. NAT sends the request to `Bing.com`. `Bing.com` returns the requested web page to the NAT server with the new NAT destination port (49212) included.

4. NAT looks at the destination port and compares it to its internal NAT table. It sees that the port is mapped to Dawn's computer with an IP address of 192.168.1.5. The router changes the destination IP and port and then sends the page back to her computer.

NAT creates its own source ports because it needs a way to identify the original requestor, and it does so by using a different source port for each one. Suppose that Jack was using `Bing.com` to search about feng shui at the same time Dawn was using `Bing.com` to search about firewalls. The NAT server would receive two answers from `Bing.com`. Without changing the source port for each request, there would not be any way for NAT to determine who should receive which response from `Bing.com`.

The NAT table now has two entries. Dawn has an IP address of 192.168.1.5, and Jack's computer has an IP address of 192.168.1.22. NAT creates different source ports for each request in the internal NAT table. When `Bing.com` returns the data on firewalls requested by Dawn, it includes the source port created by NAT. NAT then uses this information to ensure that the request is forwarded back to Dawn's computer. **See Figure 12-3.**

Most systems generate source ports from the dynamic port range of 49,152 to 65,535. Only ports that are not currently being used are selected.

✓ **FACT**

The NAT table is stored in the system's memory. If it is a router running NAT, it is stored in the router's memory. If it is a proxy server, it is stored in the server memory.

Figure 12-2 NAT Table

SOURCE IP	SOURCE PORT	DESTINATION IP	DESTINATION PORT	NAT SOURCE PORT
192.168.1.5	49155	Bing.com	80	49212

Figure 12-2. *PAT is a version of NAT that changes the source port address.*

Figure 12-3 NAT Table with Two Entries

SOURCE IP	SOURCE PORT	DESTINATION IP	DESTINATION PORT	NAT SOURCE PORT
192.168.1.5	49155	Bing.com	80	49212
192.168.1.22	49158	Bing.com	80	49213

Figure 12-3. A router running PAT can use a single outside address for many inside addresses.

NAT provides the many benefits:

Hides Internal Computers Since the computers do not have public IP addresses, they cannot be directly accessed by Internet sources.

Reduces Costs If a company does not use NAT, it would have to purchase public IP addresses for all its internal computers. This is an unnecessary expense since it is so easy to install NAT.

Extends the Lifetime of IPv4 Since companies can use a single public IP address for hundreds or thousands of internal computers, the public IPv4 address range was not depleted earlier.

Although NAT can use a single public IP address, it is also possible to use multiple public IP addresses. Consider a large network with thousands of users. A single connection to the Internet may not be enough to adequately serve all of these clients. Instead, additional connections can be added with additional public IP addresses.

Static NAT uses a single public IP address for each private address, and all connections for a private address are mapped to a single IP address. Dynamic NAT typically uses a two or more public IP addresses. Any user's request from a private IP address can be dynamically mapped to any one of the public IP addresses. One benefit of dynamic NAT is that it is able to balance the load among the different public IP addresses.

Proxy Servers

Instead of just using NAT, many organizations use proxy servers. A *proxy server* acts on behalf of the client computers on the internal network to retrieve web content from the Internet. A proxy server often includes NAT, but it does more.

Proxy servers provide three important benefits:

Caching If one user requests a page from a site, the proxy server will retrieve the page and return it to the user. It also keeps a copy of the page in its local memory, or cache. If another user then requests the same page, the proxy server retrieves the page from memory and serves it to the second user. This saves time and Internet bandwidth since the same content does not have to be retrieved repeatedly.

Filtering The proxy server can use filtering lists to restrict access to certain websites. For example, if an organization wants to ensure that employees do not access gambling sites, a filter list can contain these sites, and the proxy server will then block all access to them.

Content Checking Some proxy servers can verify that the content is valid. For

✓ **FACT**

A proxy server is not a replacement for antivirus software within a company. However, it is useful as part of a defense-in-depth security strategy.

example, the proxy server can check web pages for malicious content, such as embedded malware or malicious scripts. If the web server includes a certificate for secure HTTPS pages, the proxy server can check the certificate's validity.

In many cases, a proxy server lies between the Internet and the other computers in the intranet. This is a common configuration for many mid-sized and large organizations. **See Figure 12-4.**

The proxy server retrieves any requests that are allowed and blocks requests for pages identified in its block list.

For example, if a client wanted to access a web page on the Internet:

1. The client computer forwards the request to the proxy server.
2. The proxy server checks its internal filter.
    a. If the page is on a block list, the request is not filled. Instead, the user will usually see a web page indicating that accessing the page is against company policy.
    b. If the page is allowed, the web server will attempt to retrieve it from the Internet. It often uses the same NAT process.
3. When the web page is received, the proxy server checks the content to ensure it is valid. Suspect content can be blocked with a warning to the user that the page is suspect.

Proxy Server Filters

Some companies sell subscriptions to filter lists. These companies have web bots that constantly crawl the Web to identify content. The content is categorized, and the web pages are then added to specific lists. For example, one list might be for gambling and include all known gambling sites. Another list might be for pornography and include all known pornography sites.

Organizations can then subscribe to the different lists. These lists are added to the proxy server, and any requests to access a site on a list are blocked.

Some organizations are more proactive and create lists of only acceptable websites. If a user tries to access any website that is not on this list, access is blocked.

Figure 12-4 Proxy Server

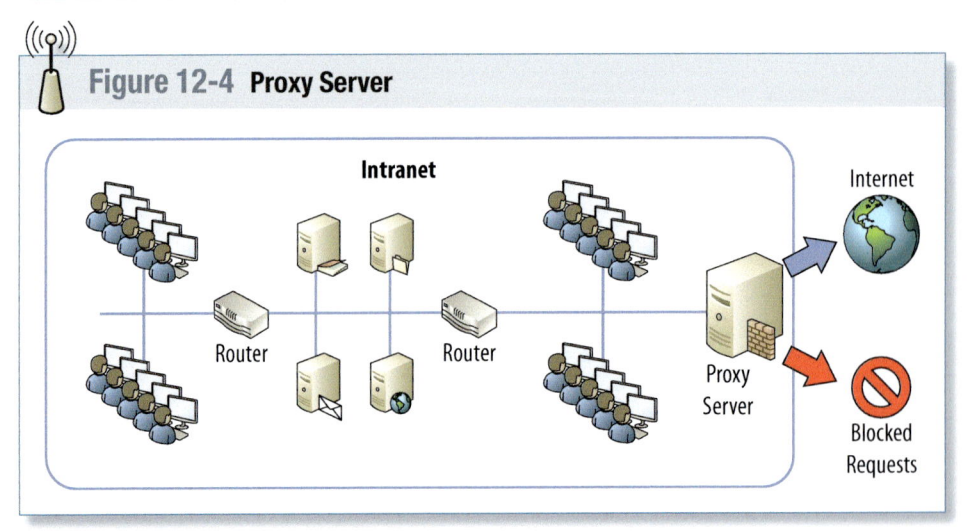

Figure 12-4. *An intranet can be connected to the Internet via a proxy server.*

4. The proxy server places valid web pages in its cache. Pages in the cache are served to other users from the cache without retrieving them from the Internet again.

5. The web page is sent to the client that originally requested it.

Client computers must be configured to use the proxy server. For example, on Windows computers, the proxy server settings are found in the Internet Options dialog box. **See Figure 12-5.**

The IP address of the proxy server is 192.168.1.251, and it is listening on port 8080.

An additional setting is "Bypass Proxy Server for Local Addresses." This ensures that requests to web servers on the internal network (the intranet) do not have to go through the proxy server.

FIREWALLS

A firewall provides protection to both networks and individual systems by controlling the traffic that can flow in or out. A host-based firewall controls the traffic on an individual host or computer. A network-based firewall controls the traffic on a network.

Firewalls have been improved over the years. The most basic firewall is simply a router with rules that define what traffic is allowed and what traffic is blocked. This is also known as a packet-filtering firewall.

Packet-Filtering Firewall A *packet-filtering firewall* filters packets based on IP addresses, ports, and some protocols. For example, to allow only HTTP traffic (which uses port 80), create a rule to allow incoming traffic on port 80. To allow traffic through a firewall from specific computers, it is possible

✓ FACT
Administrators can set the proxy server settings manually or automate the settings.

Figure 12-5 Options Dialog

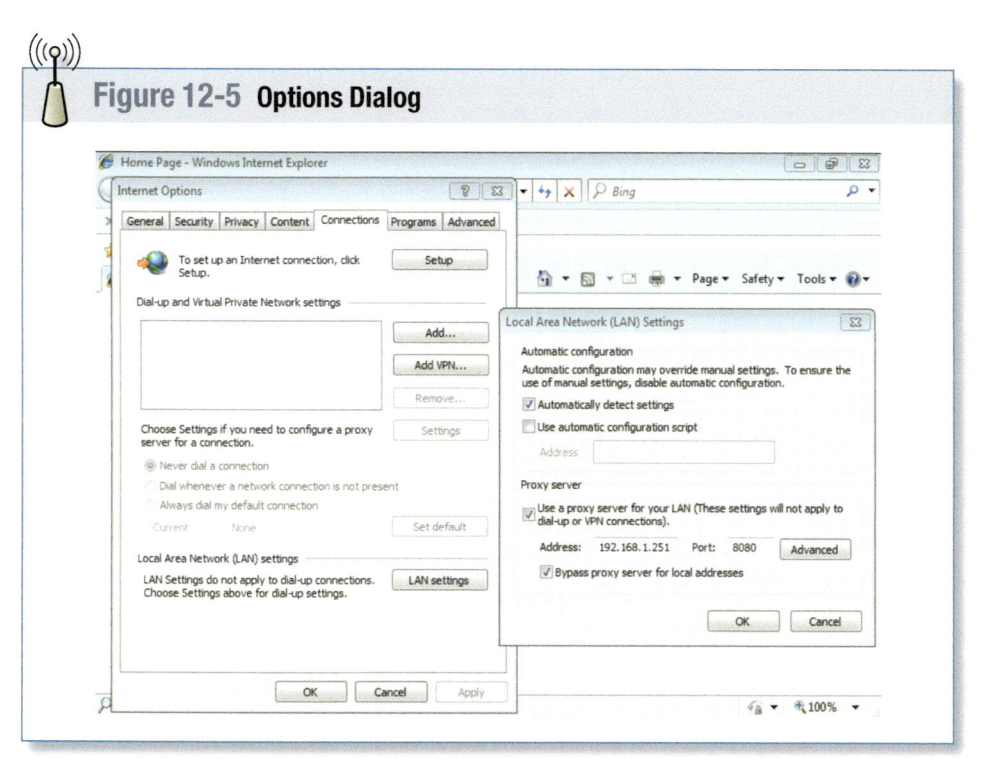

Figure 12-5. The Internet Options dialog box contains controls to configure proxy server settings.

to create rules based on their IP addresses.

Stateful Filtering *Stateful filtering* is a firewall configuration where traffic is filtered based on the state of the network connections. The firewall can examine packets in different transactions and make decisions based on connection states. Both Transmission Control Protocol (TCP) and User Datagram Protocol (UDP) traffic is analyzed. If traffic is not part of a known connection, it is blocked.

Content Filtering Some firewalls can block traffic based on the content. For example, malware is often delivered via spam embedded as a ZIP file and other types of attachments. Content filtering is often performed on e-mail servers also, to filter spam and its attachments.

Application Layer Filtering Traffic is filtered based on an application or service. The firewall has a separate component for each application protocol (such as HTTP or FTP) that it filters. These firewall components examine the traffic using that protocol to allow and block certain types of traffic. For example, HTTP `Get` commands (which enable the retrieval of documents or files) could be allowed, while `Put` commands (which post documents or files) can be blocked. In practice, application layer filters are CPU-intensive and used sparingly.

Most firewalls use an implicit deny policy. An *implicit deny policy* specifies that all traffic that has not been explicitly allowed is blocked. As an example, consider a partial listing of programs and their Windows Firewall settings on a Windows Server system. **See Figure 12-6.**

Each item that is checked is explicitly allowed. If an item is not selected, it is blocked.

Figure 12-6 Allowed Programs

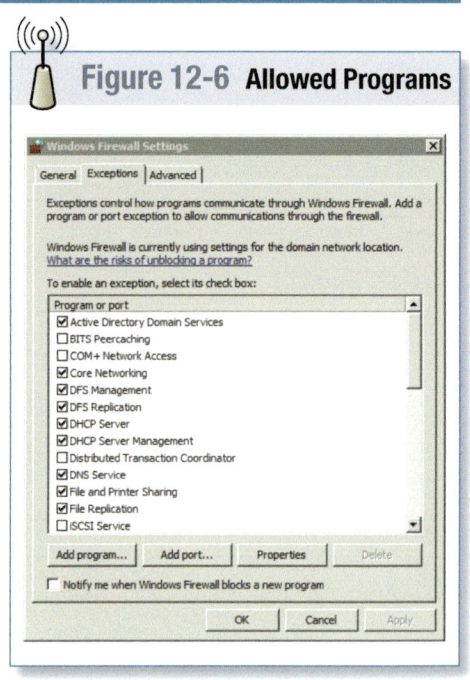

Figure 12-6. *Certain programs are allowed through the Windows Firewall.*

Windows Server Firewall

Today's Windows operating systems have the Windows Firewall built in as a host-based firewall. Following Microsoft's principle of security by default, the Windows Firewall is enabled by default.

Early Windows Firewalls

Early versions of Windows Firewall could only support rules to control inbound packets. However, since the release of Windows Vista and Windows Server 2008, it is possible to control both inbound and outbound traffic.

Current Windows Firewalls

Another feature of Windows Firewall in current Windows operating systems is the use of different rules based on where the computer is operating. For example, a Windows computer could be running on a home network, on a corporate

> **Network Discovery in Windows**
>
> Windows systems allow computers to discover each other. When network discovery is enabled, a user's computer can discover other computers on the network, and other computers can discover the user's computer. When network discovery is disabled, it prevents other computers from seeing the user's computer.
>
> Network discovery does not prevent connections. For example, a computer with network discovery disabled on a public wireless network will still be able to access the Internet by going through a known wireless router. However, network discovery does enable specific firewall rules, which makes it more difficult for other computers to discover a Windows computer running in a public network.

domain network, or on a public wireless network such as that of a coffee shop or airport. Each of these network locations has different levels of risk. Windows sometimes automatically detects this network location. At other times, the user identifies it during the first connection. Either way, Windows implements firewall rules to increase or decrease security based on the network location settings.

The different network locations are:

Public   This is a public location such as in a coffee shop or airport. Users often connect via wireless connections, and other users are completely unknown. The other users could be friendly or malicious. Attackers can try to hack into systems in a public network to steal data. Since a public network is the riskiest network location, the Windows Firewall provides the highest level of protection and helps prevent computers from being discovered on the network. Network discovery is disabled.

Private (Home/Work)   This indicates a small, protected network where other devices on the network are known and trusted, so network discovery is enabled. Users on a private network can join a homegroup. A *homegroup* is a special type of workgroup in new Windows operating systems that facilitates sharing files between computers in a home or small office network.

Domain   Computers that are joined to a domain are automatically configured for a domain network location. Administrators control these settings using domain tools.

There are two basic graphical user interfaces (GUIs) that administrators can use to manipulate the firewall in Windows Server. The basic GUI is in the Control Panel, and the second tool is the Windows Firewall with Advanced Security console, which is located in the Administrative Tools program group.

In the Control Panel view of the firewall in a Windows Server system, notice that the connection status for the domain networks is Connected. This indicates the computer is joined to a domain and that the firewall is using the settings for a domain. The firewall is On, and it is configured to block all incoming connections that have not been explicitly allowed. **See Figure 12-7.**

In the Windows Firewall with Advanced Security console in Windows Server, with the New Inbound Rule Wizard started, the inbound rules are displayed, and all the rules with a green circle are enabled to allow the traffic. The ones that are grayed out are not enabled. Note in the left pane that there are also outbound rules and connection security rules that can be manipulated. **See Figure 12-8.**

Although Windows Server firewalls include many built-in rules, administrators can also add their own rules. To launch the New Inbound Rule Wizard, click New Rule in the Actions pane.

PERIMETER NETWORKS

A *perimeter network* is an area between the Internet and an intranet that hosts servers accessible from the Internet. It provides a layer of security protection for these Internet-facing servers and isolates these servers from the internal network.

Internet-facing servers are any servers accessible from the Internet. They include web servers, mail servers, FTP servers, and more.

For example, on a perimeter network hosting a web server and a mail server, the perimeter network is between two firewalls. This is a common configuration, but there are others. **See Figure 12-9.**

> ✓ **FACT**
> A perimeter network is often called a demilitarized zone (DMZ) or a buffer zone. This is especially true when it is created with two firewalls.

> ✓ **FACT**
> Connection Security Rules use Internet Protocol Security (IPsec). IPsec can encrypt data traveling on the wire.

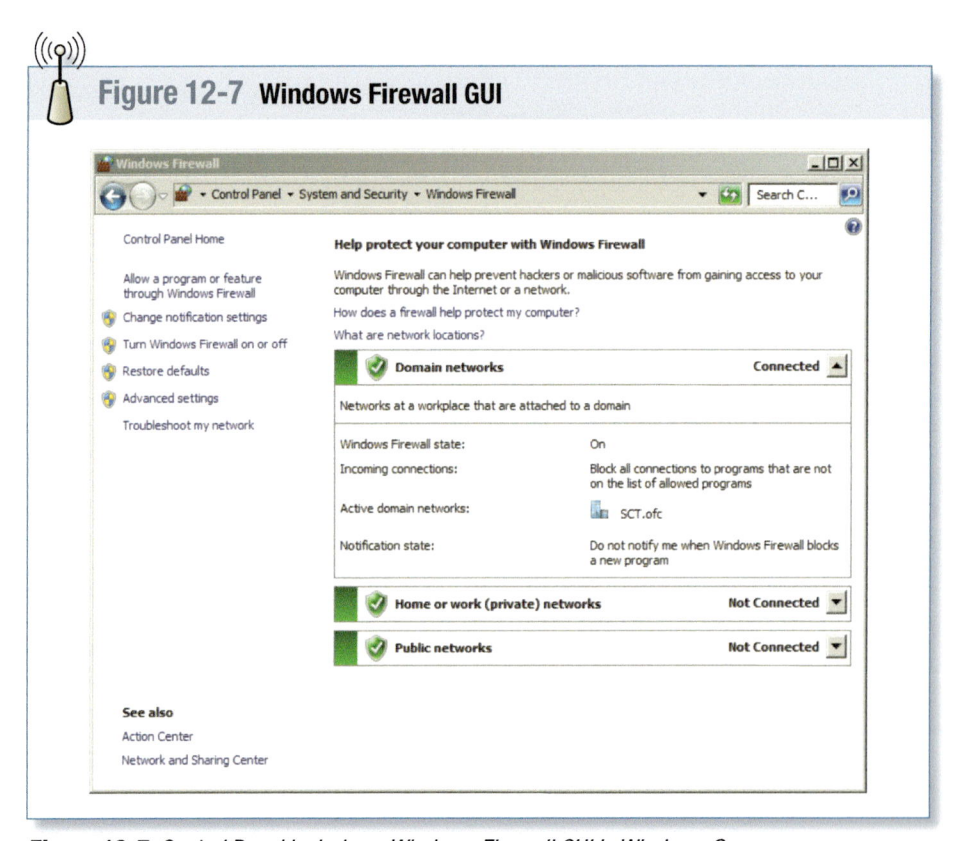

Figure 12-7. Control Panel includes a Windows Firewall GUI in Windows Server.

Figure 12-8 Windows Firewall with Advanced Security Console

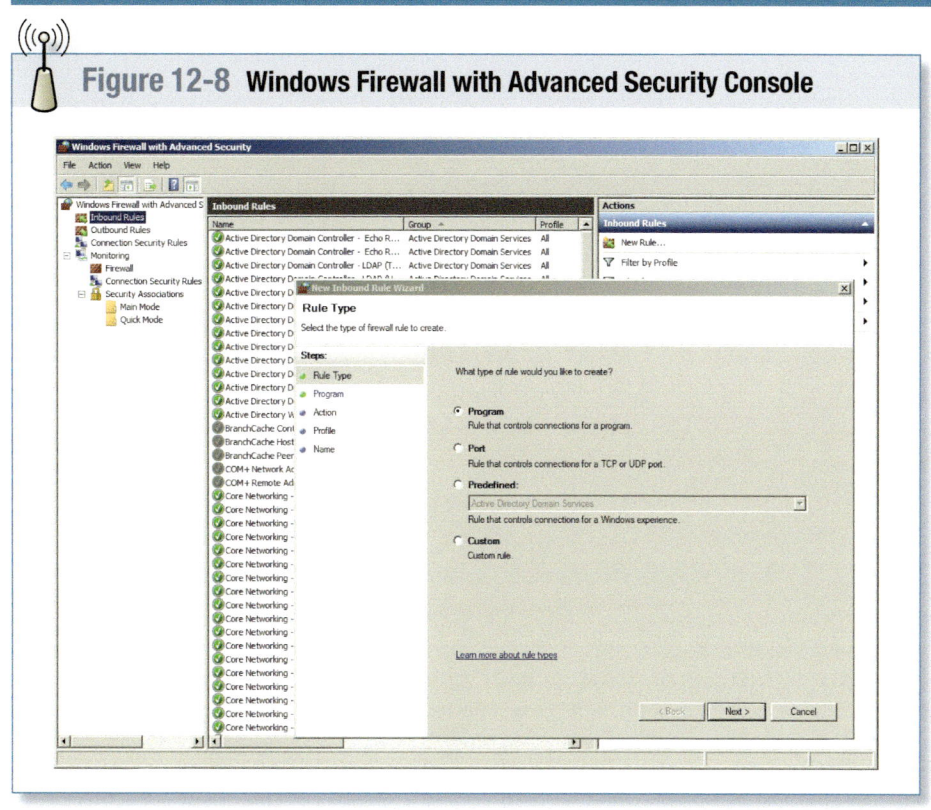

Figure 12-8. *The Windows Firewall with Advanced Security console in Windows Server provides more comprehensive control.*

Figure 12-9 Perimeter Network

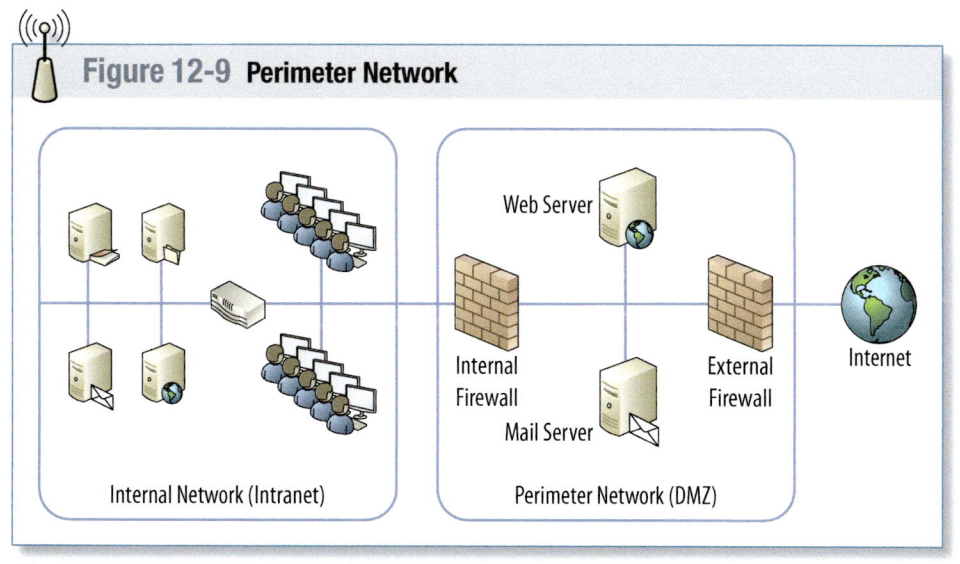

Figure 12-9. *A perimeter network is the area between two firewalls.*

An important point to realize about the perimeter network is that servers placed here are accessible from anywhere on the Internet, by anyone who has access to the Internet. However, the perimeter network does provide protection.

As an example, consider the web server. A typical web server serves web pages using HTTP on port 80 and HTTPS on port 443. The external firewall will filter traffic to this web server and can block all traffic that is not using either port 80 or port 443. This can prevent many potential attacks from reaching the server.

From a risk perspective, the perimeter network is a little safer than the Internet. However, since servers in the perimeter network are still accessible from anywhere on the Internet, there is still a significant amount of risk, especially when compared to the intranet. Additionally, if a server on the perimeter network is compromised, the internal firewall will protect resources on the intranet.

It is also possible to create a perimeter network with a single firewall. In this type of perimeter network, the mail server and web server are still isolated from both the Internet and the intranet. The firewall controls which packets can reach the perimeter network and what data can reach the intranet. **See Figure 12-10.**

Although this configuration is less expensive (since it uses only a single firewall), it is also much more complicated to configure. An administrator must configure rules to route traffic to specific network adapters. Since these rules are more complex than the rules for two firewalls, there is a greater chance of error.

A significant benefit of a two-firewall perimeter network is that it can use firewall products from two separate vendors. Although vulnerabilities might occur in any system, it is unlikely that

Figure 12-10 Perimeter Network Firewall

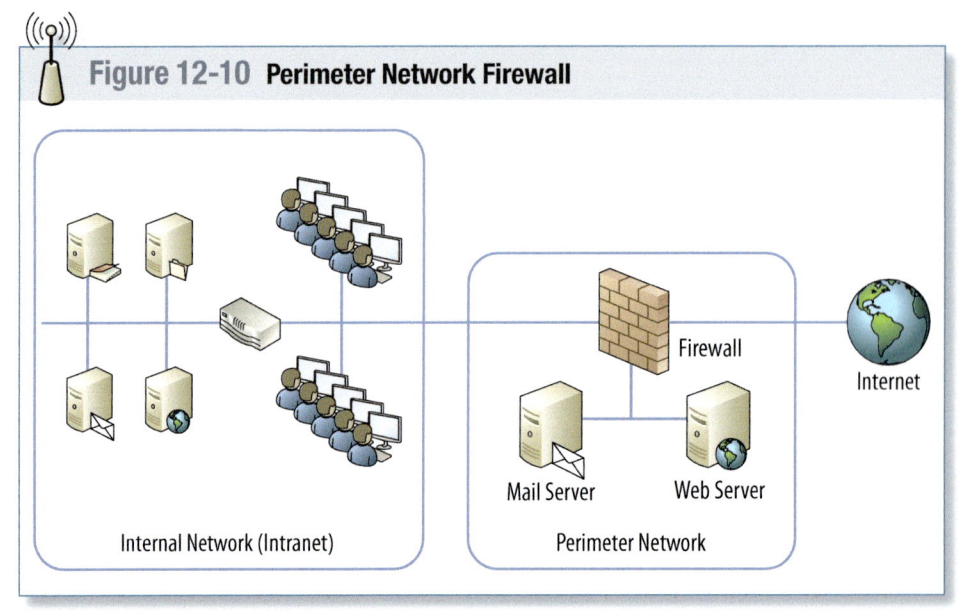

Figure 12-10. A perimeter network can use a single firewall.

both firewalls will be vulnerable at the same time. Also, although an attacker might be an expert on one of the firewalls, it is less likely that an attacker will be an expert on both at the same time.

Reverse Proxy Server

Some organizations implement reverse proxy servers to increase security and performance of web servers. A *reverse proxy server* is an additional server on the perimeter network that isolates the web servers from direct access by systems on the Internet, providing a layer of protection from Internet attackers.

When a reverse proxy server is used with a web server, the reverse proxy server receives the requests from the clients and forwards them to the web server. The web server sends the web pages back to the proxy server, and the proxy server sends them to the clients. **See Figure 12-11.**

Just as a regular proxy server can cache requests, so can the reverse proxy server. This reduces some of the load on the web server. However, since all the pages are still served over the Internet link, it does not reduce Internet usage.

Clients do not need to be configured to use a reverse proxy. Indeed, clients rarely ever know when a reverse proxy is in use, because it is transparent to the end users.

Guest Networks

Guest networks are another type of perimeter network used by larger organizations. A *guest network* is an isolated portion of the internal network that can be used by guests or visitors.

Depending on how the guest network is configured, visitors might not need to provide any credentials to access the guest network. However, their access on the network is usually

Figure 12-11 Reverse Proxy Server

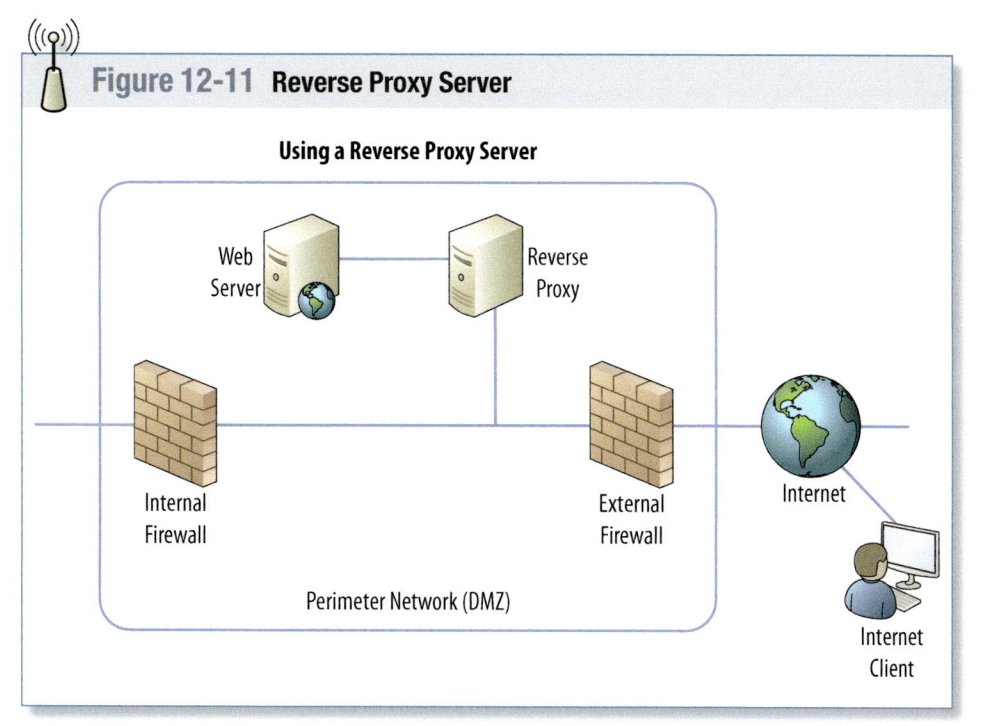

Figure 12-11. A reverse proxy server protects web servers from Internet intruders.

limited. The primary form of access that is usually granted on a guest network is Internet access.

Guest networks are also becoming popular in home wireless networks. For example, Cisco's Valet Wireless Router enables a user to create a separate password to give to visitors without compromising the primary password that is used for other connections. When the visitor leaves, the user can change the visitor password or disable visitor access. **See Figure 12-12.**

EXTRANETS

An *extranet* is an area between the Internet and an intranet that hosts resources for trusted entities. These resources are available via the Internet. An extranet is often physically the same as a perimeter network. The difference is in the intent and scope of access and the resources that are available. Specifically, an extranet is configured so that only trusted partners or customers have access to a company's resources in the extranet. These trusted partners typically need access to areas of a company's network such as private websites or databases that would not be accessible publically. This enables the company to extend access to some of its internal resources to trusted entities outside the intranet.

The difference between an extranet and a perimeter network is based on the intent. The perimeter network hosts servers that are accessible to any Internet clients from anywhere on the Internet. Extranets are available only to specific clients. **See Figure 12-13.**

For example, a boating parts company sells and ships parts to boat builders. The parts company might want some customers to be able to access their accounts, check availability of parts, place orders, and track status. To

✓ **FACT**
Extranets are often created to share data between two companies that have business relationships or partnerships.

Figure 12-12 Guest Network Interface

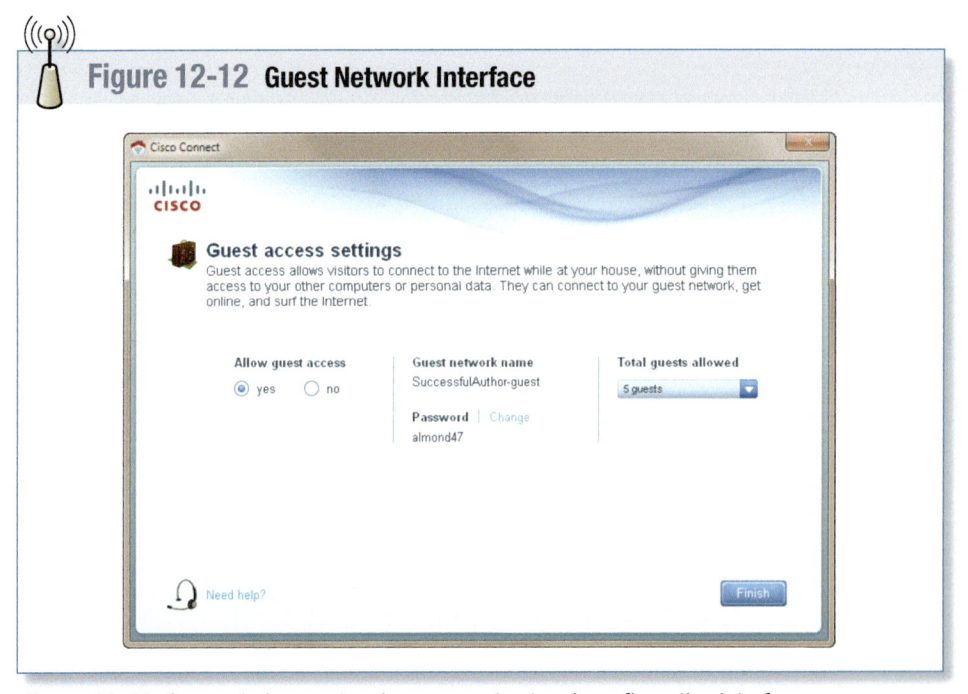

Figure 12-12. Some wireless routers have a guest network configuration interface.

do these things, it can add a web server to an extranet and restrict access to specific customers.

By allowing access to only their web server via an extranet, the company can control who is granted access to the extranet's website. This prevents unwanted users (such as competitors) from viewing information the company does not want to make public.

Figure 12-13 Extranet

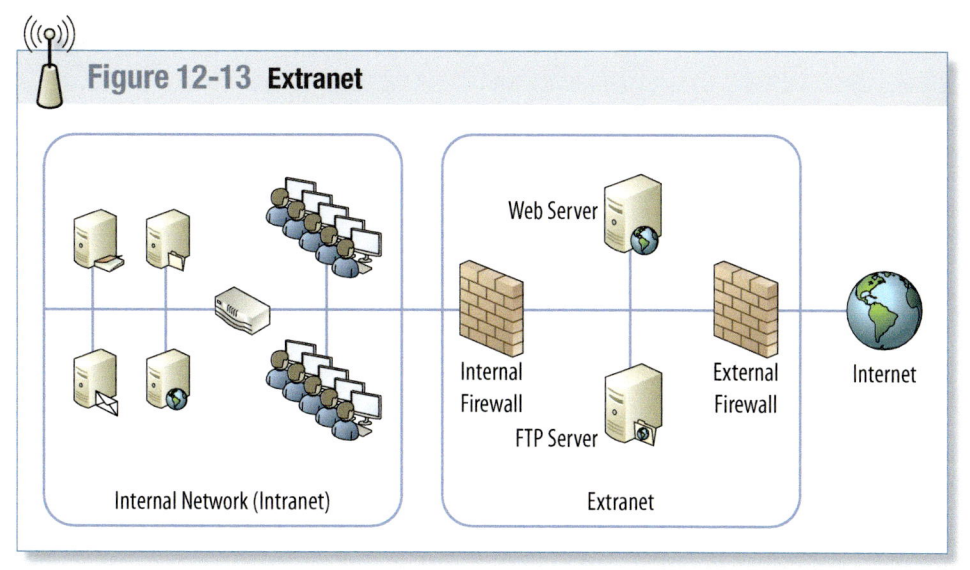

Figure 12-13. An extranet provides internal network access to specific Internet clients.

Summary

The Internet is the riskiest security zone. Any resources placed directly on the Internet are accessible from anywhere in the world and are subject to attack from anywhere in the world, as long as the attacker has access to the Internet. The intranet is an internal network and is considered the safest zone when compared to other zones. Firewalls typically separate the intranet from the Internet. Microsoft desktop and server operating systems include host-based firewalls built into the operating system. A perimeter network (also known as a DMZ) usually includes two network-based firewalls, and Internet-facing servers are placed between the two firewalls. The firewalls control traffic to and from resources in the perimeter network. Extranets are perimeter networks created to provide access to internal resources to specific trusted entities. Guest networks are perimeter networks created to provide temporary network access to visitors.

Review Questions

1. Which network security zone represents the highest risk?
   a. Extranet
   b. Internet
   c. Intranet
   d. Perimeter network

2. A proxy server translates private IP addresses to public IP addresses and translates public IP address back to private ones.
   a. True
   b. False

3. An organization wants to restrict which web pages employees can access on the Internet using company computers. What should be implemented?
   a. Firewall
   b. NAT
   c. Proxy server
   d. Reverse proxy server

4. A DMZ provides a layer of security for Internet-facing servers.
   a. True
   b. False

5. How many firewalls are used to create a perimeter network?
   a. 1
   b. 2
   c. 3
   d. 4

6. What allows computers to locate each other in a Microsoft network?
   a. Firewall
   b. Network discovery
   c. Proxy server
   d. Public network location

7. A guest network provides access to some internal resources to a business partner via the Internet. No one else receives access.
   a. True
   b. False

Wide Area Network Connectivity

Homeowners, home offices, and small offices may connect to the Internet by utilizing one of multiple methods. These range from the most basic dial-up connections to the popular broadband cable, but there are other types of connections available. It is important to understand Internet access methods and be able to identify their characteristics, such as speed and availability.

Enterprises use these and other connectivity methods to access the Internet and to connect offices via wide area network (WAN) links. WANs connect remote offices together, even when they are a significant distance apart. Some common methods used to create WANs are Integrated Services Digital Network (ISDN) connections, T1 and T3 lines in the United States, and E1 and E3 lines in Europe.

OBJECTIVES

- ▶ List various connectivity methods used in homes and SOHOs
- ▶ Identify connectivity methods used in enterprises
- ▶ Explain the primary methods for connecting to a Remote Access Service server
- ▶ Describe a RADIUS server and its purpose

CHAPTER 13

TABLE OF CONTENTS

Connectivity Methods Used
in Homes and SOHOs 248
    Dial-Up Connection 248
    DSL. 248
    Broadband Cable. 250
    Satellite. 250

Connectivity Methods in Enterprises . . . 252
    Digital Signal Lines. 253
    ISDN. 253
    T1/T3 Lines and E1/E3 Lines 254
    Ethernet WAN. 254

Remote Access Services 255
    RAS via Dial-Up 256
    RAS via a VPN 256
    Client VPNs Versus Gateway VPNs . . . 258

RADIUS 259

Summary 260

Review Questions 261

> ✓ FACT
> In rural areas, dial-up is often the only reliable method available to access the Internet.

CONNECTIVITY METHODS USED IN HOMES AND SOHOS

Users in homes and small or home offices (SOHOs) can connect to the Internet using a wide variety of methods, depending on what is available in their area. **See Figure 13-1.**

Worldwide Interoperability for Microwave Access (WiMAX) is a technology available in some cities. It provides a wireless alternative for broadband cable and DSL and gets speeds up to 40 megabits per second (Mbps).

It is possible to provide Internet sharing for each of these methods. A single router or computer connects to the Internet, and the router shares this connection among all the computers on the network. This is a little more challenging with dial-up because it is slow to start with, but it is very common with the other methods.

Dial-Up Connection

Dial-up connectivity methods use the plain old telephone service (POTS). *Plain old telephone service (POTS)* works by connecting a modem to a telephone line and then connecting to the remote network by dialing out through the phone line. To use the dial-up method, users connect to the Internet via an Internet Service Provider (ISP) or directly to a remote access server for connectivity to a company's internal network.

Compared to other technologies used today, dial-up is very slow. It has a theoretical maximum speed of 56 kilobits per second (Kbps), but typical speeds are less than 50 Kbps. However, dial-up is much cheaper than other methods, and it is available anywhere phones are available

DSL

Digital Subscriber Line (DSL) connections are popular in some urban areas. A *Digital Subscriber Line (DSL)* uses telephone lines but sends the data digitally instead of using an analog signal. One benefit is that it is possible to use the telephone for voice calls at the same time as using the telephone line for DSL access.

In a DSL installation, a user connects a router (or a single computer) to the DSL modem. A DSL splitter splits the lines so that the phone line can be used at the same time as the DSL modem. The signal travels over the telephone lines to a telephone company's central office, which then connects to the Internet. **See Figure 13-2.**

Figure 13-1 Home and SOHO Connectivity to Internet

METHOD	AVAILABILITY
Dial-up	Anywhere phone lines exist
Broadband cable	Urban/suburban areas that have cable TV
DSL	Must be close to a telephone company central office (generally within two miles)
Satellite	Widely available, but requires unobstructed view to satellite

Figure 13-1. *Internet connection options may vary from satellite in remote areas to high speed broadband connections in populated areas.*

Figure 13-2 DSL

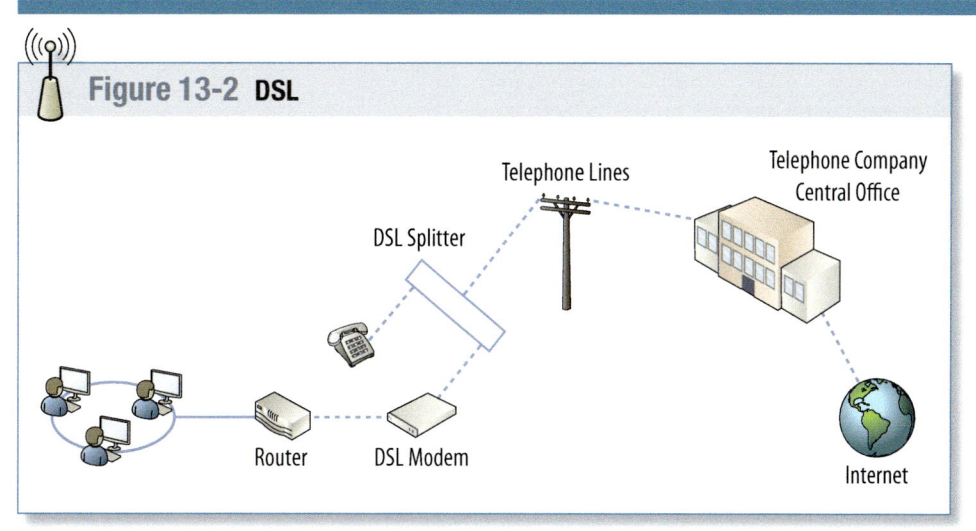

Figure 13-2. DSL provides high-speed digital access using standard telephone lines.

One limitation of DSL is that the end user must be relatively close to the central office. Improvements in recent years have extended this distance, but in general, the DSL modem must be located within two miles of the central office. The digital signal cannot travel as far on the telephone lines as an analog voice signal. Users who are closer to the central office can get higher speeds on the connection.

A DSL installation requires the addition of a DSL filter to the line, to prevent noise on the POTS line from interfering with the DSL signals.

There are many variations of DSL, which are commonly referred to as xDSL. These variations use different technologies to improve the performance. They include asymmetric DSL (ADSL), symmetric DSL (SDSL), and very-high-bit-rate DSL (VDSL). Many businesses use one of these types in what is commonly called business-class DSL.

✓ **FACT**

Compared to dial-up speeds, the DSL speed provides significant improvements.

✓ **FACT**

A central office is a nearby telephone exchange switch building that processes the digital signals.

Modems Convert Digital and Analog Signals

A modem modulates and demodulates a signal. Traditional phone lines can transmit analog signals, but they cannot transmit digital signals. Computers can understand digital data (ones and zeros), but they cannot understand analog signals.

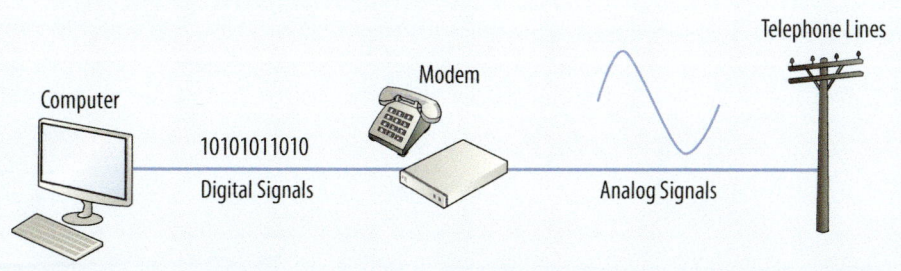

The modem modulates the analog signal to create the digital signal. In other words, it uses a specific frequency sine wave and adds variations on the analog signal to represent the digital signal. When the modem receives signals, it demodulates the signal by removing the digital data from the analog signal. It then sends the digital data to the computer.

Broadband Cable

Broadband cable provides Internet access through the same cable that provides cable TV. It is called broadband cable because the same cable transmits multiple signals across a broad frequency spectrum.

A user does not need a separate cable for every TV channel. Instead, the cable delivers all of the channels, and the user only needs to tune the TV to view the desired channel. The TV tuner strips out all the other TV channels. Similarly, the user can add a cable modem to strip out all the TV channels and leave only the Internet access signal.

The basic connectivity for broadband cable consists of a router (or a single computer) connected to the cable modem. A cable splitter splits the signal between the cable modem and the TV, with TV signals going to the TV and the Internet signal going to the cable modem. The cable company hosts ISP servers for Internet access. **See Figure 13-3.**

Speeds of broadband cable vary greatly. Although providers often advertise specific speeds, users rarely achieve them. One reason for this is that users share bandwidth with their neighbors. If several neighbors are uploading or downloading data at the same time, all the connections will be slower than if only one user was using the connection.

Some ISPs offer tiered services. For example, by paying the minimum price, a user can get speeds up to 3 Mbps. By paying a little more, the user can get up to 15 Mbps. Or the user can pay top dollar to get the top speed.

Another addition in broadband is fiber. This is similar to broadband cable, but the fiber-optic lines can carry significantly more data than traditional cable (up to 1 Gbps in some cases). This increases the capability for all users sharing the same line.

Satellite

The use of satellite Internet access is expanding, especially in rural areas where only dial-up access is available. A satellite transceiver can upload and download data from a satellite, and the satellite provides Internet access via an ISP. The satellite modem connects the transceiver to a router for a group of computers, or it can also connect to a single computer for a home user. **See Figure 13-4.**

Speeds vary wildly, in both advertised speeds and the actual speeds that users

> **✓ FACT**
> ISPs often cap speeds with bandwidth-throttling techniques. This prevents a single user from consuming bandwidth at the expense of other users.

> **✓ FACT**
> Originally, satellite access was download only and required a phone line (called terrestrial transmit) to upload data. Today, most transceivers include direct upload capability.

Figure 13-3 Broadband Cable

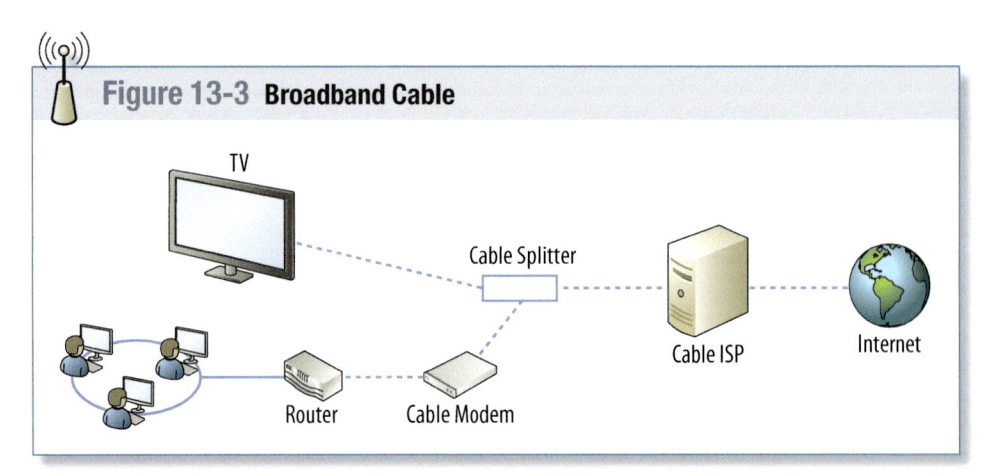

Figure 13-3. Broadband cable carries TV and Internet signals on the same cable.

Figure 13-4 Satellite Internet Access

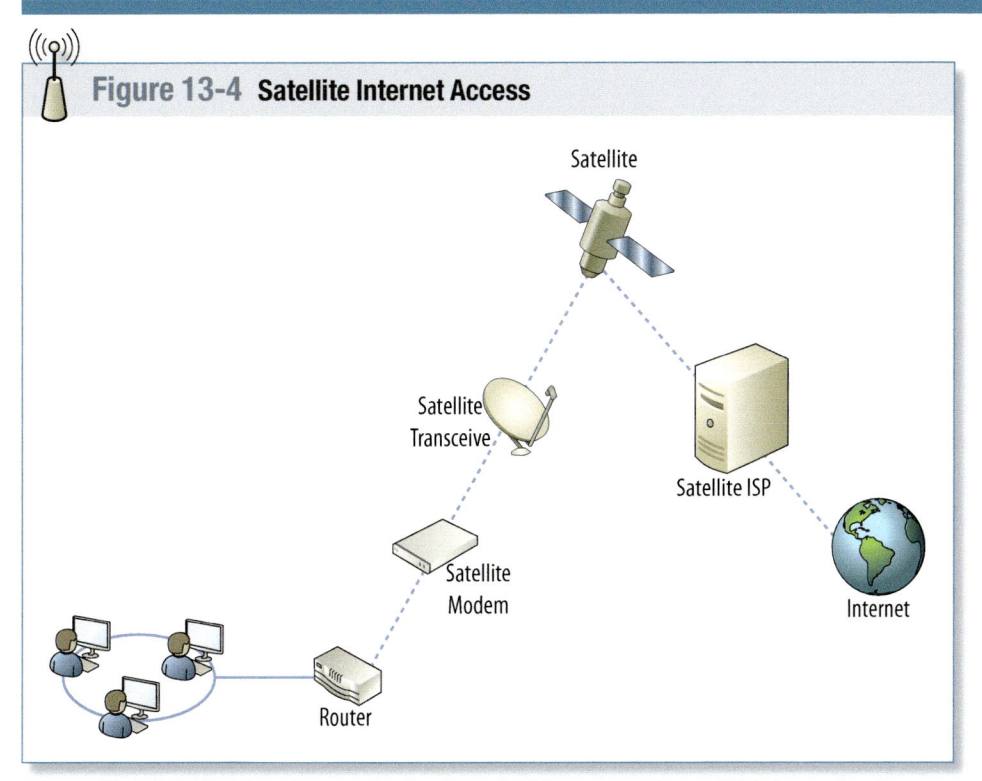

Figure 13-4. Satellite Internet access is often the only high-speed option available for rural customers.

report. Upload or uplink speeds (from the transceiver to the satellite) typically have a maximum advertised speed of 10 Mbps, although the actual uplink speeds are usually closer to 256 Kbps.

Download or downlink speeds (from the satellite to the transceiver) sometimes have maximum advertised speeds as high as 1,000 Mbps (1 Gbps). However, these quick speeds are usually available only for users willing to pay a steep premium.

Many satellites are in geostationary orbits. A geostationary orbit always appears to be in the same location from any point on the earth, even though the earth and the satellite are zipping through space at phenomenal speeds. As long as the site has a clear, unobstructed view of the satellite, satellite Internet access is possible.

One drawback to satellites is that they are susceptible to interruptions because of precipitation. Rain and snow can interfere with the signal. This can reduce speeds or completely block the signal. Another drawback is latency because of the amount of time it takes for signals to travel back and forth to the satellite. For example, when a user clicks the mouse, there are typically four trips

Bringing Internet to the Heartland

Several companies are working on bringing Internet access to rural areas using wireless technologies.

These companies have towers that send and receive the signals, and as long as users are close enough to a tower, they can connect. This business continues to grow. As more towers are added and their capabilities are increased, more and more people enjoying life in the country will be able to have broadband access similar to that of users in the city.

✓ FACT

Compared to dial-up speeds, even the slowest satellite links provide great improvement for users in rural areas.

to and from the satellite orbiting the earth before the user gets data. This can make web surfing slower, prevent some applications (such as online gaming or real-time streaming) from working, and sometimes prevent users from accessing a virtual private network (VPN).

CONNECTIVITY METHODS IN ENTERPRISES

Enterprises can connect to the Internet through any of the same methods used by homes and SOHOs. However, a bigger concern for enterprises is connecting their offices to each other.

A wide area network (WAN) is two or more local area networks (LANs) in separate geographical locations that are connected together. WAN links connect the LANs together.

For example, a WAN connecting five locations together might include a company's primary location, a regional location, and three branch offices all connected together with four WAN links. **See Figure 13-5.**

The WAN links are usually slower than the speed of the LAN. The main reason for this is that LANs are wholly owned by the organization, while WAN links often go through a service provider that charges according to the bandwidth they provide. For example, LAN speeds are often 100 or 1,000 Mbps today, but the cost of WAN link speeds that fast is rarely justified. Instead, a company will use slower WAN links that are more affordable.

There are many WAN link methods used by enterprises today. See **Figure 13-6.**

An organization often does not own these WAN links. Instead, they obtain them as leased lines. *Leased lines* are lines for which an organization contracts with a communications provider, and the provider guarantees a specific level of service identified in a service level agreement (SLA). A *service level agreement (SLA)* defines expectations for performance guaranteed by a service provider and often identifies penalties if the service fails to meet the expectations.

Similarly, communications companies do not actually run a single dedicated line between the sites but instead use their existing infrastructure to provide the service.

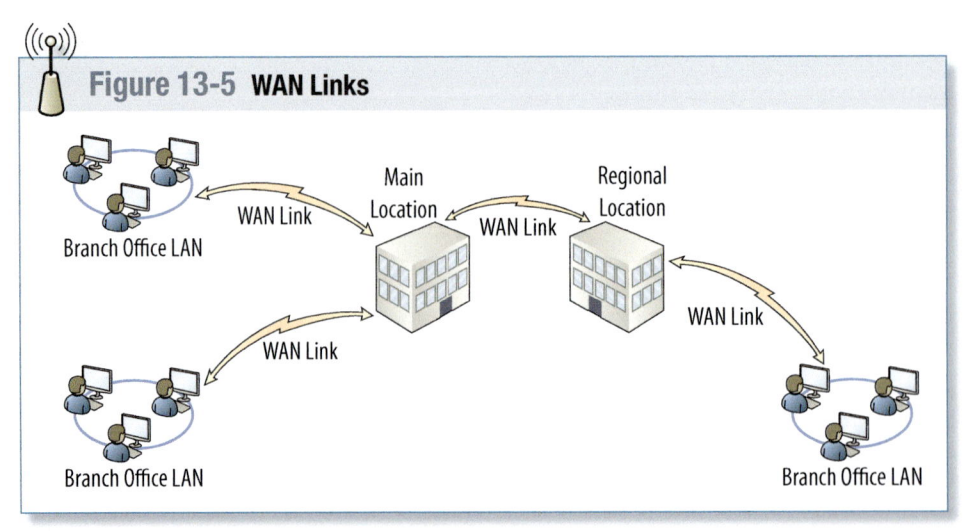

Figure 13-5. *WAN links can connect four LANs together.*

Figure 13-6 Enterprise WAN Connectivity

METHOD	SPEED	AVAILABILITY
T1	1.544 Mbps	Widely available in the United States.
T3	44.736 Mbps	Used in the United States when higher bandwidths are needed.
E1	2.048	Used in Europe.
E3	34.368 Mbps	Used in Europe.
Synchronous Optical Network (SONET)	51 to 39,813 Mbps	Fiber optic standard defining optical carrier (OC) levels from OC-1 to OC-768.
WAN DSL	Up to 24 Mbps	This is similar to DSLs used by home owners, but is instead used for WAN links as business DSL or WAN DSL. When available, it is much more affordable than T1 and T3 lines.
ISDN	128 Kbps (BRI) 1.472 Mbps (PRI)	BRI uses two B channels and one D channel. PRI uses 23 B channels and one D channel.
P2P wireless bridge	Up to 54 Mbps	Widely available but limited by line of sight.
Ethernet WAN	Up to 10 Gbps	Available in some buildings, primarily in urban areas.

Figure 13-6. Similar to residential connectivity options, many options are available for businesses to connect WAN links.

Digital Signal Lines

Many of the signaling types used with WANs are based on digital signal (DS) levels. A single DS0 channel is 64 Kbps wide, and it can carry a single digitized voice phone call. T1 and T3 lines combine multiple channels together and are often used for WAN links.

When using an enterprise technology based on the DS0, the T1 has historically been the technology most widely used.

Europe uses a different standard identified as E-carrier. E-carrier signals use time slots instead of channels, and time slots are measured differently than channels.

ISDN

Integrated Services Digital Network (ISDN) is a group of standards used for transmitting voice, data, and video. ISDN can be used as a WAN link.

An ISDN connection can be used as a WAN link to connect a main location with a branch office. The ISDN uses terminal adapters instead of modems. **See Figure 13-7.**

✓ FACT

T1 lines include an additional 8 Kbps used for overhead. In other words, a T1 is 24 * 64 Kbps (1536 Kbps) plus 8 Kbps. 1536 Kbps + 8 Kbps = 1544 Kbps or 1.544 Mbps.

Figure 13-7 ISDN

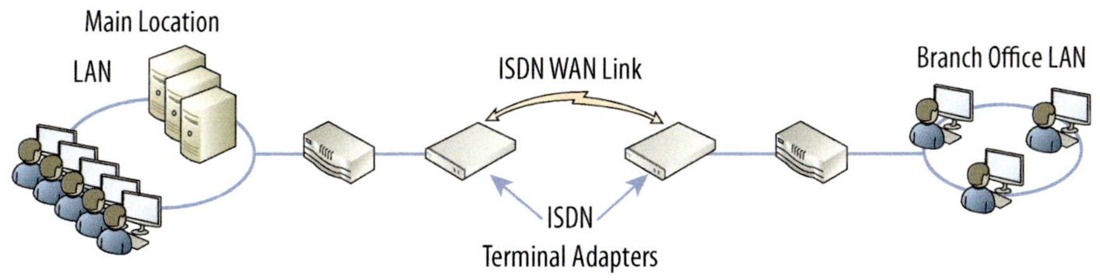

Figure 13-7. ISDN links connect offices together.

> ✓ **FACT**
> Both T1s and T3s use CSU/DSUs.

> ✓ **FACT**
> Some people use the term *ISDN modem*. However, since it does not modulate and demodulate signals, the ISDN terminal adapter is not actually a modem.

> ✓ **FACT**
> Some people refer to a PRI as a T1. However, while a PRI might be carried on a T1 (1.544 Mbps), it is not a T1 itself.

> ✓ **FACT**
> Ethernet WAN has limited availability and can only be found in larger cities.

ISDN uses bearer channels (B channels) and data channels (D channels). This is somewhat misleading since the B channels actually carry the data and the D channels provide signaling information such as caller ID, automatic number identification, and more. The B channels are 64 Kbps channels.

There are two base types of ISDN service:

Basic Rate Interface (BRI) A BRI uses two 64 Kbps B channels and one 16 Kbps D channel. This provides a 128 Kbps data link. BRIs can be used in both SOHOs and enterprises.

Primary Rate Interface (PRI) A PRI uses 23 64 Kbps B channels and one 64 Kbps D channel. This provides a 1472 Kbps data link for a total of 1.536 Mbps.

T1/T3 Lines and E1/E3 Lines

A *T1* combines 24 DS0 channels for a total of 1.544 Mbps. A *T3* combines 28 DS1 channels for a total of 44.736 Mbps. **See Figure 13-8.**

Note that the WAN link is connected with a Channel Service Unit/Data Service Unit (CSU/DSU) at each end. The CSU/DSU translates the T1 signals from the WAN link to a format that the router can accept. This is similar to how a modem translates the digital and analog signals, although the technology is different.

The cost of T1 lines has been steadily dropping in recent years. However, the availability of other cheaper and quicker methods has reduced the demand for these lines.

Europeans use E1 and E3 lines instead of T1 and T3 lines. An E1 is 2.048 Mbps, and an E3 is 34.368 Mbps. Just as T1 and T3 WAN links are terminated with CSU/DSUs, E1 and E3 lines also use CSU/DSUs.

Ethernet WAN

Although Ethernet WAN is not widely available, it provides phenomenal speeds at relatively low cost when compared to T1 lines. Speeds range from 10 Mbps to 10 Gbps.

Ethernet WANs use the same standards as Ethernet LANs, but over longer distances. They use fiber-optic cables, enabling connections to be up to 40 kilometers (km) apart.

As communications providers run more and more fiber-optic cable, the availability of Ethernet WAN connections will continue to increase.

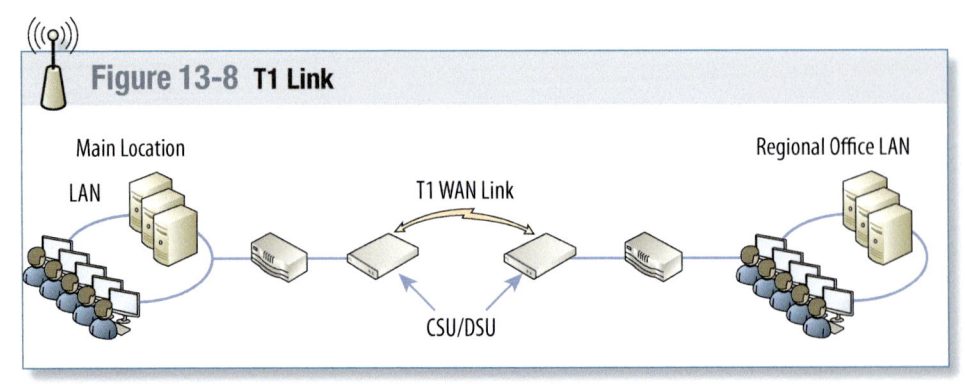

Figure 13-8. *A WAN can connect two LANs with a T1 link.*

REMOTE ACCESS SERVICES

A Remote Access Service (RAS) server provides users with the capability to access an organization's internal LAN from a remote location. Users can access a network from home while traveling or while visiting other locations such as customer sites.

Two methods that can be used for connecting to a RAS server are dial-up or a virtual private network (VPN). A virtual private network (VPN) is a network created by connecting a private network to the Internet. Administrators can configure Microsoft servers as either dial-up or VPN servers by adding the Network Policy and Access Services (NPAS) role and configuring Routing and Remote Access Services.

One of the important considerations when configuring a server to support RAS is security. Administrators only want authorized people to access the remote access server, and they want to ensure that transmissions are not intercepted. They start by using authentication and then adding encryption.

There are several different methods of authentication supported on a Microsoft RAS server. **See Figure 13-9.**

PAP *Password Authentication Protocol (PAP)* passes the password across the wire in clear text. PAP is the least secure authentication method since an attacker can intercept the password and read it.

CHAP The *Challenge Handshake Authentication Protocol (CHAP)* provides encrypted authentication. It uses Message Digest 5 (MD5) to encrypt the password instead of passing the password in clear text. Non-Microsoft clients can use CHAP.

✓ FACT
The user must provide credentials such as a username and password to authenticate their access to a RAS server. However, other methods of authentication can be used, such as a smart card.

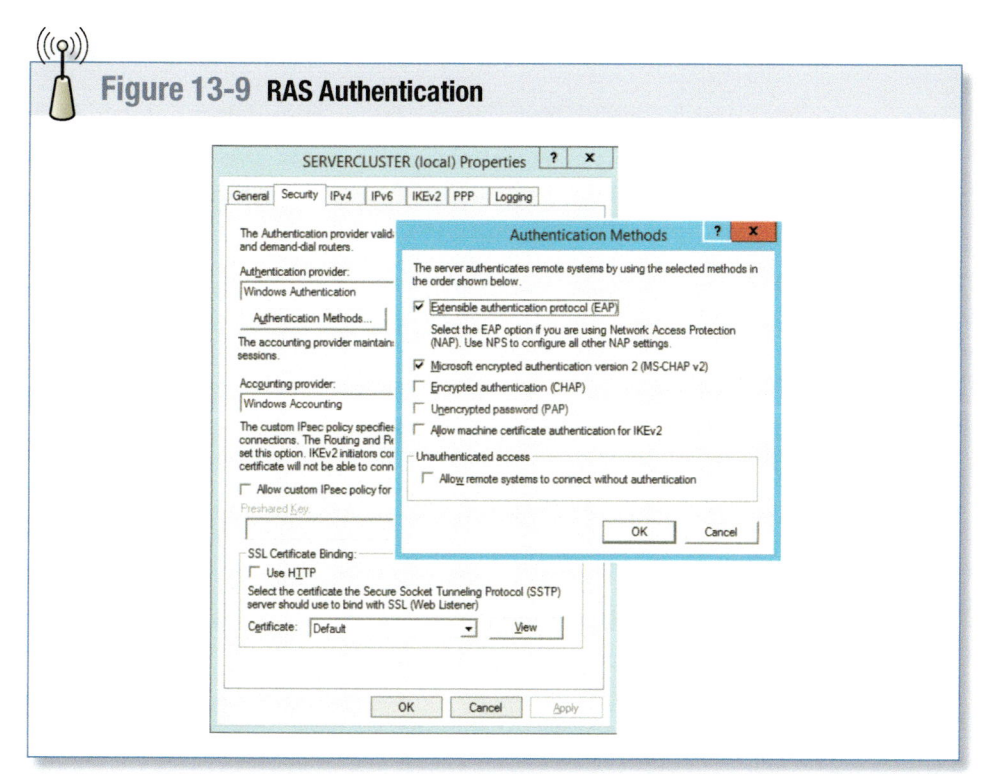

Figure 13-9. Microsoft RAS supports several authentication methods.

MS-CHAPv2 The *Challenge Handshake Authentication Protocol version 2 (MS-CHAPv2)* provides more security for the authentication process when passwords are used. One important benefit is that MS-CHAPv2 provides mutual authentication. The server authenticates back to the client before the client passes the authentication data to the server.

EAP The *Extensible Authentication Protocol (EAP)* supports additional methods including Protected EAP (PEAP) and smart cards. PEAP and smart cards use Transport Layer Security (TLS).

RAS via Dial-Up

A dial-up remote access server includes at least one modem and phone line. Any users with their own modems and phone lines can dial directly into the remote access server. However, just because users can connect does not mean they are granted access.

For example, a dial-up user connects into the remote access server, and the dial-up server challenges the user for authentication. The user must authenticate by providing credentials. The RAS server passes the credentials to a domain controller (or other identity database) to verify the authentication. Once the user is authenticated, the server checks the user account to see whether the user is authorized as a dial-in user. **See Figure 13-10.**

One of the biggest drawbacks with dial-up RAS is the speed. Dial-up modems rarely get more than 50 Kbps, which can be painfully slow. However, virtual private networks (VPNs) can be much faster.

RAS via a VPN

A VPN provides access to a private internal network over a public network such as the Internet. Administrators can configure a Microsoft server as a VPN server just as they can configure it as a dial-up server. The difference is that instead of a modem connected to a phone line, the server must have a network interface adapter that is connected to the Internet.

For example, when a RAS server is configured as a VPN server, the client first establishes a connection with the Internet and then uses a tunneling protocol to "tunnel" through the Internet. The tunnel encrypts the connection to protect the transmitted data. **See Figure 13-11.**

✓ **FACT**
Dial-up is rarely used for remote access today because of the slow speed.

Figure 13-10 Dial-Up Remote Access

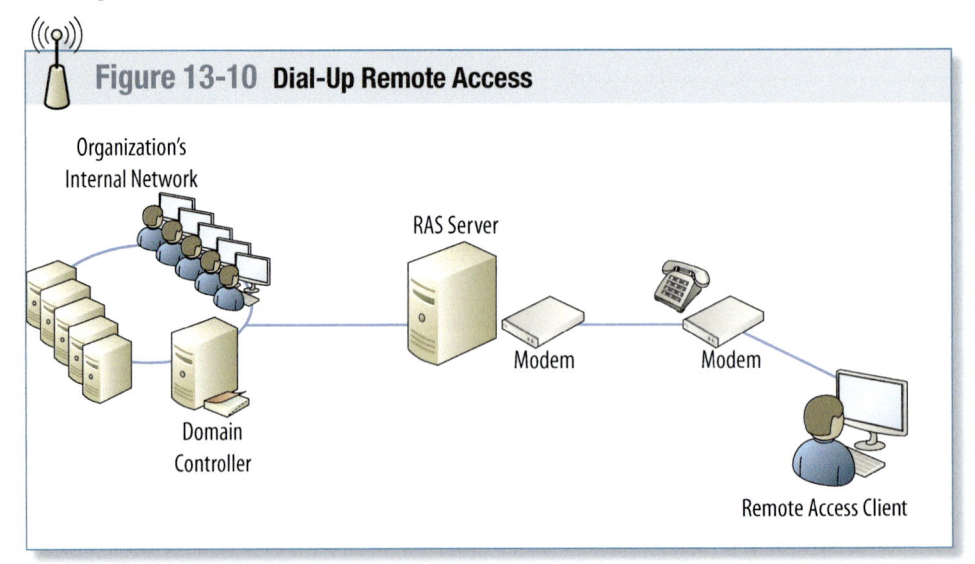

Figure 13-10. Dial-up remote access enables a computer to connect to a network.

Remote Access Versus Remote Desktop

Remote Access and Remote Desktop are two different technologies used for different purposes. Remote Access Services provide access to an entire network from a remote location outside the network. Remote Desktop provides access to a specific system from a remote location, but usually within the same network.

Administrators often use Remote Desktop to remotely access systems servers. The server is locked in a server room, and the administrator is can connect remotely from anywhere in the office.

Servers must be configured to allow Remote Desktop connections.

Remote Desktop is enabled for the server, as long as the computer is using the more secure Network Level Authentication.

Figure 13-11 Tunneling Protocols

Tunneling Protocols
- Point to Point Tunneling Protocol
- Layer 2 Tunneling Protocol
- Secure Socket Tunneling Protocol

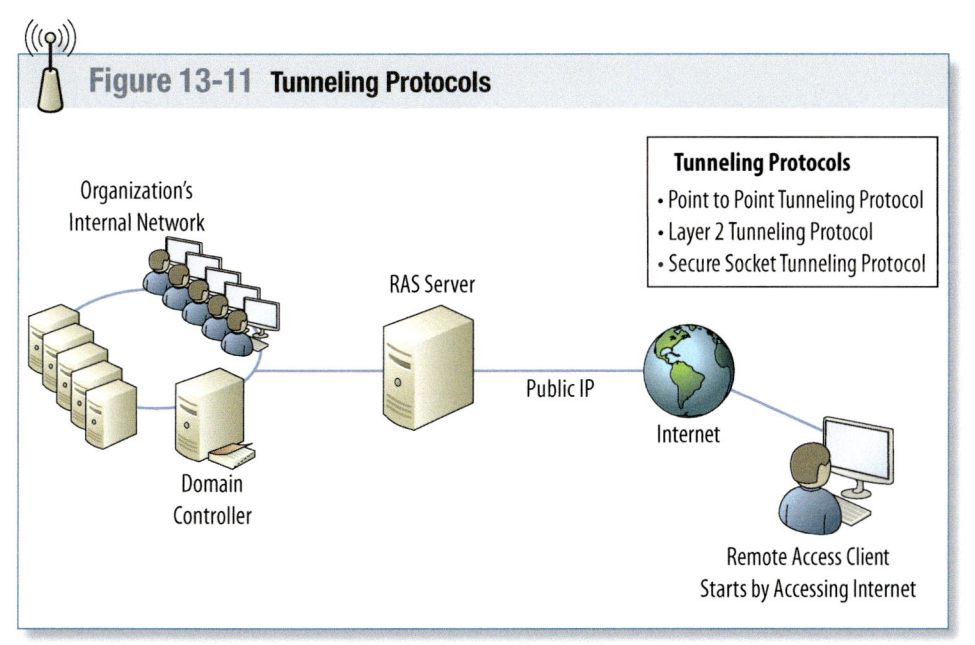

Figure 13-11. VPN uses tunneling protocols for Remote Access.

The three primary tunneling protocols are the Point-to-Point Tunneling Protocol (PPTP), the Layer 2 Tunneling Protocol (L2TP), and Secure Socket Tunneling Protocol (SSTP). These are the three VPN tunneling protocols used with Windows Server. **See Figure 13-12.**

Microsoft introduced PPTP as a VPN tunneling protocol for VPNs. PPTP is an extension of Point-to-Point Protocol (PPP), which is used for dial-up networking. PPP within PPTP offers encryption and authentication for the VPN tunnel. In addition, Microsoft uses Microsoft Point-to-Point Encryption (MPPE) to encrypt PPTP transmissions.

L2TP is a tunneling protocol created by combining the Layer 2 Forwarding (L2F) protocol and Microsoft's PPTP protocol. It is currently a standard used by many different vendors. The primary method of encrypting L2TP transmissions is with IPsec.

The one drawback with IPsec is that it cannot pass through a Network Address Translation (NAT) server. Because of how NAT translates addresses, it breaks IPsec. If the VPN server is behind a firewall that uses NAT, the administrator must decide either to step backward and use PPTP or to go forward and use SSTP.

SSTP is the newest tunneling protocol. It uses Secure Sockets Layer (SSL), which is a well-known, highly used, and respected security protocol. On the Internet, Hypertext Transfer Protocol (HTTP) combines SSL as HTTPS to encrypt the majority of encrypted web pages.

Since SSL is used so often, it is common for the SSL port (port 443) to be open on network-based firewalls. In other words, an administrator does not need to manipulate the firewall to get SSTP to work, making it a little easier to configure.

Client VPNs Versus Gateway VPNs

Administrators can also use a VPN as a WAN link. For example, they can connect a main office and a regional office via a VPN used as a WAN link. The two VPN servers create a gateway-to-gateway VPN. **See Figure 13-13.**

Users in the main office can access resources in the remote office over the VPN, and users in the remote office can access resources in the main office. A significant difference in this configuration from a client VPN is that the gateway-to-gateway VPN is transparent to the users.

Figure 13-12 VPN tunneling Protocols

VPN PROTOCOL	NAT COMPATIBILITY	PORT	COMMENTS
PPTP	Can traverse NAT	1701	Oldest of the four. Uses Microsoft Point-to-Point (MPPE) encryption.
L2TP	Cannot traverse NAT	1723	Uses Internet Protocol Security (IPsec) for encryption.
SSTP	Can traverse NAT	443	Uses Secure Sockets Layer (SSL) for encryption. Easy to configure.
IKEv2	Can traverse NAT	500	Negotiates IPsec security associations

Figure 13-12. Different VPN protocols operate in very different ways.

Figure 13-13 Gateway-to-Gateway VPN

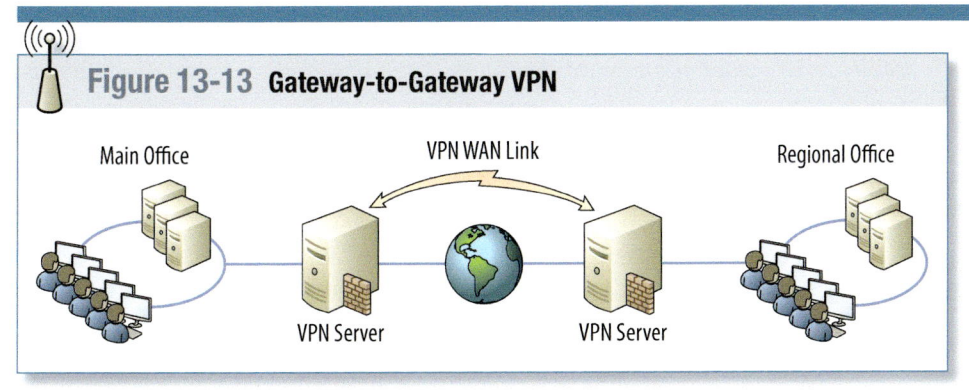

Figure 13-13. A gateway-to-gateway VPN uses a virtual private network as a WAN link.

In a client-to-gateway VPN, the user must initiate the connection. However, in a gateway-to-gateway VPN, the connection is either always on or configured as a demand-dial connection. In a demand-dial connection, as soon as the user tries to access a resource in the other LAN, the VPN servers create the connection.

RADIUS

If an organization includes multiple remote access servers, the administrators might consider the addition of a Remote Authentication Dial-In User Service (RADIUS) server. The RADIUS server provides centralized authentication and logging.

For example, the network can have multiple VPN servers that use a single RADIUS server. It is possible to have multiple VPN servers at different geographical locations such as regional offices. VPN clients can connect to any VPN server. **See Figure 13-14.**

Figure 13-14 RADIUS Server

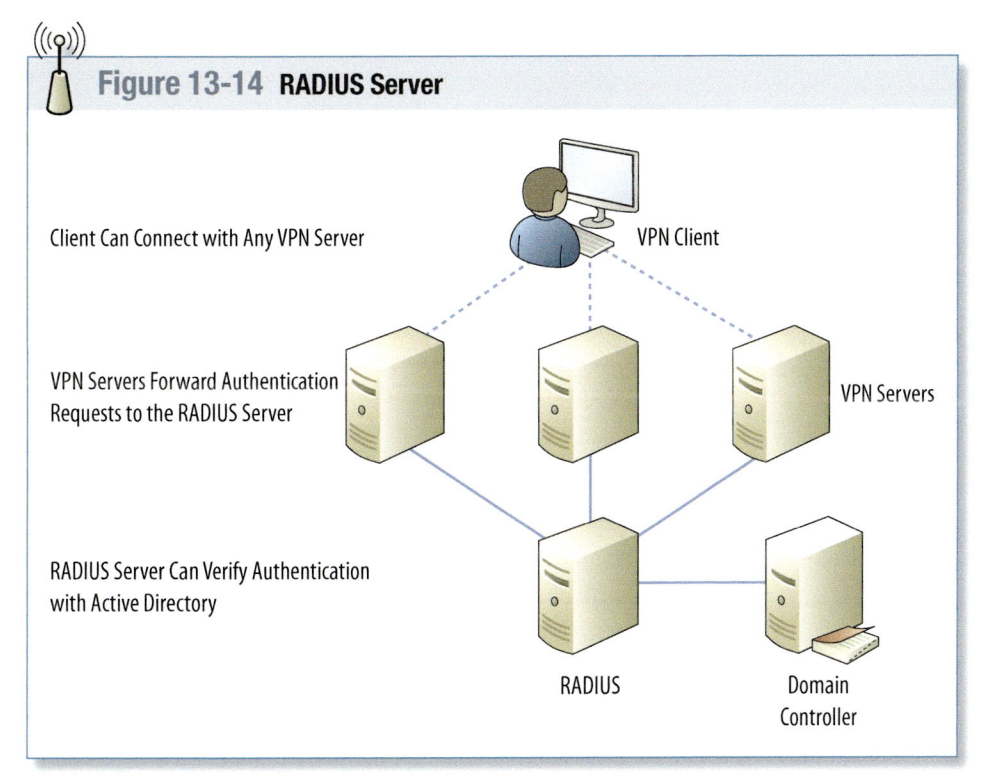

Figure 13-14. A RADIUS server can authenticate users on multiple VPN servers.

Each VPN server forwards all authentication requests to the RADIUS server. In a Microsoft domain, the RADIUS server forwards the requests to a domain controller. If the credentials are valid, the RADIUS server passes the information back to the VPN server. The VPN server then grants access.

The RADIUS server can also track all activity for each of the VPN servers with central logging. All connection attempts and usage data are logged on the RADIUS server for each of the VPN servers.

Summary

There are many different methods used to access the Internet and create wide area networks. SOHOs often use DSL, broadband cable, and many other methods for Internet access. Enterprises use ISDN, T1, and T3 (in the United States), E1 and E3 (in Europe), and other methods to create WANs. A RAS server provides end users with the ability to access an organization's network even when they are away from the organization. VPNs provide access to the private network over a public connection such as the Internet, and they use tunneling protocols to protect the connection. Common tunneling protocols include PPTP, L2TP, and SSTP. When an organization has multiple RAS servers, they can use a RADIUS server for central authentication.

Review Questions

1. How many B channels does an ISDN BRI use?
   a. 1
   b. 2
   c. 23
   d. 24

2. Which of the following connection methods requires the client to be close to the central office of a telephone company?
   a. Dial-up
   b. DSL
   c. Ethernet WAN
   d. ISDN

3. T1 and T3 lines are used in Europe.
   a. True
   b. False

4. What is the speed of a T1 link?
   a. 1.544 Mbps
   b. 2.048 Mbps
   c. 34.368 Mbps
   d. 44.736 Mbps

5. What should be added to a Windows Server 2008 server so it can be used as a VPN server?
   a. Dial-up Remote Services
   b. Routing and Remote Access Services
   c. Virtual Private Network Services
   d. Wide Area Network Services

6. Which of the following is not a valid tunneling protocol?
   a. L2TP
   b. PPTP
   c. SSTP
   d. WLTP

7. What is the primary purpose of RADIUS?
   a. Central authentication
   b. Central encryption
   c. Dial-up WAN link
   d. VPN WAN link

Troubleshooting

One of the primary reasons to study networking is to troubleshoot a network when problems occur. Many puzzle pieces must be in place when a user accesses network resources or just surfs the Internet. If any single piece is not exactly where it should be, the user will be asking for help. With a little bit of knowledge on troubleshooting, an administrator can identify the problem and fix it. This chapter contains information about key troubleshooting tools.

OBJECTIVES

- Explain the use of the command prompt
- Describe the purpose of `ipconfig`
- Identify reasons for using `ping`
- Demonstrate uses of `tracert`
- Describe various uses of `pathping`
- Describe the information shown by `netstat`
- Explain the advantages and disadvantages of installing Telnet

CHAPTER 14

TABLE OF CONTENTS

The Command Prompt 264
    Help at the Command Prompt 264
    Switches 264
    Case Sensitivity 265

TCP/IP Configuration Check
with ipconfig 265

Connectivity Troubleshooting
with ping 268

Router Identification with tracert 272

Routed Path Verification with
pathping 272

TCP/IP Statistics with netstat 275

Telnet 278

Summary 280

Review Questions 281

THE COMMAND PROMPT

Although the Windows graphical user interface (GUI) is easy to use for most end-user tasks, it does have some limitations when troubleshooting network connectivity issues. In contrast, the command prompt can be very useful in troubleshooting basic problems. The *command prompt* is a symbol on the screen that indicates the computer is ready to accept a command as input.

To launch the command prompt in just about any Windows system, open the Run dialog box and type `cmd` in the text box. Then press Enter. There is also a wealth of help available. For example, type the `Help` command to identify the available commands.

When a user types a command at the Command Prompt window (such as the `ping loopback` command), the results typically appear in the same window, in a character-based display. **See Figure 14-1.**

Help at the Command Prompt

Most commands have help available by typing in the command and adding a space, a slash (/), and then a question mark (?). For example, all the following commands will provide help:

- `ipconfig /?`
- `ping /?`
- `pathping /?`
- `tracert /?`
- `netstat /?`
- `telnet /?`

Sometimes the output can scroll past the screen before there is time to read it. To show a single page at a time, use the `More` command with the command like this:

`ipconfig /? | more`

It is also possible to redirect the output to a text file that can be read later. The following example sends the output to a text file named `config.txt`:

`ipconfig /? > config.txt`

Switches

Most commands support additional options. These options are added with

> **✓ FACT**
> The `telnet /?` command will fail if Telnet is not installed on the system.

Figure 14-1 `ping` **Results**

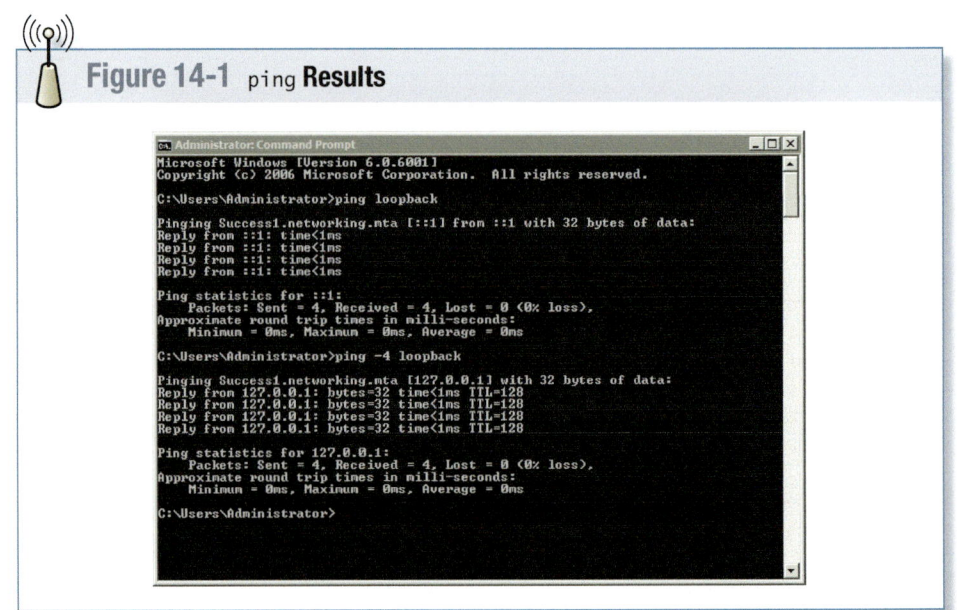

Figure 14-1. *Running* `ping` *in the Command Prompt window can display results for IPv6 or IPv4.*

switches. A *switch* is a forward slash (/) that would then be followed by the additional option. For example, entering `ipconfig` by itself gives minimal information. Entering `ipconfig /all` (using the `/all` switch) provides much more information. Entering the command with the `/? switch` shows the switches supported by the command.

Although most commands use the forward slash (/) as a switch, some commands use a hyphen (-). Most Windows commands will accept either a forward slash or a hyphen. For example, the following two commands will work the same way:

- `ipconfig /all`
- `ipconfig -all`

Case Sensitivity

With few exceptions, Windows command-prompt commands are not case-sensitive. In other words, they can be all uppercase, all lowercase, or any combination. For example, each of the following commands will provide the same results:

- `ping loopback`
- `PING LOOPBACK`
- `PiNg LoOpBaCk`

Commands are often shown with the first letter capitalized for readability. This does not mean it has to be entered that way. If a command is case-sensitive, the documentation will usually stress it.

TCP/IP CONFIGURATION CHECK WITH IPCONFIG

`ipconfig` is one of the most valuable tools for checking and troubleshooting basic TCP/IP settings.

Several pieces of key information are shown by running the `ipconfig /all` command. The `Host Name` value is the name of the computer. The `Primary DNS Suffix` value indicates that the computer joined the `network.mta` domain. The `Node Type` of `Hybrid` value indicates

Launching the Command Prompt with Administrative Permissions

Some commands require administrative permissions to run. For example, to release a Dynamic Host Configuration Protocol (DHCP) lease using the `ipconfig /release` command in Windows, the following error appears if the user has not logged on as the administrator or started the command prompt with administrative permissions:

`The requested operation requires elevation.`

If the user is logged on with the system "administrator" account, the command prompt is automatically started with administrative permissions. However, if the user is logged on with an account that is a member of the Administrators group, the command prompt does not start with administrative permissions.

The solution is to launch the command prompt with administrative permissions before executing the command. In most cases, a user can right-click a command prompt shortcut in Windows and select Run as Administrator from the context menu that appears, as shown in the following graphic.

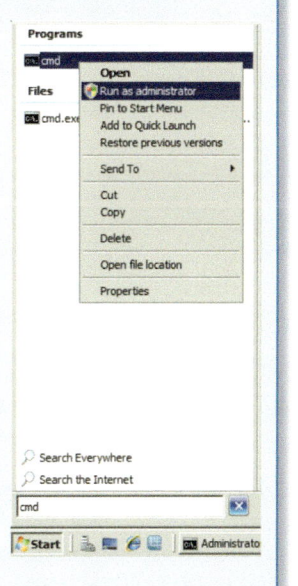

that NetBIOS names are resolved using Windows Internet Name Service (WINS) first and then broadcast. **See Figure 14-2.**

Running the `ipconfig /all` command shows the configuration of a network interface adapter on the system. Some systems may have more than one adapter, and all of the adapters will be displayed. **See Figure 14-3.**

The `Physical Address` value shows the media access control (MAC) address of the network adapter.

If the adapter is a DHCP client, `DHCP Enabled` will be listed as `Yes`, and the IP address of the DHCP server will appear, as long as the client was able to get an IP address from the DHCP server. `DHCP Enabled` is set to `No`, so an IP address for a DHCP server is not available.

`Autoconfiguration Enabled` refers to Automatic Private IP Addressing (APIPA), and it is `Yes` by default. If the system could not get an IP address from the DHCP server and `Autoconfiguration Enabled` is set to `Yes`, an IPv4 address appears that starts with `169.254`. If `Autoconfiguration Enabled` is set to `No`, then APIPA addresses are not assigned when a DHCP server cannot be reached.

A link-local IPv6 address always starts with a value ranging from `fe80` to `febf` and indicates that an IPv6 address is not assigned, but IPv6 is enabled. But remember, if an IPv6 address has been assigned the computer will have both an IPv6 unicast address and link-local address.

Use the subnet mask with the IPv4 address to determine the network ID. In Figure 14-3, the IP address of `192.168.3.10` and a subnet mask of `255.255.255.0` indicates a network ID of 192.168.3.0. The network ID must be the same as those of the other hosts on the subnetwork, including the default gateway. The default gateway and the IPv4 address share the same subnet mask.

The address of the Domain Name System (DNS) server is needed for most hostname resolutions. The same computer is the DNS server. This is evident from the IPv6 loopback address (`::1`) and the same IPv4 address (`192.168.3.10`) that is assigned to the computer. If the DNS server address information is misconfigured, there will probably be problems with name resolution.

A WINS server resolves NetBIOS names. If the network includes a WINS server, the computer configuration should include the IP address in the Primary WINS server section.

✓ **FACT**

The NetBIOS name is created from the first 15 characters of the hostname. If the first 15 characters of the hostname are not unique, duplicate NetBIOS names will result.

✓ **FACT**

An IP address starting with 169.254 means that the computer is set to get an address via DHCP, but the computer is unable to locate a DHCP server.

Figure 14-2 `ipconfig /all` **Windows IP configuration**

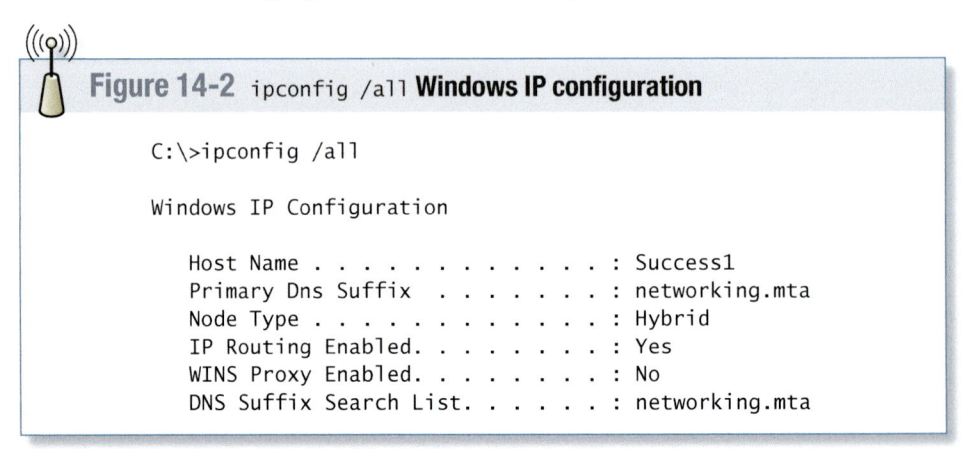

```
C:\>ipconfig /all

Windows IP Configuration

    Host Name . . . . . . . . . . . . : Success1
    Primary Dns Suffix  . . . . . . . : networking.mta
    Node Type . . . . . . . . . . . . : Hybrid
    IP Routing Enabled. . . . . . . . : Yes
    WINS Proxy Enabled. . . . . . . . : No
    DNS Suffix Search List. . . . . . : networking.mta
```

Figure 14-2. *The name networking.mta is resolved using (WINS).*

Figure 14-3 `ipconfig /all` NIC data

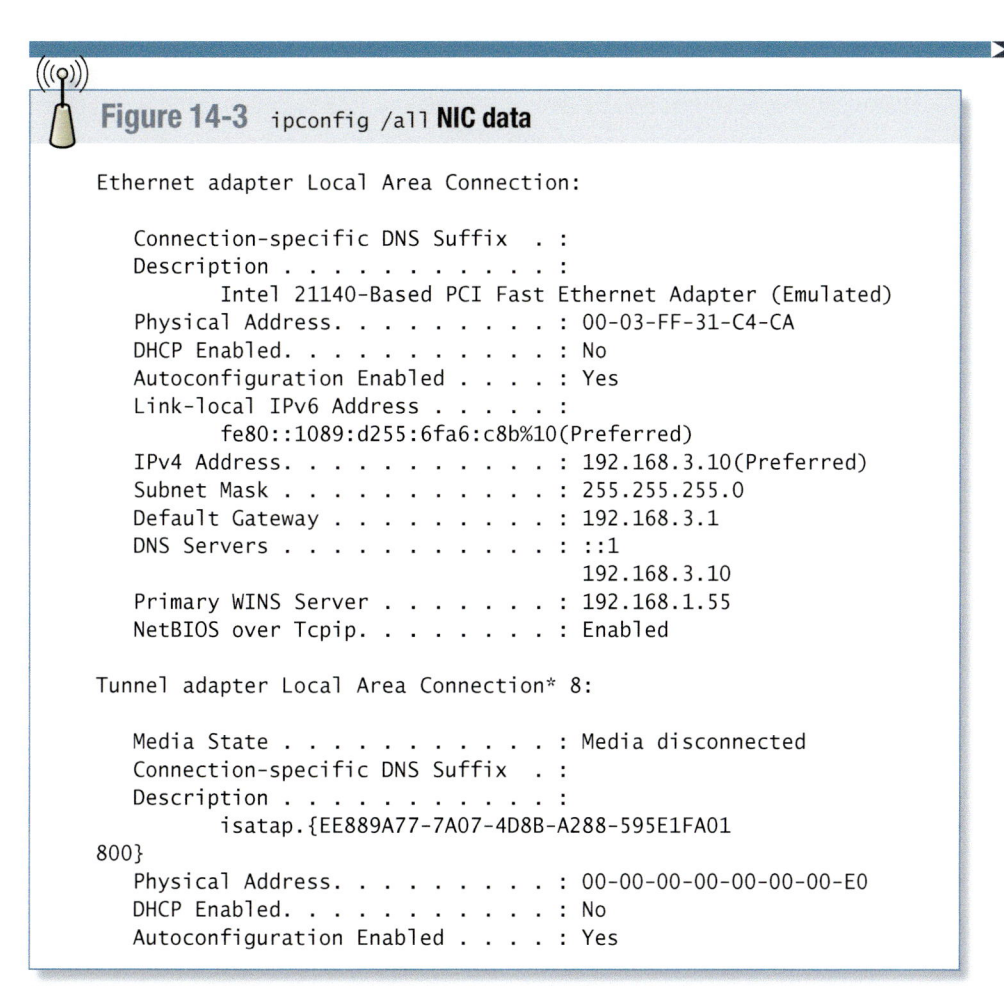

```
Ethernet adapter Local Area Connection:

    Connection-specific DNS Suffix  . :
    Description . . . . . . . . . . . :
          Intel 21140-Based PCI Fast Ethernet Adapter (Emulated)
    Physical Address. . . . . . . . . : 00-03-FF-31-C4-CA
    DHCP Enabled. . . . . . . . . . . : No
    Autoconfiguration Enabled . . . . : Yes
    Link-local IPv6 Address . . . . . :
          fe80::1089:d255:6fa6:c8b%10(Preferred)
    IPv4 Address. . . . . . . . . . . : 192.168.3.10(Preferred)
    Subnet Mask . . . . . . . . . . . : 255.255.255.0
    Default Gateway . . . . . . . . . : 192.168.3.1
    DNS Servers . . . . . . . . . . . : ::1
                                        192.168.3.10
    Primary WINS Server . . . . . . . : 192.168.1.55
    NetBIOS over Tcpip. . . . . . . . : Enabled

Tunnel adapter Local Area Connection* 8:

    Media State . . . . . . . . . . . : Media disconnected
    Connection-specific DNS Suffix  . :
    Description . . . . . . . . . . . :
          isatap.{EE889A77-7A07-4D8B-A288-595E1FA01
    800}
    Physical Address. . . . . . . . . : 00-00-00-00-00-00-00-E0
    DHCP Enabled. . . . . . . . . . . : No
    Autoconfiguration Enabled . . . . : Yes
```

Figure 14-3. *In most cases, the physical address identifies the MAC address of the NIC.*

If there is a network adapter but it is not connected, it will be listed as follows:

```
Media State . . . . . . . . . . . :
Media disconnected
```

If it does have a cable connected, check the link and activity lights on the adapter. If there are lights lit but the `Media State` indicates that the cable is disconnected, check the cabling to ensure:

- The cable is seated completely in the network interface card (NIC).
- The cable is seated completely in the wall jack.
- The cable is seated completely in the switch port. (The switch will usually be in a separate room.)
- Each of the cables is wired correctly.
- The patch cords are not damaged.

One of the simplest ways to check the wiring is to identify a known good path to the switch and use it. For example, if another computer is working, unplug the cable from that computer and plug it into the problem computer. If the problem computer now works, the fault is likely in the wiring. If it does not work, the problem is internal to the computer, and it might be necessary to replace the network adapter.

✓ **FACT**
Ethernet adapters have LED lights to indicate they are connected and have activity. Some have a single LED, and others have two LEDs.

✓ **FACT**
Before replacing hardware, always reboot the system first.

Although the `ipconfig /all` command is valuable, the `ipconfig` command has other switches. **See Figure 14-4**.

Here is one way to use the `/displaydns` and `/flushdns` switches when troubleshooting a problem connecting to another computer. The remote computer's IP address is `192.168.1.5`. However, the `ipconfig /displaydns` command shows the remote computer with a different IP address of `10.5.4.3`.

Use `ipconfig /flushdns` This should remove it from cache. After entering `ipconfig /displaydns`, if the faulty address is still in cache, this indicates that it is cached from the hosts file (not from DNS).

Try to Connect Again If the `ipconfig/flushdns` command removed the entry from the cache, try to connect to the remote computer again. If it is successful, the problem is resolved. If not, use `ipconfig /displaydns` to see what address is displayed. If it is still not the correct address (that is, `10.5.4.3` is displayed instead of `192.168.1.5`), then DNS is giving the wrong address for the computer. In other words, the problem is with the DNS server.

Use `ipconfig /registerdns` Go to the remote computer with connection problems and enter `ipconfig /registerdns`. This should correct the record in DNS.

Flush DNS and Try Again Go back to the original computer and enter `ipconfig /flushdns` to remove the cache entries. Try to connect again, and it should be successful. If not, check the cache with `ipconfig /displaydns`. If it shows the wrong address (that is, `10.5.4.3` instead of `192.168.1.5`), let the DNS administrator know.

CONNECTIVITY TROUBLESHOOTING WITH PING

`ping` is a valuable command used to check connectivity with other computers. It uses Internet Control Message Protocol (ICMP), which is the messenger service of the TCP/IP networking world.

Figure 14-4 Important ipconfig Commands

COMMAND AND SWITCH	COMMENTS
`ipconfig /release` `ipconfig /release6`	Releases an IPv4 lease (or an IPv6 lease with `release6`) obtained from a DHCP server. This does not have any effect if a system has a statically-assigned IP address instead of a DHCP-assigned IP address.
`ipconfig /renew` `ipconfig /renew6`	Renews the IPv4 lease process (or IPv6 lease process with `renew6`) from a DHCP server.
`ipconfig /displaydns`	Displays the host cache (includes names from hosts file and names resolved from a DNS server). This is useful to determine whether a name is in cache with a specific IP address.
`ipconfig /flushdns`	Removes items from the host cache. (Removes items resolved from a DNS server but not items placed in cache from the hosts file.)
`ipconfig /registerdns`	Registers the computer's name and IP address with a DNS server. This creates a host (A) record on the DNS server so that the DNS server can resolve the IP address for other computers.

Figure 14-4. *Many of the* `ipconfig` *XXXX commands will require administrator rights.*

The History of Ping

Mike Muuss wrote the original Ping program used with UNIX systems while studying radar and sonar in 1983. Sonar sends echo signals out, and the reply sounds like "ping," so he called his program Ping. According to Muuss, Dave Mills decided that Ping was actually an acronym (PING) that stood for Packet InterNet Groper.

ping can run with an IP address or a hostname. However, when it runs with a hostname, the first step in the process is that ping resolves the hostname to an IP address. For example, a basic ping command can be used to check connectivity with a server named dc1 on a network. **See Figure 14-5.**

Two important points occurred with the ping command.

First, the computer named dc1 was resolved to the IP address of 192.168.1.112. If name resolution did not work, an error would appear:

```
Ping request could not find host
dc1. Please check the name and try
    again.
```

Second, the ping command sent four packets to the server named dc1 and received four packets back. This reply verifies that the computer named dc1 is operational and able to respond to the ping request. If the server was not operational or not able to respond to the ping request, there would be a different response. **See Figure 14-6.**

Figure 14-5 Successfully pinging a computer

```
C:\>ping dc1

Pinging dc1 [192.168.1.112] with 32 bytes of data:
Reply from 192.168.1.112: bytes=32 time=1ms TTL=128
Reply from 192.168.1.112: bytes=32 time=1ms TTL=128
Reply from 192.168.1.112: bytes=32 time=1ms TTL=128
Reply from 192.168.1.112: bytes=32 time=1ms TTL=128

Ping statistics for 192.168.1.112:
    Packets: Sent = 4, Received = 4, Lost = 0 (0% loss),
Approximate round trip times in milli-seconds:
    Minimum = 1ms, Maximum = 1ms, Average = 1ms
```

Figure 14-5. Successful pings verify a valid end to end connection.

Figure 14-6 Unsuccessfully pinging a computer

```
C:\>ping dc1

Pinging dc1 [192.168.1.112] with 32 bytes of data:
Request timed out.
Request timed out.
Request timed out.
Request timed out.
Ping statistics for 192.168.1.112:
    Packets: Sent = 4, Received = 0, Lost = 4 (100% loss),
```

Figure 14-6. Even a failed ping can identify that name resolution is working properly.

Notice that even though the requests timed out, name resolution still worked. The `ping` command provides a reliable method to test name resolution. `ping` assumes the name is a hostname, so it attempts hostname resolution methods first (such as DNS).

It is also important to realize that just because a "Request timed out" response appears does not necessarily mean that the other computer is not operational. Secure networks and secure computers often have firewall rules blocking Internet Control Message Protocol (ICMP). If ICMP is blocked, the `ping` command will fail even when the computer is operational.

Other error messages that might result from the `ping` command include:

Destination Host Unreachable A destination unreachable reply usually indicates a problem with routing. The local computer might not be configured with the correct default gateway, or a router between the two might be misconfigured or faulty.

TTL Expired in Transit The time to live (TTL) value starts at 128 on Windows systems. It is decremented each time the `ping` passes through a router (also called a hop). If the TTL value is lower than the number of routers the `ping` must pass through to reach its destination, the `ping` packet is discarded. However, it is very rare that a `ping` will need to go through 128 routers, unless there is a problem with routing.

For an example of how to use the `ping` command to troubleshoot a system, consider a network with several systems on two subnetworks separated by a router. Imagine that Sarah is unable to connect with the server named FS1 and she asks for help. An administrator can use the `ping` command to check for several different situations. **See Figure 14-7.**

1. Enter **ping localhost** or **ping 127.0.0.1**.

   Using `ping` for the localhost or the loopback address (`127.0.0.1`) can verify that TCP/IP is functioning correctly on Sarah's local system. There should be four successful replies. Use **ping -4 localhost** or **ping -6 localhost** to check IPv4 or IPv6, respectively.

Figure 14-7 ping **Connectivity Test**

MikePC1
192.168.1.5/24

Mike

Default Gateway
192.168.1.1/24

SarahPC1
192.168.1.8/24

Sarah

Router

FS1
192.168.3.10/24

Default Gateway
192.168.3.1/24

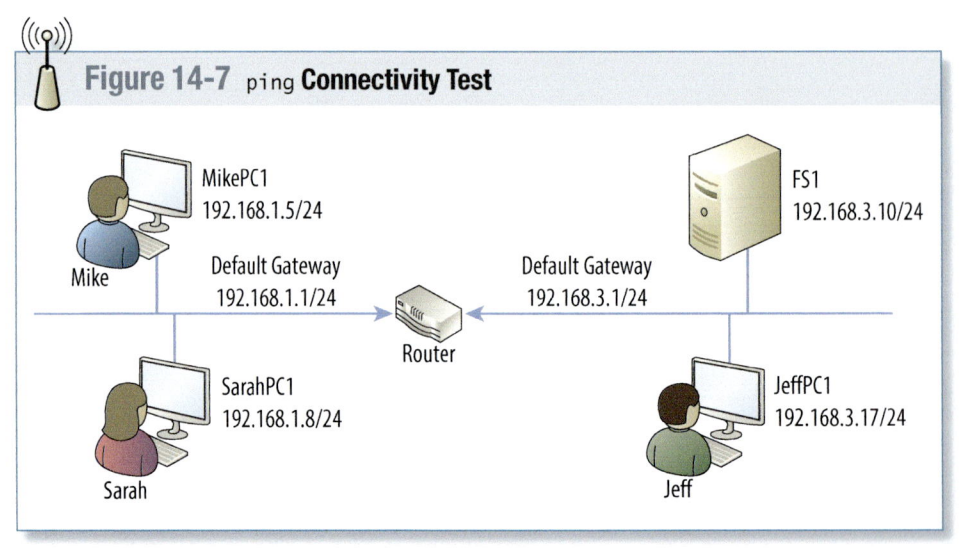

JeffPC1
192.168.3.17/24

Jeff

Figure 14-7. *The* `ping` *program can test connectivity.*

2. Enter **ping 192.168.1.8**.

   This will determine if the network adapter itself is functioning. Sending a ping command to 127.0.0.1 will only test the software in the host, but a ping to the administrator's own IP address will test the hardware as well.

3. Enter **ping 192.168.1.5**.

   This checks connectivity through a switch (or a hub) but not the router. Sending a ping to any other computer with the same network ID will also work. If these ping commands fail, the problem is on this side of the router.

4. Enter **ping 192.168.1.1**.

   This sends a ping command to the default gateway. Use `ipconfig` to determine the IP address of the default gateway.

5. Enter **ping 192.168.3.1**.

   This is the far side of the router. If this ping is successful, it indicates the router is successfully routing traffic. If it fails, but sending a ping to 192.168.1.1 (the default gateway for 192.168.1.1) is successful, it indicates the router is causing the connectivity problem and may be misconfigured or faulty.

6. Enter **ping 192.168.3.10**.

   This sends a ping command to the IP address of the server named FS1. If this succeeds, it indicates that the server is up and operational. Remember, though—if it fails, it could be because the server is blocking ICMP traffic.

7. Enter **ping fs1**.

   The first step of the ping should be to resolve the name fs1 to the IP address of 192.168.3.10. If ping cannot resolve the name, the problem is with name resolution. The primary name resolution methods to check are DNS, the host cache, and the hosts file.

Other switches are supported by ping. See Figure 14-8.

Figure 14-8 Some ping Switches

SWITCH	COMMENTS
-4 Ping fs1 -4	Forces the use of an IPv4 address instead of IPv6.
-6 Ping fs1 -6	Forces the use of an IPv6 address instead of IPv4.
-t Ping fs1 -t	Continues the ping process until it is stopped. Press Ctrl+C to stop the pings.
-a Ping -a 192.168.1.5	Resolves IP addresses to hostnames. This requires that DNS has reverse lookup zones and associated pointer records, which are both optional. In other words, it might not work but does not indicate a problem.
-w	This changes the timeout from the default of one second to five seconds (5,000 milliseconds).
Ping 192.168.1.5 -w 5000	In cases when a computer is heavily loaded or under an attack, ping may fail with a timeout even when it is operational and ICMP is not blocked.

Figure 14-8. An administrator can use many different ping switches in order to identify different issues.

ROUTER IDENTIFICATION WITH TRACERT

If a network includes multiple routers, the `tracert` (pronounced as "trace route") command can trace the path a packet takes through these routers. The `tracert` command can verify the path throughout an entire network.

`tracert` is similar to `ping` in that it checks connectivity. However, it also includes information on all routers between the computer running the program and the destination computer.

The `tracert` command also uses ICMP. Although this normally works well, the results may be incomplete if ICMP is blocked.

For example, say the `tracert` command is sent from a home computer to the computer hosting the Microsoft.com website. A few of the lines shown in the result indicate that the request timed out. This is not because the path is faulty but instead because ICMP is being blocked. **See Figure 14-9.**

The `tracert` command identifies round-trip times for each hop listed in milliseconds (ms). Three different times are listed as `tracert` sends three separate probe requests by default for each hop. Shorter times indicate the trip is faster than longer times. The round-trip times are progressively longer for each additional hop, because those routers are farther away.

The program also lists the names of the routers when it can identify them. If `tracert` cannot identify the name of the router, it just lists the IP address.

If the path between two systems is not working and `tracert` fails to complete, it is possible to use the output to determine the location of the problem. For example, the path was successful up to the 17th step. This indicates the 17th router from the source computer. The problem could be one of three things:

- The routing information on the 17th router is incorrect. This prevents the data from reaching the 18th router.
- The 18th router is faulty.
- ICMP is blocked on the 18th router.

Additional switches also work for the `tracert` command. **See Figure 14-10.**

ROUTED PATH VERIFICATION WITH PATHPING

`pathping` is a combination of `ping` and `tracert`. It starts by checking the route between the two computers similar to how `tracert` does so. It then uses `ping` to check for connectivity at each router.

The `pathping` program sends each router 100 echo request commands, and it expects to receive 100 echo replies back. It then calculates the percentage of data loss based on what it receives. For example, if it receives 100 replies, there is 0% packet loss. However, if it receives only 95 replies, there is 5% packet loss.

Consider the output of a `pathping` command. In the first part of the `pathping` process, it checks the path similar to `tracert`. Lines 1 through 13 represent routers identified as hops. After the program calculates the path, it then starts calculating the statistics by measuring loss. By default, the calculation process takes five minutes. **See Figure 14-11.**

The last few hops are similar to hop 8 and are not listed. They are showing 100% loss since 100 packets were sent

---

✓ **FACT**

Although the primary troubleshooting value of `tracert` is on internal networks, it can also display the routing path to computers on the Internet.

✓ **FACT**

Attackers often use ICMP to launch attacks. It is common for Internet systems to block ICMP traffic to protect against these attacks.

✓ **FACT**

The round-trip times are recalculated for each hop. Additional packets are sent for each router to calculate the round-trip time for that router.

Figure 14-9 Output of tracert command

```
C:\>tracert microsoft.com

Tracing route to microsoft.com [207.46.232.182]
over a maximum of 30 hops:

  1     3 ms     <1 ms    <1 ms   [192.168.1.1]
  2    10 ms     8 ms      9 ms   10.10.184.1
  3    11 ms    11 ms     10 ms   68.10.14-77
  4    14 ms    10 ms     13 ms   172.22.48.33
  5    12 ms     9 ms      9 ms   nrfkdsrj02-ge600.0.rd.hr.cox.net
                                  [68.10.14-17]
  6    16 ms    16 ms     54 ms   ashbbprj02-ae4.0.rd.as.cox.net
                                  [68.1.1.232]
  7    15 ms    15 ms     17 ms   209.240.199.130
  8    17 ms    22 ms     18 ms   ge-3-1-0-0.blu-64c-1a.ntwk.msn.net
                                  [207.46.47.29]
  9    16 ms    17 ms     19 ms   ge-7-0-0-0.blu-64c-1b.ntwk.msn.net
                                  [207.46.43.113]
 10    41 ms    78 ms     40 ms   xe-0-1-3-0.ch1-16c-1b.ntwk.msn.net
                                  [207.46.46.151]
 11    44 ms    40 ms     51 ms   xe-7-0-0-0.ch1-16c-1a.ntwk.msn.net
                                  [207.46.43.146]
 12    93 ms    90 ms     92 ms   ge-3-1-0-0.co1-64c-1a.ntwk.msn.net
                                  [207.46.46.118]
 13    95 ms    93 ms     93 ms   ge-2-3-0-0.co1-64c-1b.ntwk.msn.net
                                  [207.46.35.151]
 14    95 ms    95 ms     94 ms   ge-0-1-0-0.wst-64cb-1b.ntwk.msn.net
                                  [207.46.43.185]
 15    93 ms    94 ms     94 ms   ge-4-3-0-0.tuk-64cb-1b.ntwk.msn.net
                                  [207.46.46.162]
 16   142 ms    96 ms     97 ms   ten2-4.tuk-76c-1b.ntwk.msn.net
                                  [207.46.46.23]
 17   107 ms   181 ms    101 ms   po16.tuk-65ns-mcs-1b.ntwk.msn.net
                                  [207.46.35.142]
 18     *         *         *     Request timed out.
...
Trace complete.
```

Figure 14-9. The `tracert` command includes information on all routers from the host computer to the remote computer.

Figure 14-10 Some tracert Switches

SWITCH	COMMENTS
-4 tracert -4 microsoft.com	Forces the use of an IPv4 address instead of IPv6.
-6 tracert -6 microsoft.com	Forces the use of an IPv6 address instead of IPv4.
-d tracert -d microsoft.com	Suppresses IP address to name resolution. Only the IP addresses are listed.

Figure 14-10. tracert -4 and tracert -6 will force the use of IPv4 and IPv6 respectively.

Figure 14-11 Output of pathping command

```
C:\Users\Dar>pathping microsoft.com

Tracing route to microsoft.com [207.46.197.32]
over a maximum of 30 hops:
  0  Laptop.hr.cox.net [192.168.1.114]
  1  [192.168.1.1]
  2  10.10.184.1
  3  68.10.14-77
  4  172.22.48.33
  5  nrfkdsrj02-ge600.0.rd.hr.cox.net [68.10.14-17]
  6  ashbbprj02-ae4.0.rd.as.cox.net [68.1.1.232]
  7  209.240.199.130
  8  ge-3-1-0-0.blu-64c-1a.ntwk.msn.net [207.46.47.29]
  9  xe-0-1-3-0.ch1-16c-1a.ntwk.msn.net [207.46.46.169]
 10  ge-3-1-0-0.co1-64c-1a.ntwk.msn.net [207.46.46.118]
 11  ge-1-0-0-0.wst-64cb-1a.ntwk.msn.net [207.46.43.163]
 12  ge-7-1-0-0.cpk-64c-1b.ntwk.msn.net [207.46.43.228]
 13  ten3-4.cpk-76c-1a.ntwk.msn.net [207.46.47.197]
 14   *         *         *
Computing statistics for 325 seconds...
             Source to Here   This Node/Link
Hop  RTT    Lost/Sent = Pct  Lost/Sent = Pct  Address
 0                                            [192.168.1.114]
                                0/ 100 =  0%  |
 1   3ms     0/ 100 =  0%       0/ 100 =  0%  [192.168.1.1]
                                0/ 100 =  0%  |
 2   - -   100/ 100 =100%     100/ 100 =100%  10.10.184.1
                                0/ 100 =  0%  |
 3   11ms    0/ 100 =  0%       0/ 100 =  0%  68.10.14-77
                                0/ 100 =  0%  |
 4   - -   100/ 100 =100%     100/ 100 =100%  172.22.48.33
                                0/ 100 =  0%  |
 5   14ms    0/ 100 =  0%       0/ 100 =  0%
                                              nrfkdsrj02-ge600.0.rd.hr.cox.net
                                              [68.10.14-17]
                                0/ 100 =  0%  |
 6   25ms    0/ 100 =  0%       0/ 100 =  0%
                                              ashbbprj02-ae4.0.rd.as.cox.net
                                              [68.1.1.232]
                                0/ 100 =  0%  |
 7   21ms    0/ 100 =  0%       0/ 100 =  0%  209.240.199.130
                              100/ 100 =100%  |
 8   - -   100/ 100 =100%       0/ 100 =  0%
                                              ge-3-1-0-0.blu-64c-1a.ntwk.msn.net
                                              [207.46.47.29]
 . . .
Trace complete.\
```

Figure 14-11. The tracert command only sends 3 echo-requests to each router, but pathping sends 100 by default.

and 100 packets were lost. Again, this is likely because the routers are blocking ICMP, not because there is actual data loss.

On a large network with many routers, the `pathping` command can be useful to help identify whether there is any data loss at specific routers. It could be that the routers simply have too much traffic for their capacity. Offload some of the traffic to another subnet or increase the capacity of the router.

There are some additional switches for the `pathping` command. **See Figure 14-12.**

TCP/IP STATISTICS WITH NETSTAT

The `netstat` (short for "network statistics") command displays information on any TCP/IP connections on a computer. It can show all the connections, ports, and applications involved in network connections. It can also check TCP/IP statistics.

Several common commands incorporate `netstat`. **See Figure 14-13.**

A basic listing can show open ports for a computer running on a network without any browser sessions opened. With a few web pages open in a browser, the number of open ports can easily fill a page. **See Figure 14-14.**

The `Local Address` indicates the local computer (with an IP address of `192.168.1.114`) and is in the format of *IP address: port*. The `Foreign Address` indicates the name or IP address of the remote computer. The `State` column indicates the state of the connection.

Following are some of the common states of a connection:

ESTABLISHED Indicates that a TCP session is established

LISTENING Indicates that the system is ready to accept a connection

CLOSE_WAIT Indicates that the system is waiting for a final packet from the remote system to close the connection

Some connection states are described in RFC 793 with a hyphen, but `netstat` displays them with an underscore. For example, RFC 793 uses `CLOSE-WAIT`, but `netstat` displays `CLOSE_WAIT`.

Running the `netstat` command can display something that looks suspicious. For example, the `Foreign Address`

✓ **FACT**

Switches can be combined. For example, the `netstat -ano` command combines the output of the -a, -n, and -o switches.

Figure 14-12 Some pathping Switches

SWITCH	COMMENTS
-4 pathping -4 microsoft.com	Forces the use of an IPv4 address instead of IPv6.
-6 pathping -6 microsoft.com	Forces the use of an IPv6 address instead of IPv4.
-n pathping -n microsoft.com	Suppresses IP address to name resolution. Only the IP addresses are listed.
-q pathping -q 50 microsoft.com	Changes the number of queries per hop. By default, 100 queries per hop are used.

Figure 14-12. `pathping -q` *changes the number of ICMP echo requests.*

Figure 14-13 Common netstat Commands

COMMAND	COMMENTS
Netstat -a	Shows all connections and listening ports.
Netstat -b	Shows connections that all applications are using to connect on the network (including the Internet if the client is connected to the Internet).
Netstat -e	Shows Ethernet statistics.
Netstat -f	Shows fully qualified domain names (FQDNs).
Netstat -n	Shows addresses and port numbers in numerical form.
Netstat -o	Includes the process that owns the connection.
Netstat -p protocol Netstat -p TCP	Shows connections for specific protocols, including IP, IPv6, ICMP, ICMPv6, TCP, TCPv6, UDP, and UDPv6. For example, netstat -p TCP shows connections for TCP only.
Netstat -r	Shows the routing table. This is the same routing table displayed by the route print command.
Netstat -s	Shows statistics for the protocols running on the system. This includes packets received, packets sent, errors, and more.
Netstat interval Netstat 15	Redisplays the statistics after waiting the interval period. The interval is specified in seconds, as in netstat 15 to wait 15 seconds before executing the netstat command again.

Figure 14-13. Typing netstat ? and then pressing Enter will display this list.

Figure 14-14 Output of netstat command

```
C:\Users\Dar>netstat

Active Connections

  Proto  Local Address              Foreign Address          State
  TCP    192.168.1.114:135          WIN7-PC:49766            ESTABLISHED
  TCP    192.168.1.114:1030         WIN7-PC:49767            ESTABLISHED
  TCP    192.168.1.114:1060         MYBOOKWORLD:microsoft-ds
                                                             ESTABLISHED
  TCP    192.168.1.114:2078         beta:http                ESTABLISHED
  TCP    192.168.1.114:3389         Server08R2:56080         ESTABLISHED
  TCP    [fe80::41f0:f763:5451:198a%10]:135  Darril-PC:50506
                                                             ESTABLISHED
```

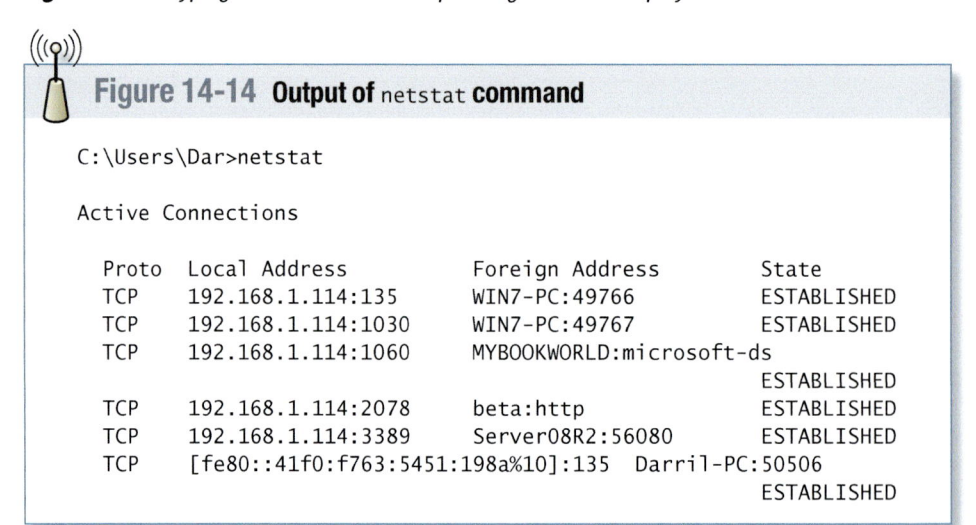

Figure 14-14. The netstat command shows the active TCP sessions.

✓ FACT

netstat can be useful in detecting spyware and malware. If the applications are unknown, they might be malicious.

of beta:http. The netstat -b command can identify the application or process using the port. The netstat-b command is one of the commands that must be run from an administrator prompt. **See Figure 14-15.**

With a little information about ports, an administrator can use the output of the netstat command, the names of the applications, and the port numbers to determine what each of the ports is doing.

Figure 14-15 Using `netstat -b` to identify applications and processes

```
C:\>netstat -b

Active Connections

  Proto  Local Address          Foreign Address          State
  TCP    192.168.1.114:135      WIN7-PC:49766            ESTABLISHED
 RpcSs
 [svchost.exe]
  TCP    192.168.1.114:1030     WIN7-PC:49767            ESTABLISHED
 [spoolsv.exe]
  TCP    192.168.1.114:1060     MYBOOKWORLD:microsoft-ds ESTABLISHED
 Can not obtain ownership information
  TCP    192.168.1.114:2078     beta:http                ESTABLISHED
 [OUTLOOK.EXE]
  TCP    192.168.1.114:3389     Server08R2:56080         ESTABLISHED
 CryptSvc
 [svchost.exe]
```

Figure 14-15. `netstat -b` *must be run from the administrator prompt.*

Port 135 Port 135 is used for NetBIOS and Remote Procedure Calls (RPCs) in Windows systems. This shows an IPv4 connection (the first line) with another computer named `Win7-PC` in the network.

Port 1030 This is being used by the print spooler service (`spoolsv.exe`).

Port 1060 This port is being used to connect to a network drive (named `MYBOOKWORLD`) that is mapped to the system as an additional drive.

Port 2078 This is being used by Microsoft Outlook for a connection to the Internet.

Port 3389 CryptSvc is short for the Cryptographic Services service. Port 3389 is the port used by Microsoft for Remote Desktop Services (RDS). Combined, they indicate an RDS session is established with a remote computer named `Server08R2`.

That still may not be enough information if the application looks suspicious. Use the following steps to get more information about any of these connections:

1. Enter **netstat** at the command prompt.

2. Review the listing and determine whether there are ports to investigate more.

   Note the port number in the `Local Address` column. For example, an administrator might want to investigate the `beta:http` line, which shows port 2078.

3. Enter **netstat -ano** at the command prompt.

   This provides a more detailed listing, including the process ID (PID). Look for the line with the selected port number. The following code snippet shows the line for this port:

   ```
   Proto  Local Address
   Foreign Address  State       PID
     TCP    192.168.1.114:2078
   65.55.11.163:80  ESTABLISHED 5356
   ```

   The `PID` column shows a PID of `5356` for port `2078`.

4. Launch Task Manager by pressing the Ctrl+Shift+Esc keys at the same time.

5. Select the Processes tab.

6. Click View, and then click Select Columns.

7. Select the PID (Process Identifier) box. Click OK.

8. Look for the entry with the selected PID. **See Figure 14-16.**

    Notice that the `Image Name` value (the process) is `Outlook`.

9. Launch the Performance Monitor by searching for **perfmon** and pressing Enter.

10. Launch the Resource Monitor by right-clicking Monitoring Tools and selecting Resource Monitor.

11. Look for the PID in the CPU, Disk, Network, and Memory sections.

    This provides additional information on the process such as how much resources the process is consuming. **See Figure 14-17.**

It is possible to perform more advanced searches to narrow down the source of connections. The goal of these steps is not to make someone a master at identifying all the resources that an open port might be using, but instead to show some of the possibilities. It provides a chance to dig into the system and learn a little more about it.

TELNET

Telnet is a lesser used tool for troubleshooting. It is not installed on Windows

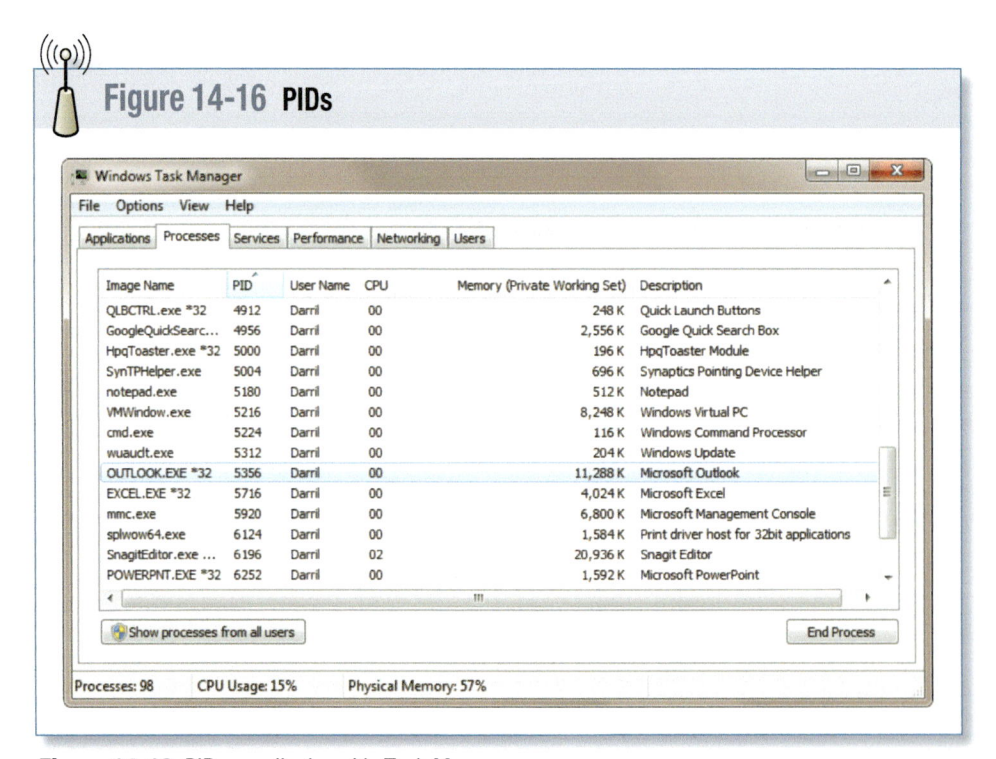

Figure 14-16. PIDs are displayed in Task Manager.

Figure 14-17 Resource Monitor

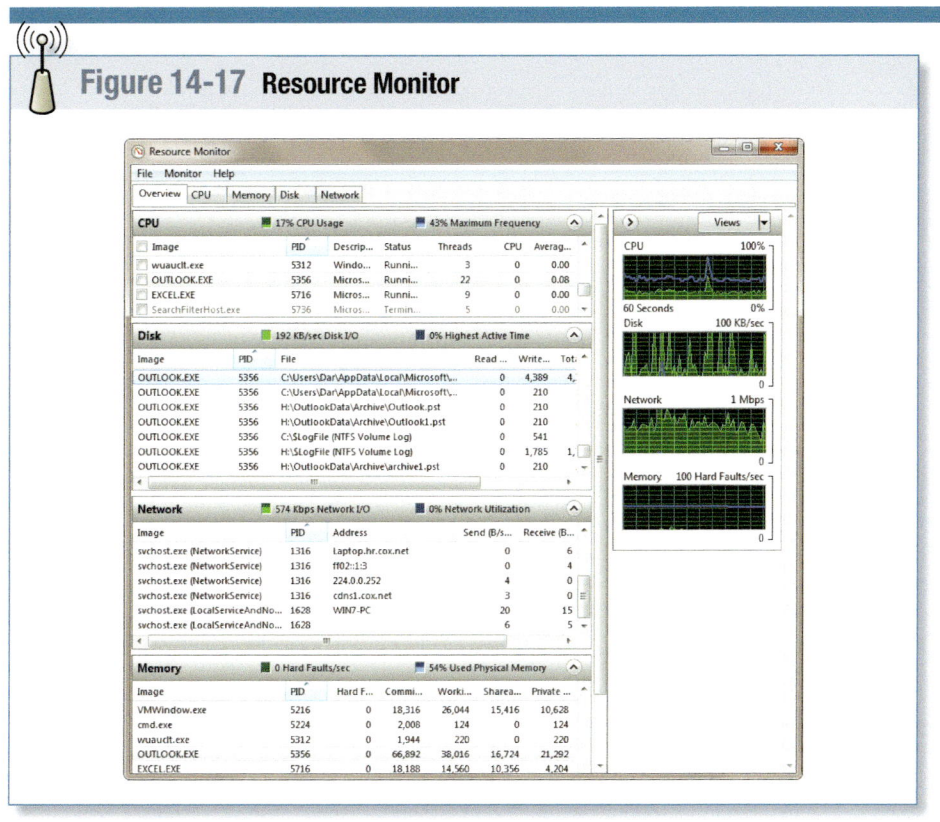

Figure 14-17. Users can view resource usage in Resource Monitor.

systems by default because of security risks, but it can be added. Attackers often use Telnet to check for open ports on a system that has Telnet enabled, so it is more secure to keep Telnet disabled unless it is needed.

When Telnet is installed on client and server computers, it is possible to connect a Telnet client to a Telnet server. It provides a command-line interface that enables a user to run Telnet commands from the Telnet client that are executed on the Telnet server. Commands include command-line programs, shell commands, and scripts.

Many programs that use Telnet can be configured to encrypt the traffic with Secure Shell (SSH). This ensures that attackers are not able read the traffic.

After Telnet is installed, enter **telnet /?** at the command prompt to display the output of a help file.

To start a Telnet session from a Telnet client, use the following command:
`Telnet TelnetServerName`

✓ **FACT**

One of the risks with Telnet is that commands go across the network in clear text. An attacker with a sniffer can capture the traffic and easily read it.

✓ **FACT**

Telnet presents significant risks to computers in a network. Do not install Telnet on a computer in a production environment unless it is actually needed.

> **Summary**
>
> Many different methods are available for troubleshooting network and connectivity problems using tools that run from the command prompt. Useful commands include `ipconfig`, `ping`, `tracert`, `pathping`, and `netstat`. By using these tools, an administrator can check the configuration of a system and check its interoperability on a network.

Review Questions

1. Which switch can display help for a command?
   a. ?
   b. /?
   c. /?Help
   d. /hlp

2. Which one of the following commands can be used to determine the MAC address of a NIC?
   a. ipconfig
   b. ipconfig /all
   c. netstat
   d. telnet

3. An administrator wants to verify that a computer can connect with the default gateway. It has an IP address of 192.168.1.1. What command should the administrator use?
   a. 192.168.1.1 Netstat
   b. 192.168.1.1 Ping
   c. netstat 192.168.1.1
   d. ping 192.168.1.1

4. To ensure that the ping command only uses IPv4, one must use the /ip4 switch.
   a. True
   b. False

5. The ping command has been used to check connectivity with a server named dc1 and the following error has been returned: "Ping request could not find host dc1. Please check the name and try again." What does this error mean?
   a. The default gateway is not configured.
   b. DHCP is down.
   c. Name resolution did not work.
   d. A router is not configured properly.

6. The ping command has been used to check connectivity with a server named fs1 and the following error has been returned: "Destination Host Unreachable." What is the most likely reason for this error?
   a. DHCP is down.
   b. DNS is not configured.
   c. Name resolution did not work.
   d. A router is not configured properly.

7. Packet loss can be measured by using the tracert command.
   a. True
   b. False

8. An administrator can use netstat to view TCP/IP statistics, including the number of packets that have been sent and received.
   a. True
   b. False

9. What is the command to view all open ports including known applications on a Windows Server 2008 system?
   a. ipconfig /all
   b. netstat
   c. netstat -b
   d. pathping

10. The Telnet client is installed by default on Windows systems.
    a. True
    b. False

Network Fault Tolerance

Increased availability refers to technologies that enable users to continue accessing a resource despite the occurrence of a disastrous hardware or software failure. A number of availability mechanisms are at an administrator's disposal, including shadow copies and offline files, disk redundancy solutions, application availability solutions, and server clustering.

OBJECTIVES

- ▶ Describe various malware that can threaten a network
- ▶ Explain how the failure of a hard disk drive can result in losing data
- ▶ Identify ways to ensure applications will not stop running on a network
- ▶ List various methods for keeping a server up and running
- ▶ Describe several methods of disaster recovery

CHAPTER 15

TABLE OF CONTENTS

Malware 284
    Denial of Service 284
    Man in the Middle 284
    Malware 285
    Buffer Overflow 286
    Social Engineering 286
    Wireless Threats 286
    Mitigation Techniques 287

Availability 288
    Data Availability 288
    Application Availability 292
    Server Availability 294

Disaster Preparation 301

Summary 302

Review Questions 303

MALWARE

Understanding the nature of the threats against a network is an essential part of building an effective strategy against them. The relationship between the attackers and the defenders is a type of arms race, with most of the attackers creating ever-different variations on a few familiar themes.

Denial of Service

The busier a service is, the longer it takes for customers to get their orders. This axiom applies as easily to networking as it does to retail. When a server is busy processing thousands of incoming requests, performance degrades and all of the clients suffer. A *denial of service (DoS) attack* is an attempt to overwhelm a server or an application with incoming traffic.

In the simplest form of DoS attack, an attacker can use the `ping` utility to send an endlessly repeating stream of Internet Control Message Protocol (ICMP) messages to a server and bring it to a near halt. This is why the default configuration on many firewalls blocks the ICMP Echo Request messages that `ping` generates.

A *smurf attack* is one particularly sneaky form of ICMP-based DoS attack that involves flooding a network with `ping` messages sent to the network's broadcast address. These messages are also spoofed—the source address field contains the IP address of the computer that is the intended victim. This way, all of the computers receiving the broadcast will send their responses to the victim, flooding its in-buffers.

A DoS attack does not have to use ICMP messages, however. All a determined attacker has to do is find an open port on a server, and it is not difficult to bombard it with some kind of traffic. Some attacks even use the type of traffic the server is designed to accept. For example, an attacker can flood a web server with incoming web requests, excluding the requests generated by legitimate users. In this case, the server administrator cannot just close off port 80 in the firewall, because that would exclude all of the incoming traffic. The counter for this type of attack would be to discover where the flood is coming from and block it by its IP address.

A single computer can only transmit so many packets, limiting the potential effectiveness of an individual DoS attack against a large server farm. However, attackers who distribute a type of malware called a Trojan horse can take control of other peoples' computers without them knowing it. These remote controlled computers are called bots or zombies, and the hoard of zombies under one attacker's control is called a botnet. The attacker can use the zombies to generate DoS messages from hundreds of systems at once, overwhelming even the most robust application. This is called a distributed denial of service (DDoS) attack.

Man in the Middle

A *man in the middle (MITM) attack* is one in which the attacker interposes himself or herself between two individuals who think they are communicating with each other. The attacker receives the messages from each party in the transaction and relays them to the other party, but not without reading them (or even modifying them) first.

The attacker in a MITM attack must be able to receive all of the communications generated by both of the other parties, to remain a convincing intermediary. While undetected, MITM attackers can use their access to obtain sensitive information, such as passwords and shared keys.

Detection of an MITM attack is sometimes possible through analysis of the latency periods between message transmissions and receipts. The sudden appearance of communication delays between two parties not attributable to other causes can indicate the presence of an intermediary. The best defense against MITM attacks is mutual authentication, such as that performed by public key infrastructure (PKI) systems. In fact, these authentication systems were invented in part to counter this type of attack.

An FTP bounce attack is a variation on the MITM attack that involves the use of the `Port` command in the FTP protocol to gain access to ports in another computer that are otherwise blocked. After connecting to an FTP server, the attacker uses the commands within the FTP protocol, but directs them at a different computer. At one time, many FTP implementations could do this, but most developers have since closed this potential exploit.

Malware

Malware is a generic term referring to any software that has a malicious intent, whether obvious or obscure. Some types of malware are relatively benign and intended only to generate business for the distributor, while others are deliberate attempts to cause damage with no rational motive. The most common types of malware are:

Virus A virus is a type of program that replicates by attaching itself to an executable file or a computer's boot sector and performs a specified action—usually some form of damage—at a prearranged time. Viruses do not spread through a network by themselves. They require a user to run the infected program and load it into memory. Viruses typically replicate through removable media, such as USB flash drives.

Worm A worm is a program that replicates itself across a network by taking advantage of weaknesses in computer operating systems. Unlike a virus, a worm can replicate across a network without any user activity. Some worms do nothing more than consume network bandwidth, while others can damage files, generate spam e-mail, or install a backdoor program, turning the target computer into a zombie.

Trojan Horse A trojan horse is a non-replicating program that appears to perform an innocent function, but that in reality has another, more malicious, purpose. One common tactic is to insert code into a free game or other application, which turns the computer into a server by opening up specific ports to incoming traffic. This enables an attacker on the Internet to take control of the computer without the owner's knowledge and use it as a zombie for any purpose, including initiating distributed attacks against other targets.

Spyware Spyware is a hidden program that gathers information about computer activities and sends it to someone on the Internet. *Adware* is a relatively harmless type of spyware that tracks the Internet sites that have been visited for the purpose of sending targeted advertisements to the user, while others are more dangerous. For example, some spyware can record keystrokes and

> **✓ FACT**
> The term "virus" is often used as a catchall for various types of malware, including worms, Trojan horses, and spyware. The primary characteristic that distinguishes a virus is its capability to replicate itself.

other usage data to capture passwords and other sensitive information. Spyware is usually something the user downloads unknowingly, by clicking on a link in an e-mail or on a web page.

Buffer Overflow

A *buffer* is an area of computer memory designed to hold incoming data as it is being processed. A *buffer overflow* is a condition in which a program sends too much data to a buffer, and it spills over into an area of memory intended for another purpose. The results of a buffer overflow depend on the application and the operating system, but they can include error messages, data corruption, or even a system crash.

Ordinarily the result of a programming error, some people take advantage of inherent weaknesses in operating systems or applications and write code designed to deliberately cause buffer overflows to occur. These attackers can deliver the code to the target system in any number of ways, including viruses, worms, and Trojan horses.

Social Engineering

Sometimes the easiest way for an attacker to obtain sensitive information is simply to ask for it. *Social engineering* is the term used to describe a practice in which a friendly attacker contacts a user by telephone, mail, or e-mail and pretends to be an official of some sort. That attacker then gives some excuse for needing the user's password or other confidential information.

Most people are reasonably helpful by nature, and when someone asks for a favor they can easily perform, they usually do it. This is particularly true in a corporate environment, where a call from an unknown person in another department is not unusual. An attacker with a friendly nature and a convincing story can often compel users to give up all sorts of valuable information.

Another, more refined form of this tactic is called phishing. *Phishing* consists of sending out an official looking e-mail or letter to users that points them to a website containing a form asking for personal information. The e-mail might appear to be from a bank, a credit company, or a government agency, and it might contain a request for help or a threat of some inconvenient action if the user does not comply with the instructions provided. The letter and website are, of course, bogus, and the confidential information the users supply goes right to the attacker.

Wireless Threats

Wireless networks have their own specialized security protocols, because of the specialized nature of the threats against them. As with other threats, the continued development of greater security technologies drives certain people to constantly search for new weaknesses they can exploit.

Some of the most common threats against wireless networks are:

War Driving Virtually all wireless LANs today use some form of encryption, but that was not always the case. In the early days of Wi-Fi, many people left their networks unprotected, making it possible for unauthorized users to connect to them and access their files or use their Internet connection. *War driving* is the process of cruising around a neighborhood with a scanner, looking for unprotected wireless networks to which one can connect.

War Chalking *War chalking* is a practice associated with war driving in which the people discovering an unprotected network leave a mark on a wall or gatepost indicating its presence, so that future drivers can find it.

Cracking *Cracking* is the process of penetrating an encryption protocol by discovering its cryptographic key. An encryption protocol has not been invented that cannot be cracked; it is just a question of how long it will take and how much computing power can be devoted to it. All of the encryption protocols that wireless LANs use—WEP, WPA, and WPA2—are crackable with enough time and effort. The process basically consists of locating a wireless network, using a packet sniffer program to capture some of its traffic, and then analyzing the contents of the packets to discover the keys used to encrypt them. Wireless network cracking tools are freely available on the Internet, so an attacker does not even have to possess the expertise needed to write them.

Rogue Access Point A routing access point is an unauthorized wireless access point connected to a network. It is arguably the greatest possible security hazard for a wireless network administrator, because its perpetrators are often innocents. A user wanting the convenience of wireless laptop access in the office purchases an inexpensive access point and plugs it into the network with no security enabled. This enables anyone in the area to access the wired network without anyone's knowledge.

Evil Twin An *evil twin* is an unauthorized wireless access point deliberately configured to closely mimic an authorized one. Users fooled by the impersonation connect to the access point, which provides the attacker with access to the packets and the data inside them.

Mitigation Techniques

Learning about the threats might be the first step in combatting them, but then the administrator has to devise a strategy for fighting back, or for preventing them in the first place. Threat mitigation is an ongoing process, in which both sides continue to learn. However, there are several standard mitigation techniques that all network administrators should keep in mind, including:

Training and Awareness In many cases, successful attacks are the result of user error. A person clicks the wrong link, opens the wrong e-mail, or executes the wrong file, and the door admitting the intruder to the network is opened. Educating users about the potential dangers and what not to do when confronted with them is the best way to protect the network from intrusion.

Patch Management Attacks are often possible because of weaknesses in applications or operating systems, which intruders have learned to exploit. Software developers are constantly discovering new weaknesses, and they release patches and updates to close the security holes that result. Keeping all of the computers on the network updated must be an essential part of an administrator's security regimen. In addition, all computers should be equipped with appropriate anti-malware software, which also must be updated on a regular basis.

Policies and Procedures For a network to run efficiently, there must be policies in place that govern what administrators

> ✓ FACT
>
> A packet sniffer is an application that intercepts and captures packets as they are transmitted over a network. Sniffers are legitimate tools for network administrators, but they are also valuable weapons for attackers.

and users should and should not do. Published policies can prevent many security breaches before they happen, and proper procedures can enable users to recognize security problems when they happen and take appropriate action.

Incident Response Policy should dictate how administrators respond to security-related events, and all threats and attacks should be carefully documented. A history of occurrences can provide evidence of an escalation of tactics, indicating that attackers are targeting the organization specifically.

AVAILABILITY

Keeping data and services available for use is one of the primary functions of the network administrator. There are a variety of strategies and technologies administrators can use to guard against faults and failures that can interrupt user productivity.

Data Availability

As the computer component with the most moving parts and the closest physical tolerances, hard disk drives are more prone to failure than virtually any other element of a data network.

When a hard disk drive fails, the data stored on it obviously becomes unavailable. Depending on the type of data stored on the disk and the availability technologies implemented on the computer and the network, the effect of a disk failure on user productivity can be transitory, temporary, or permanent.

Data Backup

The simplest and most common type of data availability mechanism is the disk backup—that is, a copy of a disk's data stored on another medium. The traditional medium for network backups is magnetic tape, although other options are now becoming more prevalent, including portable hard drives and online backups.

While backups are designed to protect data against the outright loss caused by a drive failure, they cannot be considered a high-availability mechanism, because in the event of a failure, administrators must restore the data from the backup medium before it is again accessible to users. Depending on the nature of the failure and what files are lost, data might be unavailable for hours (or even days) before the drive is replaced and the restoration completed.

Despite these drawbacks, however, regular backups are an essential part of any network maintenance program, even when other availability mechanisms are in place.

Shadow Copies

Backups are primarily designed to protect against major losses, such as drive failures, computer thefts, and natural disasters. However, the loss of individual data files is a fairly common occurrence on most networks, typically caused by accidental deletion or user mishandling. For backup administrators, the need to locate, mount, and search backup media just to restore a single file can be a regular annoyance. However, Windows Server includes a feature called Shadow Copies that can make recent versions of data files highly available to end users.

Shadow Copies is a mechanism that automatically retains copies of files on a server volume in multiple versions from specific points in time. When users accidentally overwrite or delete files, they can access the shadow copies to restore earlier versions. This feature is specifically designed to prevent administrators from having to load backup media to restore individual files for users. Shadow Copies is a file-based fault-tolerance mechanism that does not provide protection against disk failures, but it does protect against the minor disasters that inconvenience users and administrators on a regular basis.

It is only possible to implement Shadow Copies for an entire volume, not for specific shares, folders, or files. After an administrator configures Shadow Copies, the system creates shadow copies of each file on the storage volume selected at scheduled times. Shadow copies are block-based; the system only copies the blocks in each file that have changed since the last copy. This means that, in most cases, the shadow copies take up far less storage space than the originals.

Once the server begins creating shadow copies, users can open previous versions of files on the selected volumes, either to restore those that they have accidentally deleted or overwritten, or to compare multiple versions of files as they work.

To open a shadow copy of a file, a user must select it in any File Explorer window and open its Properties sheet. On the Previous Versions tab, all of the available copies of the file appear, along with their dates. **See Figure 15-1.**

After selecting one of the file versions, the user clicks a tab to open it.

Open Open launches the application associated with the file type and loads the file from its current location. The user cannot modify the file without saving it to another location.

Copy Copy copies the file to the system's Clipboard, so that the user can paste it to any location.

Figure 15-1 Previous Versions Tab

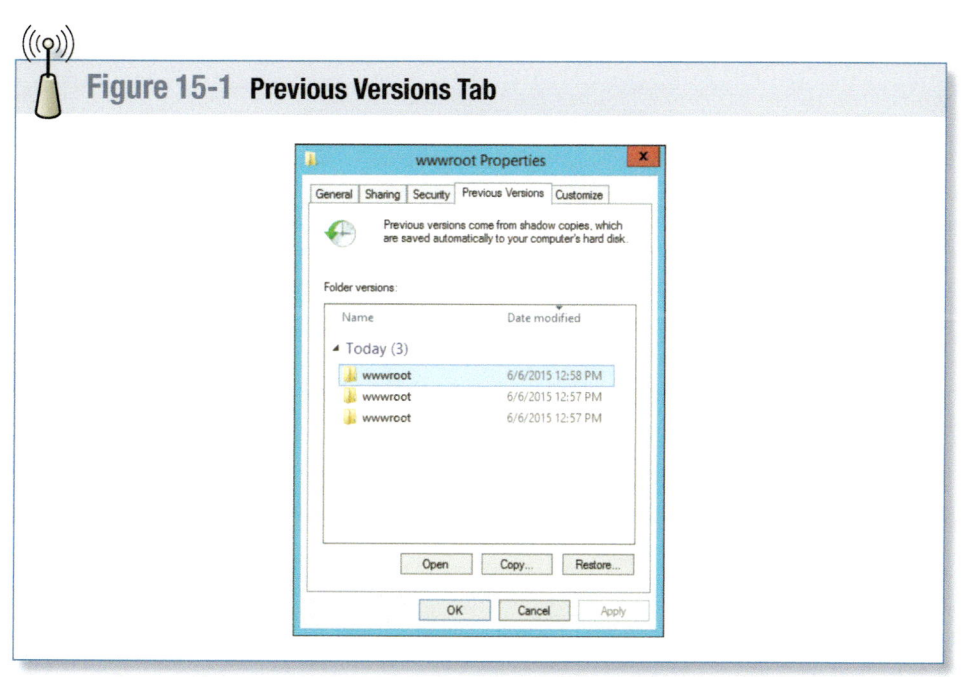

Figure 15-1. The Previous Versions tab provides users with access to Shadow Copies.

Restore Restore overwrites the current version of the file with the selected previous version. Once this is done, the current version of the file is lost and cannot be restored.

Offline Files

Offline Files is a mechanism that individual users can employ to maintain access to their server files, even if the server becomes unavailable. Offline Files works by copying server-based folders that users select for offline use to a workstation's local drive. The users then work with the copies, which remain accessible whether the workstation can access the server or not. No matter what the cause (be it a drive malfunction, a server failure, or a network outage), the users can continue to access the offline files without interruption.

When the workstation is able to reconnect to the server drive, a synchronization procedure replicates the files between server and workstation in whichever direction is necessary. If the user on the workstation has modified the file, the system overwrites the server copy with the workstation copy. If another user has modified the copy of the file on the server, the workstation updates its local copy. If there is a version conflict (such as when users have modified both copies of a file), the system prompts the user to specify which copy to retain.

Although an effective availability mechanism, primary control of Offline Files rests with the user, not the administrator, making it a less than reliable solution for an enterprise network. Administrators can configure server shares to prevent users from saving offline copies, but they cannot configure a server to force a workstation to save offline copies.

To use Offline Files, the user of the client computer must first activate the feature, using the Windows Control Panel. The Control Panel interface also provides settings that enable the user to specify how much local disk space is allotted for offline files, when synchronization events should occur, and whether the system should encrypt the locally stored offline files.

Once Offline Files is enabled, the workstation user can right-click any folder on a server share and select Always Available Offline or Make Available Offline from the context menu. The workstation will then perform an initial synchronization that copies the contents of the folder to the local disk.

From this point on, Offline Files works automatically for that selected folder. If the user is working on a protected file when access to the server share is interrupted, the system indicates that the server is no longer available, but there is no interruption to the file access, because the user is actually working with the offline copy. When the server share is once again available, the system notifies the user and offers to perform a synchronization.

Disk Redundancy

Shadow Copies and Offline Files are both effective mechanisms for maintaining data availability. However, for mission-critical data that must be continuously available, using a mechanism controlled by administrators (not end users), there is no better solution than to have redundant disks containing the same data online at once.

Storage technologies such as disk mirroring and Redundant Array of Independent Disks (RAID) enable servers to maintain multiple copies of data files online at all times, so that in the event of a disk failure, users can still access their files.

Disk redundancy is the most common type of high availability technology currently in use. Even organizations with small servers and modest budgets can benefit from redundant disks, by installing two or more physical disk drives in a server and using the disk mirroring and RAID-5 capabilities built into Windows Server. For larger servers, external disk arrays and dedicated RAID hardware products can provide more scalability, better performance, and a greater degree of availability.

Generally speaking, when planning for high availability, administrators must balance three factors: fault tolerance, performance, and expense. The more fault tolerance required for the data, the more expensive it is to achieve it, and the more likely the possibility of degraded performance as a result of it.

Disk mirroring is the simplest form of disk redundancy, and it typically does not have a negative effect on performance, as long as a disk technology that enables the computer to write to both disks at the same time is used, such as Small Computer System Interface (SCSI) or serial ATA (SATA). However, mirroring reduces by half the amount of storage space realized from the array, which is less efficient than parity-based RAID technologies, in this respect, and more expensive in the long run. However, the price per gigabyte of hard disk storage continues to fall, which makes disk mirroring a viable high-availability solution.

Parity-based RAID is the most commonly used high-availability solution for data storage, primarily because it is far more scalable than disk mirroring and enables more storage space to be realized from hard disks. In fact, the more physical disks added to the array, the greater the percentage of space that can be devoted to actual storage.

For example, a RAID 5 array that consists of three 250 GB disks yields 500 GB of available space—that is, 66% of the total storage. This is because, for each bit on the three disks, one must be devoted to the parity data that provides the fault tolerance. However, adding a fourth 250 GB disk to the array (for a total of 1,000 GB) results in 750 GB of storage, or 75%, because there is still only one out of each four bits devoted to parity information. In the same way, a five-disk array will realize 80% of its total as usable storage.

Windows Server supports RAID 5 as its only parity-based data availability mechanism. While RAID 5 is more scalable than disk mirroring (because it is always possible to add another drive to the array) and more economical (because it provides more usable storage), administrators must accept a performance hit.

Because RAID 5 distributes the parity information among the drives in the array, disk write performance is not heavily affected, as it is with RAID levels 3 and 4, which use dedicated parity disks. However, the administrator must consider the processing burden required to calculate the parity information. The RAID 5 implementation included with Windows Server uses the system's processor and memory resources to perform all of the parity calculations, and depending on the amount of I/O traffic the server must handle, this additional processor burden can be significant.

When using only the data availability mechanisms built into Windows Server, consider carefully whether to use disk

> ✓ FACT
>
> Disk duplexing is strictly a hardware modification that adds a further degree of fault tolerance to a mirrored disk environment. By mirroring disks connected to different host adapters, the system can continue to function, even if one of the host adapters fails. This is admittedly a far less likely occurrence than a hard drive failure, but fault tolerance is all about planning for unlikely occurrences.

mirroring (and pay a larger price per gigabyte of usable storage) or RAID 5 (and possibly pay more for a server with sufficient memory and processor speed to handle the parity calculations).

The alternative to the Windows Server RAID 5 mechanism is to purchase a hardware-based RAID solution, either in the form of a host adapter card that must be installed in the server, or an external disk array, which is connected to the server. The downside to this solution, obviously, is the additional cost.

The advantages to hardware-based RAID are improved performance and flexibility. These solutions include dedicated hardware that performs the parity calculations, so there is no additional burden placed on the computer's own resources. Third-party products often provide additional RAID capabilities as well, such as RAID 6, which maintains two copies of the parity data, enabling the array to survive the failure of two drives without service interruption or data loss. Some third-party RAID implementations also include hybrid RAID technologies, such as mirrored stripe sets or other proprietary combinations of striping, mirroring, and/or parity.

When selecting a data high-availability solution, consider questions like the following:

- How much data is there to protect?
- How critical is the data is to the operation of the enterprise?
- How long of an outage can the organization comfortably endure?
- How much can the organization afford to spend?

None of these high-availability mechanisms is intended to be a replacement for regular system backups. For document files that are less than critical, or files that see only occasional use, it might be more economical to keep some spare hard drives on hand and rely on backups. If a failure occurs, it is possible to replace the malfunctioning drive and restore it from the most recent backup, usually in a matter of hours. However, if access to server data is critical to the organization, the expense of a RAID solution might be seen as minimal, when compared to the lost revenue or even more serious consequences of a disk failure.

Application Availability

High availability is not limited to data. Applications also must be available for users to complete their work.

Application availability can mean different things, depending on the nature of the applications the users run. Applications can be classified into three categories: client-run, client/server, and distributed. These describe the server interaction needed by each one. Keeping each of these application types available to users at all times requires different strategies and mechanisms, some of which are simply a matter of setting configuration parameters, while others call for additional hardware and an elaborate deployment. As with data availability, the solutions must depend on how important the applications are to the organization and how much it can afford to spend.

When a standalone application is installed on a client workstation, it will obviously continue to run, even if the computer is disconnected from the network for any reason. However, an application might still require access to the network at least occasionally to download updates needed to keep it current. Network connectivity aside, a standalone application can also be rendered unavailable by a local disk failure or an accidentally deleted executable or other application file.

Client/server applications are more liable to be rendered unavailable than client-run applications, because they require both the client and server components to function properly, and the two must also be able to communicate. As a result, damage to either component or any sort of network outage can prevent the application from functioning. Distributed applications are even more sensitive, because they have more components that can go wrong.

For example, a web-based application that accesses a database requires three separate application components (the web browser, the web server, and the database server) to function properly on three different computers. There are also two network connections (browser-to-web server and web server-to-database server) that must be operational as well. If any one of these components or connections fails, the application becomes unavailable.

Increasing Application Availability

An application running on workstations (whether it is a standalone application or the client half of a client/server application) is left to the responsibility of the end users. If a user deletes the wrong file or damages the computer in some other way, he or she might disable the application.

One way of protecting workstation applications and ensuring their continued availability is to run them using Remote Desktop Services (RDS). With RDS, users can access applications that are actually running on a server, rather than their own workstations. Because the server administrators have ultimate control over the application files, they can protect them against accidental deletion and other types of damage.

In addition, deploying applications using RDS makes it far easier to replace a workstation that has experienced a catastrophic failure of some type. For example, if a user's workstation experiences a complete hard disk failure, the applications on the computer are obviously rendered unavailable. The user's productivity stops.

Replacing or repairing the computer and installing all of the applications the user needs can be a lengthy process, all during which the user sits idle. If the applications are deployed on RDS servers, an administrator can supply the user with a temporary workstation containing just the operating system. The user can then run his or her applications from the RDS servers and get back to work almost immediately.

Virtualization (such as that provided by Hyper-V in Windows Server) is another way of increasing application availability. Administrators can move virtual machines from one computer to another relatively easily, in the event of a hardware failure, or when the applications running on the virtual machines require more or different hardware resources.

Application Resilience

Application resilience refers to the capability of an application to maintain its own availability by detecting outdated, corrupted, or missing files and automatically correcting the problem.

There are a number of methods administrators can use to make applications resilient, most of which are integrated into application deployment tools.

Application Availability Using Group Policy

Administrators can use Group Policy to deploy application packages to computers and/or users on the network. When an administrator assigns a software package to a computer, the client installs the package automatically when the system boots. When an administrator assigns a package to a user, the client installs the application when the user logs on to the domain, or when the user invokes the software by double-clicking an associated document file.

Both of these methods enforce a degree of application resilience, because even if the user manages to uninstall the application, the system will reinstall it during the next startup or domain logon. This is not a foolproof system, however. Group Policy will not recognize the absence of a single application file, as some other mechanisms do.

Application Availability Using Windows Installer

Windows Installer is the component in Windows Server that enables the system to install software packaged as files with an .msi extension. One of the advantages of deploying software in this manner is the built-in resiliency that Windows Installer provides to the applications.

When an .msi package is deployed (either manually or using an automated solution, such as Group Policy or System Center Configuration Manager), Windows Installer creates special shortcuts and file associations that function as entry points for the applications contained in the package. When a user invokes an application using one of these entry points, Windows Installer intercepts the call and verifies the application to make sure that its files are intact and all required updates applied before executing it.

Implementing this resilience is up to the developers of the application. In organizations that create their own software in-house, the server administrators should ensure that the application developers are aware of the capabilities built into Windows Installer, so that they can take advantage of them.

Server Availability

Server clustering can provide two forms of high availability on an enterprise network. In addition to providing fault tolerance in the event of a server failure, it can provide network load balancing for busy applications. As mentioned earlier, high availability is typically a matter of using redundant components to ensure the continued functionality of a particular network resource. A database stored on a RAID array remains available to users, even if one of the hard disk drives fails. An application installed on an RDS server is readily available to users from any workstation.

Servers themselves can suffer failures that render them unavailable. Hard disks are not the only computer components that can fail, and one way of keeping servers available is to equip them with redundant components other than hard drives. The ultimate in fault tolerance, however, is to have entire servers that are redundant, so that if anything goes wrong with one computer, another one can take its place almost immediately.

In Windows Server, this is known as a failover cluster. *Network load balancing* is a type of clustering, useful when a web server or other application becomes overwhelmed by a large volume of users, in which multiple identical servers (also known as a server farm) are deployed and the user traffic is distributed evenly among them.

Hardware Redundancy

Servers contain the same basic components as workstations: processors, memory, hard disks, and so forth. In many cases, what differentiates a server from a workstation (apart from the size, speed, and expense of the individual components) is the inclusion of secondary hardware that provides enhanced or redundant functions.

Servers often have faster processors than workstations, or multiple processors. They also tend to have more, and larger, hard drives installed. One of the byproducts of this extra hardware is additional heat. There are few things that can kill a computer faster than uncontrolled heat generation, and many server computers are equipped with cooling systems that include multiple redundant fans that can run at various speeds.

When the computer's motherboard senses a temperature rise inside the case, it can switch the fans to a higher speed and, in some cases, warn an administrator of the problem. Because there are multiple cooling fans in most servers, this variable speed capability can also compensate for the failure of one or more fans.

Another critical component to server operation that is known to be fallible is the computer's power supply. The power supply connects to an AC power source and provides all of the components in the computer with the various voltages they need to operate. A power supply can fail because of factors such as power surges or excessive heat, and when it does, the computer stops. To prevent this, some servers have redundant power supplies, so that the system can continue running even if one power supply fails.

There are other types of specialized hardware often found in servers as well. Some of these (such as hot-swappable drive arrays) are designed to enable administrators to make repairs while the server is still running.

Failover Clustering

A *failover cluster* is a collection of two or more servers that perform the same role or run the same application and appear on the network as a single entity.

Windows Server includes a Failover Cluster Management console that enables administrators to create and configure failover clusters after setting up an appropriate environment. Before creating a failover cluster in Windows Server, an administrator must install the Failover Clustering feature.

Failover Cluster Requirements

Failover clusters are intended for critical applications that must keep running. If an organization is prepared to incur the time and expense required to deploy a failover cluster, Microsoft assumes that the organization is prepared to take every possible step to ensure the availability of the application. As a result, the recommended hardware environment for a failover cluster calls for an elaborate setup, including the following:

Duplicate Servers The computers that will function as cluster nodes should be as identical as possible in terms of memory, processor type, and other hardware components.

Shared Storage All of the cluster servers should have exclusive access to shared storage, such as that provided by a Fibre Channel or iSCSI storage area network (SAN). This shared storage will be the location of the application data, so that all of the cluster servers have access to it. The shared storage can also contain the witness disk. A *witness disk* holds the cluster configuration database. This, too, should be available to all of the servers in the cluster.

Redundant Network Connections Connect the cluster servers to the network in a way that avoids a single point of failure. Each server can be connected to two separate networks or a single network can be built using redundant switches, routers, and network adapters.

Shared storage is a critical aspect of server clustering. In a pure failover cluster, all of the servers must have access to the application data on a shared storage medium. However, there cannot be two instances of the application accessing the same data at the same time. For example, if a database application is running on a failover cluster, only one of the cluster servers is active at any one time. If two servers accessed the same database file at the same time, they could modify the same record simultaneously, causing data corruption.

In addition to the hardware recommendations, the cluster servers should use the same software environment, which consists of the following elements:

Operating System All of the servers in a cluster must be running the same edition of the same operating system.

Application All of the cluster servers must run the same version of the redundant application.

Updates All of the cluster servers must have the same operating system and application updates installed.

Active Directory All of the cluster servers must be in the same Active Directory domain, and they must be either member servers or domain controllers. Microsoft recommends that all cluster servers be member servers, not domain controllers. Do not mix member servers and domain controllers in the same failover cluster.

Before creating a failover cluster, connect all of the hardware to the networks involved and test the connectivity of each device. Ensure that every cluster server can communicate with the other servers and with the shared storage device.

Failover Cluster Configuration Validation

The Failover Cluster Management console is included with Windows Server as a feature. After installing the feature, an administrator can start to create a cluster by validating the hardware configuration. The Validate a Configuration Wizard performs an extensive battery of tests on selected computers, enumerating their hardware and software resources and checking their configuration settings. If any elements required for a cluster are incorrect or missing, the wizard lists them in a report.

Failover Cluster Creation

After validating the cluster configuration and correcting any problems, the administrator can create the cluster. A failover cluster is a logical entity that exists on the network, with its own name and IP address, just like a physical computer.

After creating the cluster, the administrator uses the Failover Cluster Management console to specify the applications the cluster will manage. If a server fails, the selected applications are immediately executed on another server to keep them available to clients at all times.

Network Load Balancing

Network load balancing (NLB) differs from failover clustering because its primary function is not fault tolerance, but rather the more efficient support of heavy user traffic.

If an Internet website experiences a sudden increase in traffic, the web server could be overwhelmed, causing performance to degrade. To address the problem, administrators can add another web server that hosts the same site, but how can they ensure that the incoming traffic is split equally between the two servers? Network load balancing (NLB) is one possible answer.

In a failover cluster, only one of the servers is running the protected application at any given time. In network load balancing, all of the servers in the cluster are operational and able to service clients. The *NLB cluster* itself, like a failover cluster, is a logical entity with its own name and IP address. Clients connect to the cluster, rather than the individual computers, and the cluster distributes the incoming requests evenly among its component servers.

Because all of the servers in an NLB cluster can actively service clients at the same time, this type of cluster is not appropriate for database and e-mail applications, which require exclusive access to a data store. NLB is more appropriate for applications that have their own data stores, such as web servers. A website can easily be replicated to multiple servers on a regular basis, enabling each computer to maintain a separate copy of the data it provides to clients.

NLB ClusterCreation

To create and manage NLB clusters on a Windows Server computer, the Network Load Balancing feature must first be installed. This feature also includes the Network Load Balancing Manager console. After creating the NLB cluster itself, administrators can add servers to and remove them from the cluster as needed.

The process of implementing an NLB cluster consists of the following tasks:

- Creating the cluster
- Adding servers to the cluster
- Specifying a name and IP address for the cluster
- Creating port rules that specify which types of traffic the cluster should balance among the cluster servers

After creating the NLB cluster, administrators can add servers to it at will by using the Network Load Balancing Manager console from any computer. As each server is added, the Network Load Balancing service automatically incorporates it into the cluster. The Network Load Balancing feature must be installed on each server before it can be added to the cluster, but administrators can manage the cluster from any server running the Network Load Balancing Tools feature.

Once the cluster is operational, the only modification on the part of the clients is that they connect to the cluster, using the name or IP address specified in the New Cluster wizard, not to the individual servers in it. In the case of a web server, switching from a single server to an NLB cluster would mean changing the Domain Name System (DNS) record for the web server's name to reflect the IP address of the cluster, not the address of the original web server. Users would still direct their browsers to the same URL, but the DNS would resolve that name to the cluster IP address instead of to an individual server's IP address.

The servers in an NLB cluster continually exchange status messages with each other, known as heartbeats. *Heartbeats* enable an NLB cluster to check the availability of each server. When a server fails to generate five consecutive heartbeats, the cluster initiates a process called convergence. *Convergence* stops an NLB cluster from sending clients to a server that is not functioning or offline. When the offending server is operational again, the cluster detects the resumed heartbeats and again performs a convergence, this time to add the server back into the cluster. These convergence processes are entirely automatic, so administrators can take a server offline at any time, for maintenance or repair, without disrupting the functionality of the cluster.

This type of server clustering is intended primarily to provide a scalable solution that can use multiple servers to handle large amounts of client traffic. However, NLB clustering provides fault tolerance as well. If a server in the cluster should fail, an administrator can simply remove it from the cluster, and the other servers will take up the slack.

The main thing to remember when managing this type of cluster is that when making any changes to a clustered application on one server, the administrator must change the other servers in the same way. For example, if an administrator adds a new page to a website, someone must update all of the servers in the cluster with the new content.

RDS Servers Load Balancing

Windows Server supports the use of network load balancing for terminal servers in a slightly different manner. For any organization with more than a few RDS clients, multiple servers are required. Network load balancing can ensure that the client sessions are distributed evenly among the servers.

One problem inherent in the load balancing of RDS servers is that a client can disconnect from a session (without terminating it) and be assigned to a different server when he or she attempts to reconnect later. To address this problem, the Remote Desktop Services role includes the Remote Desktop Connection Broker role service, which maintains a database of client sessions and enables a disconnected client to reconnect to the same RDS server.

The process of deploying Remote Desktop Services begins with the creation of an RDS server farm.

RDS Server Farm

To create a load-balanced terminal server farm, the Remote Desktop Services role with the Remote Desktop Session Host role service must be installed on at least two Windows Server computers. The Remote Desktop Connection Broker role service must also be installed on one computer. The computer running Remote Desktop Connection Broker can be (but does not have to be) one of the RDS servers. The RDS computers are subject to the following requirements:

- The RDS computers must be running the same version of Windows Server.
- The RDS servers and the computer running Remote Desktop Connection Broker must be members of the same Active Directory domain.
- The RDS servers must be configured identically, with the same installed applications.
- Clients connecting to the RDS server farm must run Remote Desktop Connection (RDC) version 5.2 or later.

When the Remote Desktop Connection Broker role service is installed, the system installs the Remote Desktop Connection Broker. The server must then be added to an RDS server farm and repeated on all of the other RDS servers.

To automate the configuration process, administrators can apply these settings to an organization unit (OU) using Group Policy. The Remote Desktop Connection Broker settings are located in the Computer Configuration/Policies/Administrative Templates/Windows Components/Remote Desktop Session Host/RD Connection Broker node of a group policy object. **See Figure 15-2**.

When RDS servers are configured to use the load-balancing capability built into Remote Desktop Connection Broker, the client connection process proceeds as follows:

1. A client attempts to connect to one of the RDS servers in the server farm.
2. The RDS server sends a query to the Remote Desktop Connection Broker server, identifying the client attempting to connect.
3. The Remote Desktop Connection Broker searches its database to see if the specified client is already in an existing session.

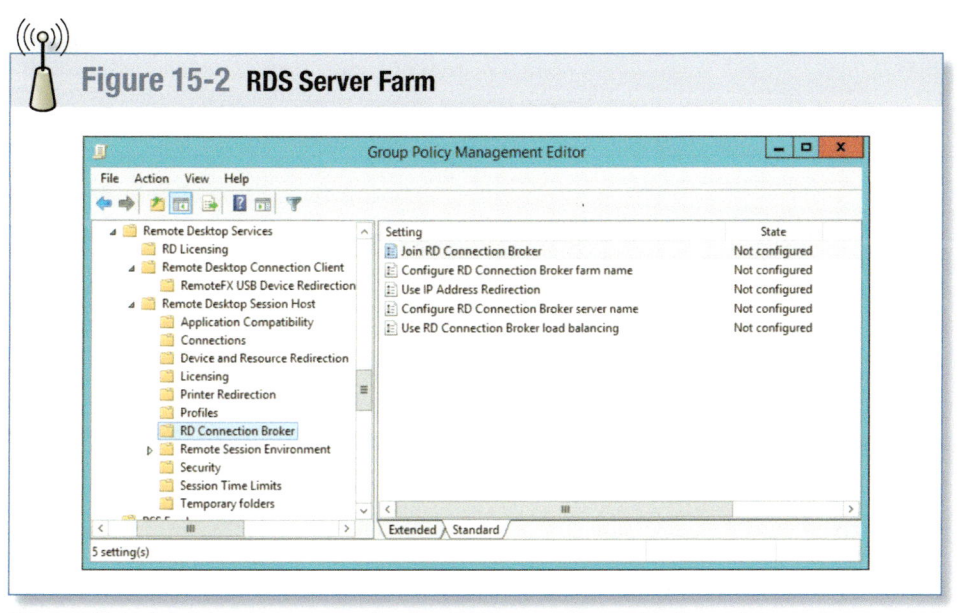

Figure 15-2. Administrators can configure an RDS server farm using Group Policy.

4. The Remote Desktop Connection Broker server sends a reply to the RDS server, instructing it to redirect the connection to one of the servers in the farm. If a session already exists for the client, the Remote Desktop Connection Broker server redirects it to the RDS server running that session. If a session does not exist for the client, the Remote Desktop Connection Broker server redirects it to the RDS server with the fewest sessions.

5. The RDS server forwards the client connection to the computer specified by the Remote Desktop Connection Broker server.

DNS Round Robin

While Remote Desktop Connection Broker is an effective method for keeping the sessions balanced among RDS servers, it does nothing to control which server receives the initial connection requests from clients on the network. To balance the initial connection traffic among the RDS servers, an NLB cluster can be used, as described earlier in this lesson, or another, simpler load-balancing technique called DNS Round Robin can be used.

When a client connects to a server, the user typically specifies the server by name. The client's computer then uses the DNS to resolve the name into an IP address, and then uses the address to establish a connection to the server. Under normal circumstances, the DNS server has one resource record for each hostname. The resource record equates the name with a particular IP address, enabling the server to receive requests for the name and respond with the address. Therefore, when the DNS server receives a request for a given name, it always responds with the same IP address, causing all clients to connect initially to the same server.

In the case of an RDS server, this is undesirable, even if there is a Remote Desktop Connection Broker server in place. In the *DNS Round Robin* technique, multiple resource records are created using the same name, with a different server IP address in each record. When clients attempt to resolve the name, the DNS server supplies them with each of the IP addresses in turn. As a result, the clients are evenly distributed among the servers.

For example, to distribute traffic among five web servers using DNS Round Robin, an administrator should create five host (A) resource records in the contoso.com zone, all with the name www. Each resource record should have the IP address of one of the five servers. By doing this, the administrator is essentially creating a cluster called www.contoso.com, although there is no software entity representing that cluster, as there is in NLB. Once the DNS Round Robin feature is activated on the DNS server, incoming name resolution requests for the www.contoso.com name use each of the five IP address associated with that name in turn.

DNS Round Robin is a simple mechanism, and it enables administrators to create a basic NLB cluster without installing additional services or performing elaborate configurations. However, when compared with the Network Load Balancing service, DNS Round Robin has several disadvantages.

The main disadvantage of DNS Round Robin is that the DNS server has no connection with the servers in the cluster. This means that if one of the servers fails, the DNS server will continue trying to send clients to it, until an administrator deletes that server's resource record. The DNS server also has no conception of how the clients make use of the servers in the cluster.

For example, some clients might connect to a web server cluster and view one web page, while others might download huge video files. The DNS servers balance the incoming client connection requests among the servers in the cluster evenly, regardless of the burden on each server.

Another disadvantage is that DNS servers cache resource records for reuse by subsequent clients. When a client on the Internet connects to a DNS Round Robin cluster, its DNS server caches only one of the resource records, not all of them. Therefore, any other clients using the DNS server will all connect to the same server, throwing off the distribution of cluster addresses.

Generally speaking, the Network Load Balancing service is far superior to DNS Round Robin in its capabilities, but it's much more difficult to implement.

DISASTER PREPARATION

Disaster recovery is a set of policies and technologies that enable an organization to maintain its essential IT services after a disaster—natural or otherwise—threatens them. Disaster-recovery mechanisms can include:

Preventive Measures Preventive measures are used to minimize the possibility of disasters.

Preventive measures can include the redundant hardware technologies such as duplicate power supplies, hard disk drives, and other components, as well as uninterruptible power supplies and fire prevention systems. On the software side, anti-virus and anti-malware products can prevent users from falling prey to outside intrusion, as can data encryption mechanisms.

Failover clusters and NLB installations can provide for rapid disaster recovery in the event of a server failure. This is particularly true when the servers are installed in different locations, in case of fire or other disasters. The increasing popularity of cloud-based servers and services also functions as a disaster-prevention mechanism, because the technology is stored offsite in secure locations, and service providers typically contract for a specified level of service.

Discovery Measures Discovery measures are used to report the occurrence (or possible occurrence) of disasters to administrators.

Discovery measures can include network management products such as System Center Configuration Manager, as well as burglar alarms and other notification systems.

Corrective Measures Corrective measures are used to mitigate the effects of disasters and restore systems to operational status.

Corrective measures can include data-recovery mechanisms such as offsite backups, but the most important part of a disaster-recovery plan is a comprehensive set of policies and instructions, so that everyone involved knows what to do in the event of a disaster.

Summary

Several different methods can be used to protect against malware, provide increased availability and fault tolerance, and prepare for disasters. Disaster preparation includes preventative measures, discovery measures, and corrective measures.

Review Questions

1. In addition to the application data, which of the following is stored on a failover cluster's shared storage?
   a. The Failover Cluster management console
   b. The Shadow Copies database
   c. The TS Session Broker database
   d. The witness disk

2. Which of the following Windows Server features enables users to access files that they have accidentally overwritten?
   a. Offline files
   b. Parity-based RAID
   c. Shadow Copies
   d. Windows Installer 4.0

3. Which of the following server-clustering solutions requires a storage area network?
   a. DNS Round Robin
   b. A failover cluster
   c. A network load balancing cluster
   d. A terminal server farm

4. To use Shadow Copies, the feature must be enabled at which of the following levels?
   a. The file level
   b. The folder level
   c. The server level
   d. The volume level

5. How many heartbeat messages must a server in an NLB cluster miss before it is removed from the cluster?
   a. 1
   b. 5
   c. 50
   d. 500

6. Which of the following is not a requirement to use the Remote Desktop Connection Broker role service?
   a. All of the RDS computers must be configured identically.
   b. All of the RDS computers must be running the same Windows version.
   c. The RDS computers must all be member of the same Active Directory domain.
   d. The Remote Desktop Connection Broker role service must be installed on one of the terminal servers.

7. Which of the following statements is true about DNS Round Robin load balancing?
   a. DNS Round Robin requires the creation of multiple resource records containing the same hostname.
   b. To use DNS Round Robin, a cluster resource record specifying the cluster name must be created.
   c. When one of the servers in the cluster fails, the Windows Server DNS server performs a convergence and disables the resource record for that server.
   d. When the DNS server receives a name resolution request for a cluster, it replies with all of the resource records for that cluster name.

Management and Administration

Managing and administering a network requires many different tools, skills, and techniques. It is important to understand some of the fundamental tools and tasks that make future administrative chores easier, such as network management systems and documenting a network.

OBJECTIVES

- ▶ Describe the purpose of SNMP
- ▶ Explain the process of creating a subnet
- ▶ Identify several advantages for using network monitoring software
- ▶ Create a network process flowchart
- ▶ List several types of documentation that network administrators often use

CHAPTER 16

TABLE OF CONTENTS

SNMP Communications 306
    SNMP Agents. 306
    SNMP Versions 307

Subnet Calculations 307
    Network Address Subnetting. 308
    Subnetting Between Bytes 309
    Binaries and Decimals 310
    IP Address Calculation with the
    Subtraction Method 311

Network Monitoring Software 311

Network Process Flowcharts 312

Network Documentation 312
    Cable Diagrams 314
    Network Diagrams 315
    Hardware Configurations 315
    Change Management. 315
    Baselines 317

Summary 319

Review Questions 319

SNMP COMMUNICATIONS

When an application or an operating system experiences a problem, it usually generates an error message. These error messages are easy to monitor by reviewing logs, but receiving error messages from other network components (such as routers and switches) can be more difficult.

A hardware router does not have a screen on which it can display error messages, but it does usually have an administrative interface that is accessible through a remote connection. However, even with this capability, it is difficult for an administrator who is responsible for dozens or hundreds of devices to monitor them all. In this case, it is possible to arrange for many networking devices to supply administrators with information about their status.

Network management products are designed to provide administrators with a comprehensive view of network systems and processes, using a distributed architecture based on a specialized management protocol, such as the Simple Network Management Protocol (SNMP) or the Remote Monitoring (RMON) protocol.

The *Simple Network Monitoring Protocol* is a TCP/IP application layer protocol and query language that specially equipped networking devices use to communicate with a central console. Many of the networking hardware and software products on the market (including routers, switches, network adapters, operating systems, and applications) are equipped with SNMP agents.

SNMP Agents

An *SNMP agent* is a software module that is responsible for gathering information about a device and delivering it to a computer that has been designated as the network management console. The agents gather specific information about the network devices and store them as managed objects in a *management information base* (*MIB*). At regular intervals, the agents transmit their MIBs to the console using SNMP messages, which are carried inside User Datagram Protocol (UDP) datagrams. The agents use UDP port 161 and the management console uses port 162.

The network management console processes the information that it receives from the agents in SNMP messages and provides the administrator with a composite picture of the network and its processes. The console software can usually create a map of the interconnections between network devices, as well as display detailed log information for each device. In the event of a serious problem, an agent can generate a special message called a *trap*, which it transmits immediately to the console, causing it to alert the administrator of a potentially dangerous condition. In many cases, it is possible to configure the console software to send alerts to administrators in a variety of ways, including e-mails and text messages.

In addition to network reporting capabilities, network management products can provide other functions as well:

- Software distribution and metering
- Network diagnostics
- Network traffic monitoring
- Report generation

Network management products are available with a wide range of capabilities, ranging from relatively modest open source packages to extremely

---

✓ **FACT**

The language used by hardware and software manufacturers to identify SNMP-capable devices is not consistent. But whenever a network interface adapter, switch, router, access point, or other device is purported to be managed, or is claimed to have network management capabilities, this generally means that the device includes an SNMP agent.

complex and expensive, commercial products.

Deploying a network management system is a complex undertaking that is intended for administrators of large networks that cannot possibly monitor all of their network devices individually. To use a product like this effectively, for example, administrators must be sure that all of the equipment they purchase when designing and building a network supports the network management protocol they intend to use. However, products like these can greatly simplify the tasks of network administrators, and they can often bring serious problems to the administrator's attention before they cause serious outages.

SNMP Versions

The first version of the SNMP standard (which the Internet Engineering Task Force (IETF) published in 1988 as RFC 1065, RFC 1066, and RFC 1067) provides the protocol's basic functionality, but is hampered by shortcomings in security. SNMPv1 messages contain no protection other than a community string, which functions as a password, and which the systems transmit in clear text.

SNMPv2, released in 1993, adds some improvements in functionality. Version 2 also includes a new security system that many people criticized as being overly complex. Such was the resistance to this system that an interim version appeared, called SNMPv2c, which consists of SNMP version 2 without the new security system, and with the old version 1 community string instead.

In 2002, the IETF published an SNMP standard with a workable security solution, which became version 3, and was ratified as an Internet standard. SNMPv3 includes all of the standard security services administrators have come to expect, including authentication, message integrity, and encryption. Many network management products that implement SNMPv3 also include support for the earlier, unprotected versions, such as SNMPv1 and SNMPv2c.

SUBNET CALCULATIONS

Originally, IP addresses were divided into three classes, in which the network IDs are 8, 16, and 24 bits, respectively. **See Figure 16-1.**

Figure 16-1 Classful IP Addresses

CLASS	FIRST NUMBER	RANGE OF IP ADDRESSES	CLASSFUL SUBNET MASK	EXAMPLE
Class A	1 to 126	1.0.0.0 to 126.255.255.254	255.0.0.0	10.80.1.15
Class B	128 to 191	128.0.0.0 to 191.255.255.254	255.255.0.0	172.16.32.15
Class C	192 to 223	192.0.0.0 to 223.255.255.254	255.255.255.0	192.168.1.5

Figure 16-1. IPv4 addresses starting with 127 are reserved for traffic not intended to leave a computer.

It might at first seem odd that the IP address classes are defined as they are. After all, there are no private networks that have 16 million hosts on them, so it makes little sense even to have Class A addresses. In the early days of IP, no one worried about the depletion of the IP address space, so the wastefulness of the classful addressing system was not an issue.

Eventually, this wastefulness was recognized, and the designers of the protocol developed a system for subdividing network addresses by creating subnets on them. A subnet is simply a subdivision of a network address that administrators can use to represent a part of a larger network, such as one LAN on an internetwork or the client of an ISP. Thus, a large ISP might have a Class A address registered to it, and it might allocate sections of that network address to its clients in the form of subnets. In many cases, a large ISP's clients are smaller ISPs, which in turn supply addresses to their own clients.

To understand the process of creating subnets, it is important to understand the function of the subnet mask. TCP/IP systems at one time recognized the class of an address simply by examining the values of the first three bits. Today, however, when administrators configure the TCP/IP client on a computer, they assign it an IPv4 address and a subnet mask. Simply put, the subnet mask specifies which bits of the IP address are the network identifier and which bits are the host identifier. For a Class A address, for example, the default subnet mask value is 255.0.0.0.

When expressed as a binary number, a subnet mask's 1 bits indicate the network identifier, and its 0 bits indicate the host identifier. A mask of 255.0.0.0 in binary form is as follows:

11111111 00000000 00000000 00000000

This mask indicates that the first 8 bits of a Class A IP address are the network identifier bits and the remaining 24 bits are the host identifier. The default subnet masks for the three main address classes are as follows:

▶ Class A: 255.0.0.0
▶ Class B: 255.255.0.0
▶ Class C: 255.255.255.0

Network Address Subnetting

If all the IP addresses in a particular class used the same number of bits for the network and host identifiers, there would be no need for a subnet mask. The value of the first byte of the address would indicate its class. However, administrators can create multiple subnets, using a single address of a given class, by applying a different subnet mask. For example, with a Class B address, the default subnet mask of 255.255.0.0 would allocate the first 16 bits for the network identifier and the last 16 bits for the host identifier. However, using a mask of 255.255.255.0 with a Class B address allocates an additional 8 bits to the network identifier, borrowed from the host identifier. The third byte of the address thus becomes part of the network identifier. **See Figure 16-2.**

By subnetting in this way, it is possible to create up to 256 subnets using that one address, with up to 254 network interface adapters on each subnet. Using Classless Inter-Domain Routing (CIDR) notation, an IP address of 131.107.67.98/24 would use a subnet mask of 255.255.255.0 and would therefore indicate that the network is

Figure 16-2 Multiple Subnets

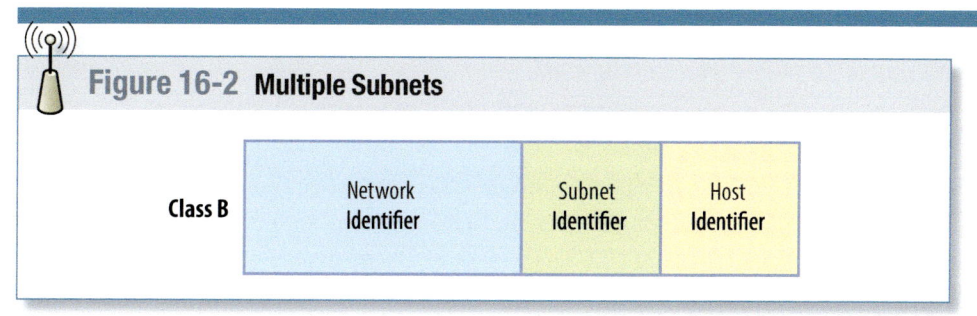

Figure 16-2. Changing the subnet mask makes it possible to create multiple subnets out of one network address.

using the classful network address 131.107.0.0, and that the interface is host number 98 on subnet 67. A large corporate network might use this scheme to create a separate subnet for each of its LANs.

Subnetting Between Bytes

To complicate matters further, the boundary between the network identifier and the host identifier does not have to fall between two bytes. An IP address can use any number of bits for its network address, and more complex subnet masks are required in this type of environment. For example, suppose there is a Class C network address of 192.168.65.0 that is to be subnetted. There are already 24 bits devoted to the network address, and it is obviously not possible to allocate the entire fourth byte as a subnet identifier, or there would be no bits left for the host identifier. It is possible, however, to allocate part of the fourth byte. By using 4 bits of the last byte for the subnet identifier, there are 4 bits left for the host identifier. To do this, the binary form of the subnet mask must appear as follows:

11111111 11111111 11111111 11110000

The decimal equivalent of this binary value is 255.255.255.240 because 240 is the decimal equivalent of 11110000. This leaves a 28-bit subnet identifier and a 4-bit host identifier, which means that it is possible to create up to 16 subnets with 14 hosts on each one. Figuring out the correct subnet mask for this type of configuration is relatively easy. Figuring out the IP addresses that must be assigned to workstations is more difficult. To do this, the 4 subnet bits must be incremented separately from the 4 host bits. Once again, this is easier to understand when looking at the binary values. The 4-bit subnet identifier can have any one of the following 16 values:

0000 0001 0010 0011 0100 0101 0110 0111 1000 1001 1010 1011 1100 1101 1110 1111

Each one of these subnets can have up to 14 workstations, with each host identifier having any one of the same values except for 0000 and 1111. Thus, to calculate the value of the IP address's fourth byte, combine the binary values of the subnet and host identifiers and convert them to decimal form. For example, the first host (0001) on the second subnet (0001) would have a fourth-byte binary value of 00010001, which in decimal form is 17. Thus, the IP address for this system would be 192.168.65.17 and its subnet mask would be 255.255.255.240.

The last host on the second subnet would use 1110 as its host identifier, making the value of the fourth byte

00011110 in binary form, or 30 in decimal form, for an IP address of 192.168.65.30. Then, to proceed to the next subnet, increment the subnet identifier to 0010 and the host identifier back to 0001, for a binary value of 00100001, or 33 in decimal form. Obviously, the IP addresses used on a network like this do not increment normally. The numbers 31 and 32 cannot be used because they represent the broadcast address of the second subnet and the network address of the third subnet, respectively. They must be computed carefully to create the correct values.

Binaries and Decimals

Part of the difficulty in calculating IP addresses and subnet masks is converting between decimal and binary numbers. The easiest way to do this, of course, is to use a calculator. Most scientific calculators work with binary as well as decimal numbers and can usually convert between the two. However, it is also useful to be able to perform the conversions by hand.

To convert a binary number to a decimal, assign a numerical value to each bit, starting at the right with 1 and proceeding to the left, doubling the value each time. The values for an 8-bit number are therefore as follows:

| 128 | 64 | 32 | 16 | 8 | 4 | 2 | 1 |

Then line up the values of the 8-bit binary number with the eight conversion values, as shown here:

| 128 | 64 | 32 | 16 | 8 | 4 | 2 | 1 |
| 1 | 1 | 1 | 0 | 0 | 0 | 0 | 0 |

Finally, add together the conversion values for the 1 bits only:

| 128 | +64 | +32 | +0 | +0 | +0 | +0 | +0 | =224 |
| 1 | 1 | 1 | 0 | 0 | 0 | 0 | 0 | |

Therefore, the decimal equivalent of the binary value 11100000 is 224.

At times it might be necessary to convert decimal numbers into binaries. To do this, use the same basic process in reverse, by subtracting the conversion values from the decimal to be converted, working from left to right. For example, to convert the decimal number 202 into binary form, subtract the conversion value 128 from 202, leaving a remainder of 74. Because it was possible to subtract 128 from 202, put a value of 1 in the first binary bit as follows:

| 128 | 64 | 32 | 16 | 8 | 4 | 2 | 1 |
| 1 | 0 | 0 | 0 | 0 | 0 | 0 | 0 |

Then subtract 64 from the remaining 74, leaving 10, so the second binary bit has a value of 1 also:

| 128 | 64 | 32 | 16 | 8 | 4 | 2 | 1 |
| 1 | 1 | 0 | 0 | 0 | 0 | 0 | 0 |

It is not possible to subtract 32 or 16 from the remaining 10, so the third and fourth binary bits are 0:

| 128 | 64 | 32 | 16 | 8 | 4 | 2 | 1 |
| 1 | 1 | 0 | 0 | 0 | 0 | 0 | 0 |

It is possible to subtract 8 from 10, leaving 2, so the fifth binary bit is a 1:

| 128 | 64 | 32 | 16 | 8 | 4 | 2 | 1 |
| 1 | 1 | 0 | 0 | 1 | 0 | 0 | 0 |

It is not possible to subtract 4 from 2, so the sixth binary bit is a 0, but 2 can be subtracted from 2, so the seventh bit is a 1. There is now no remainder left, so the eighth bit is a 0, completing the calculation as follows:

| 128 | 64 | 32 | 16 | 8 | 4 | 2 | 1 |
| 1 | 1 | 0 | 0 | 1 | 0 | 1 | 0 |

Therefore, the binary value of the decimal number 202 is 11001010.

> ✓ FACT
> A number of software tools are available that can simplify the process of calculating IP addresses and subnet masks for complex subnetted networks.

IP Address Calculation with the Subtraction Method

Manually calculating IP addresses by using binary values can be a slow and tedious task, especially if there are hundreds or thousands of computers on the network. However, when the subnet mask for the network is known and the relationship between the subnet and host identifier values is understood, IP addresses can be calculated without having to convert them to binary values.

To calculate the network address of the first subnet, begin by taking the decimal value of the octet in the subnet mask that contains both subnet and host identifier bits and subtracting it from 256. For example, with a network address of 192.168.42.0 and a subnet mask of 255.255.255.224, the result of 256 minus 224 is 32. The network address of the first subnet is therefore 192.168.42.0 and the second is 192.168.42.32. To calculate the network addresses of the other subnets, repeatedly increment the result of the previous subtraction by itself. For example, if the network address of the second subnet is 192.168.42.32, the addresses of the remaining six subnets are as follows:

192.168.42.64
192.168.42.96
192.168.42.128
192.168.42.160
192.168.42.192
192.168.42.224

To calculate the IP addresses in each subnet, repeatedly increment the host identifier by one. The IP addresses in the first subnet are therefore 192.168.42.1 to 192.168.42.30 and the addresses of the second are 192.168.42.33 to 192.168.42.62. The 192.168.42.31 and 192.168.42.63 addresses are omitted because these addresses have binary host identifier values of 11111, which are broadcast addresses. The IP address ranges for the subsequent subnets are as follows:

192.168.42.65 to 192.168.42.94
192.168.42.97 to 192.168.42.126
192.168.42.129 to 192.168.42.158
192.168.42.161 to 192.168.42.190
192.168.42.193 to 192.168.42.222
192.168.42.225 to 192.168.42.254

NETWORK MONITORING SOFTWARE

A protocol analyzer is one of the most powerful tools for learning about, understanding, and monitoring network communications. A *protocol analyzer*—sometimes called a packet sniffer—captures a sample of the traffic passing over the network, decodes the packets into the language of the individual protocols they contain, and allows an administrator to examine them in minute detail. Some protocol analyzers can also compile network traffic statistics, such as the number of packets using each protocol, and the number of collisions that are occurring on the network.

Using a protocol analyzer to capture and display network traffic is relatively easy, but interpreting the information that the analyzer presents and using it to troubleshoot the network requires a detailed understanding of the protocols running on the network. However, there is no better way to acquire this type of knowledge than to examine the actual data transmitted over a live network. However, it should be noted that permission must be obtained to do this at work.

> ✓ FACT
>
> Protocol analyzers are useful tools in the hands of experienced network administrators, but they can also be used for malicious purposes. In addition to displaying the information in the captured packets' protocol headers, the analyzer can also display the data carried inside the packets. This can sometimes include confidential information, such as unencrypted passwords and personal correspondence. If possible, do not permit users to run protocol analyzers unsupervised.

A protocol analyzer is typically a software product that runs on a computer connected to a network. On an Ethernet network that uses hubs, protocol analyzers work by switching the network interface adapter they use to access the network into promiscuous mode. *Promiscuous mode* sets a network interface adapter to read and process all the traffic that is transmitted over the network, not just the packets that are addressed to it. This means that the system can examine all of the traffic transmitted on the network from one computer.

On today's networks, however, switches are more common than hubs, and as a result, capturing traffic for the entire network is more difficult. Because switches forward incoming unicast traffic only to their intended recipients, a protocol analyzer connected to a standard switch port only has access to one computer's incoming and outgoing traffic, plus any broadcasts transmitted over the local network segment.

To capture all of the traffic transmitted on a switched network, plug the computer running the protocol analyzer into a switch that supports port mirroring. Switches that support port mirroring have a special port to which the switch sends all incoming traffic.

One of the most commonly used protocol analyzers is the Microsoft Network Monitor application, mostly because it is available as a free download from the Microsoft website. There are many other protocol analyzer products available for Windows, UNIX, and Linux. The analyzers for Windows are all graphical and provide varying capabilities. For UNIX and Linux, both commercial and open source protocol analyzers are available, some of which are character-based (such as `tcpdump`), while others are graphical (such as Wireshark). There are also some dedicated hardware products that are essentially special-purpose computers with the analyzer software already installed.

NETWORK PROCESS FLOWCHARTS

One way to understand networking protocols more fully is to document their functions using flowcharts. A *flowchart* is a systematic diagram of the activities performed and the decisions made by a computer during a specific protocol task.

For example, a Dynamic Host Configuration Protocol (DHCP) server leases IP addresses using a process by which it broadcasts DHCPDISCOVER messages and waits for replies from clients on the network. When a client replies, a transaction commences in which the client and server exchange additional messages called DHCPOFFER, DHCPREQUEST, and DHCPACK.

Documenting the transaction in paragraph form can be wordy and confusing, but a flowchart can help to illustrate the process. See **Figure 16-3.**

Many tools are available that can be used to create flowcharts digitally, including Microsoft Visio.

NETWORK DOCUMENTATION

When maintaining a home or small office network, it is sometimes possible for an administrator to work "on the fly," dealing with issues as they arise, solving problems as they happen, and keeping all of the details about the network in his or her head. Beyond a four- or five-node network, however, this method becomes increasingly unmanageable. Administrators begin to forget some of the details, things slip

Figure 16-3 Flowcharts

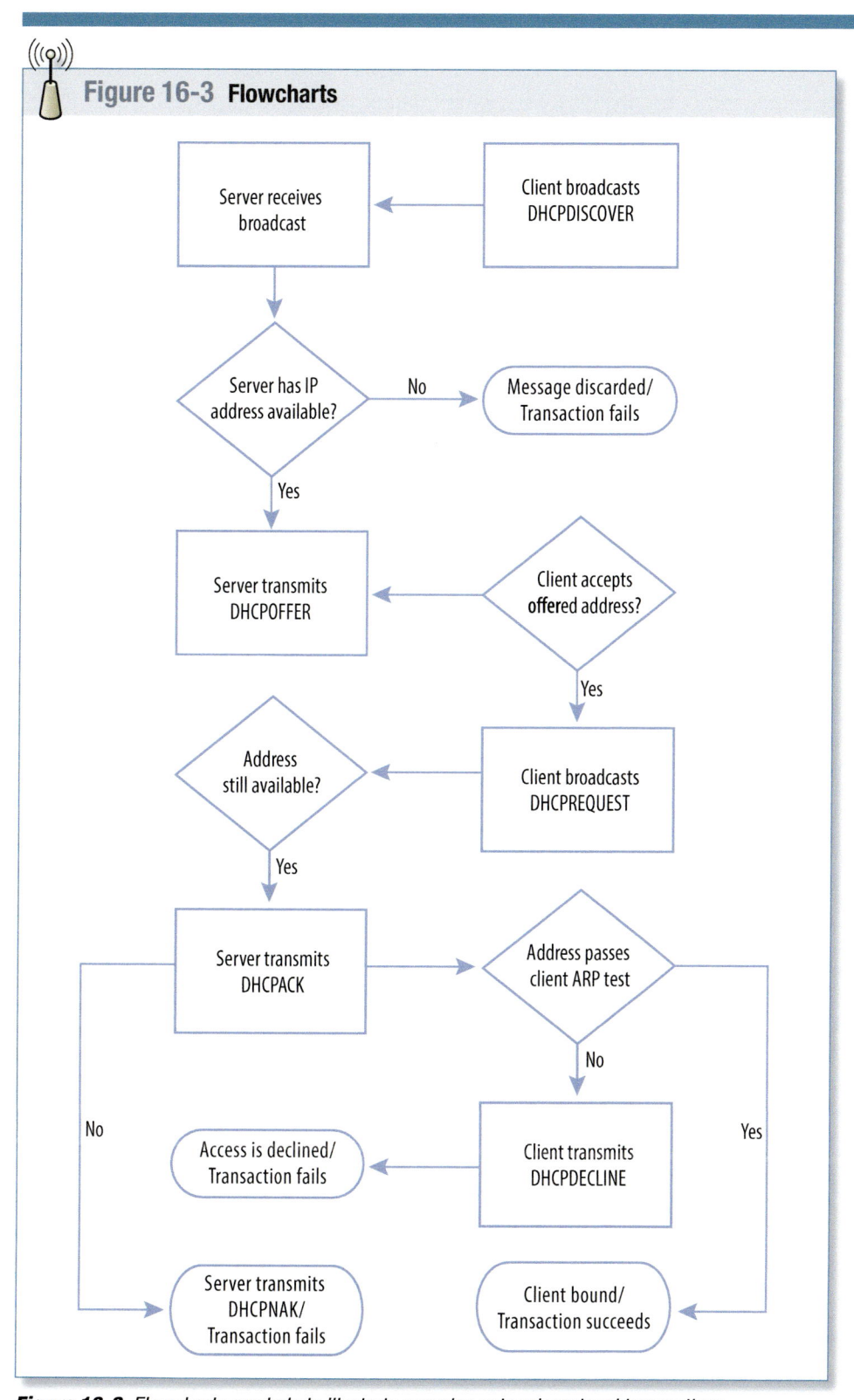

Figure 16-3. Flowcharts can help to illustrate complex network protocol transactions.

by that they should have remembered, and they find themselves repeating tasks unnecessarily.

In truth, this network management philosophy is impractical and unprofessional for even the smallest network. Documentation is a critical part of any network management plan, and the time to start thinking about it is well before installing the network hardware. The planning phase of the network must also be documented, so that the people who have to work on it later know what has been done.

There are many types of documentation that network administrators use and maintain. How and in what form administrators choose to create these documents is a matter of personal preference and company policy, but the important factor is that everyone involved in the network management and administration processes knows where the documentation is and can access it.

Some of the most important types of network documentation are described in the following sections.

Cable Diagrams

Documentation of a network's cable installation is particularly important, both because much of it is probably hidden from view, and because the organization probably had an outside contractor install it. The purpose of having this documentation is so that if something goes wrong with a cable run, or the network must be expanded, it is known where the existing hardware is and it will not be necessary to needlessly poke holes in walls and lift ceilings tiles.

In many cases, the best way to ensure that administrators have all of the documentation they need is to begin with the classic questions posed by journalists. In the case of a cable installation, ask the following:

- Who installed it? If the cables were installed by an outside contractor, maintain contact information for them and copies of the original contract. Every aspect of the arrangement should be documented; no oral agreements.

- What was installed? The documentation should include a complete list of all the hardware used in the cable installation, including the bulk cable and all connectors, wall plates, patch panels, and other components. Save receipts and invoices attesting to the rating of the cable components. If a contractor agrees to use CAT6 hardware throughout the installation, they should provide documentation to prove it.

- Where was it installed? A wiring schematic or cabling diagram is essential to the document collection. Cable installers may be required to document the exact path of every cable run through walls, floors, and ceilings. The best way to accomplish this is to obtain a copy of the original plan or blueprint for the site and add the cable runs to it. Documents should also record the numbers assigned to each cable end and connector.

- When was it installed? For warranty purposes and to track conformance to ever-changing standards, record when the cables were installed, especially if different parts of the network were installed at different times.

- How was it installed? It is essential to record the decisions made during the cable installation process, such as whether the pinouts conform to the T568A or T568B standard. This enables the administrators to ensure that future cabling work in the network conforms to the same standards.

Finally, it is important for all cable diagrams and other network documents to reflect the as-built condition of the installation, especially when work is performed by outside contractors. *As-built* is a term that refers to the documentation of all changes made to the original plans during the network installation process. Cable installers frequently encounter surprises when they look into walls, ceilings, and other hidden spaces, so any changes they make to the plan as a result of these surprises must be included in the final diagrams and documentation.

Network Diagrams

The terminology is not always consistent, but a network diagram differs from a cable diagram in that its intention is to illustrate the relationships between the network components. It is not usually drawn to scale, and does not necessarily include architectural elements of the site, such as walls, ducts, and fixtures.

What a network diagram does have is a representation of every device and component on the network and all the connections between them. This means that the diagram includes not only computers, but all of the switches, routers, access points, wide area network (WAN) devices, and other hardware components that make up the network infrastructure. **See Figure 16-4.**

A number of software tools can be used to create network diagrams. In most cases, these products use generic icons to represent network hardware components, but there are packages that provide genuine depictions of specific products, enabling administrators to create a realistic diagram of the racks in a data center, for example.

Hardware Configurations

When a new software product is released, and it looks as though it might be necessary to upgrade the hardware in a network's computers to run it, how do administrators know for sure which computers need the upgrade and which can already support it? There are various tools that can inventory the hardware in computers, but how many of them can specify whether a system has memory slots free, or room for an expansion card?

The best way to keep track of a network's computers and their configurations is to document them personally. Large enterprise networks typically assign their own identification numbers to their computers and other hardware purchases, as part of an asset management process that controls the entire lifecycle of each device, from recognition of a need to retirement or disposal.

The record for each device should contain all available information about each one, including the original documentation for the computer, an inventory of its internal components, and detailed information about its software configuration. This way, anyone seeking to upgrade or troubleshoot the computer can find out what is inside without having to travel to the site and open the case.

The record for each computer should also document any changes that administrators make to it, whether to the hardware or software, so that the information is continually updated.

Change Management

A properly documented network also has written policies regarding how things are supposed to be done. When an

316   Introduction to Network Technologies

Figure 16-4 Network Diagram

Domain Registrar   ISP   Mail Server

IP Phone

Router

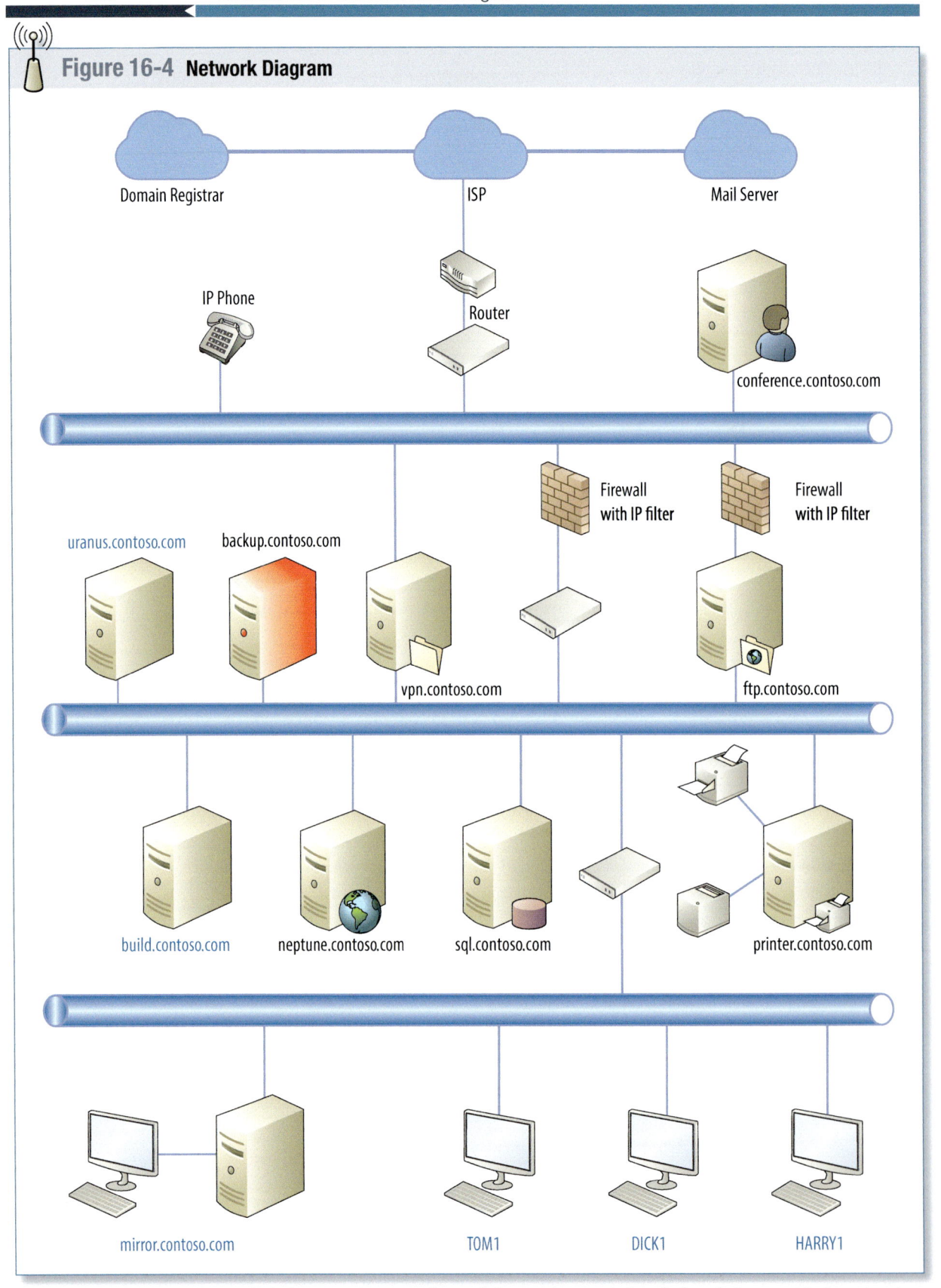

conference.contoso.com

uranus.contoso.com   backup.contoso.com

vpn.contoso.com

Firewall with IP filter   Firewall with IP filter

ftp.contoso.com

build.contoso.com   neptune.contoso.com   sql.contoso.com   printer.contoso.com

mirror.contoso.com   TOM1   DICK1   HARRY1

Figure 16-4. A network diagram includes computers as well as switches, routers, and other devices that connect to the network.

administrator troubleshoots a computer, for example, and in doing so replaces a hard drive, there should be more to it than taking the drive out of the box and installing it in the computer. There should be a change management policy that leads the administrator through all the ancillary tasks related to the hard drive replacement. *Change management* is the practice of identifying and implementing necessary changes to a computer system.

For example, the administrator might have to update the parts inventory to show one less drive in stock; check the warranty status of the failed drive and, if necessary, file a claim; update the computer's record with the serial number and characteristics of the new drive; rebuild the user's local data from backups; and any number of other related tasks.

The same sort of documents should be on file for network-related tasks, including expansions and upgrades, so that the policies used to build the network in the first place are maintained throughout its lifecycle.

Baselines

One of the basic principles of network management is to observe and address any changes that might occur in the performance of a system, be it a computer or a network. Administrators do this by comparing the system's performance levels at various times. A *baseline* is the starting point for comparisons of varying performance levels of a system.

Administrators can use a variety of tools and criteria to measure performance. Microsoft Windows includes a Performance Monitor tool that can display performance counters. *Counters* are measurements of hundreds of different system and network performance characteristics. **See Figure 16-5.**

✓ FACT

Although its functionality has remained largely the same, Performance Monitor has gone by various names in different versions of Windows. In Windows XP, it is System Monitor. In Windows Vista, the tool is called the Reliability and Performance Monitor. In Windows 7 and Windows 8/8.1, it is just Performance Monitor.

Figure 16-5 Performance Monitor

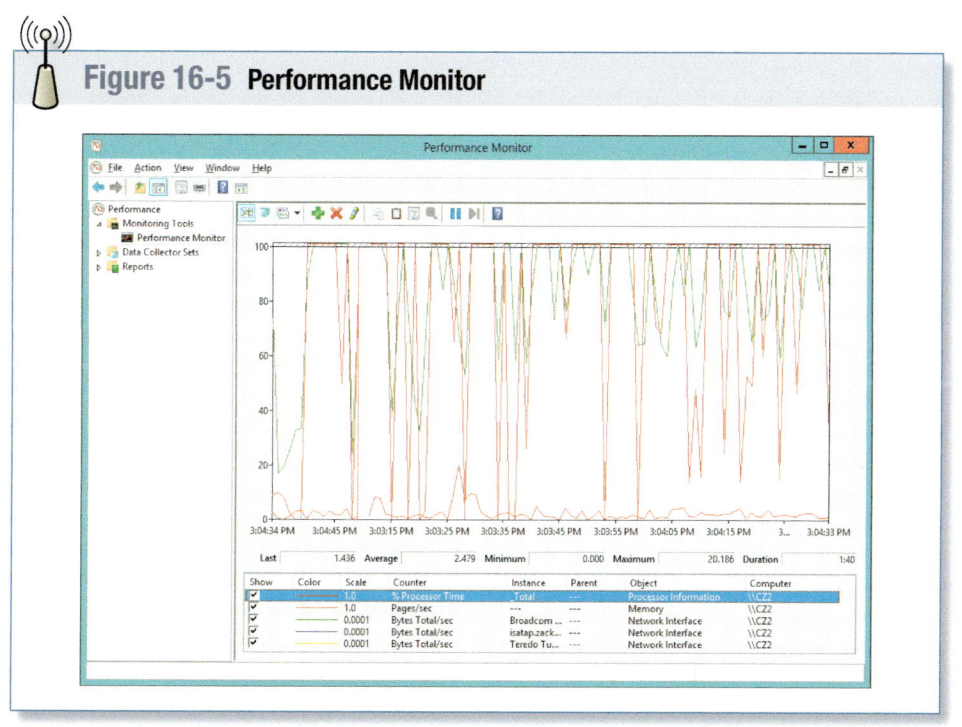

Figure 16-5. The Windows Performance Monitor application displays system and network performance characteristics in real time.

In addition to displaying performance data in real time, Performance Monitor can also capture data to log files over extended periods of time. A *data collector set* is the name for the data logs captured by the Performance Monitor. To capture an effective baseline, an administrator can capture data over the course of several hours, days, or even longer.

What is most important is documenting both the exact testing procedure and the initial results of the tests, which will function as a baseline for future comparisons. These documents should become part of the permanent record for the system.

Then, at regular intervals, administrators repeat the tests using the same tools and the same procedures, and compare the results to the previous ones. If there are major discrepancies between the new results and the earlier ones, make an effort to determine why. This basic technique can help to identify trends in performance that enable administrators to address problems before they become severe.

Summary

It is important to understand several different elements of the network management and administration processes, including IP address calculations and network documentation.

Review Question

1. **Which of the following subnet mask values would be used when configuring a TCP/IP client with an IPv4 address on the 172.16.32.0/19 network?**

   a. 255.224.0.0
   b. 255.240.0.0
   c. 255.255.224.0
   d. 255.255.240.0
   e. 255.255.255.240

Glossary

10Base2: A second version of the standard published in 1982, called DIX Ethernet II; also called Thin Ethernet, this version added a second Physical layer specification, calling for RG-58 coaxial cable.

10Base5: The first Ethernet standard, it described a network that used RG-8 coaxial cable in a bus topology up to 500 meters long, with a transmission speed of 10 Mbps; also called Thick Ethernet.

5-4-3 Rule: A rule stating that an Ethernet network can have as many as five cable segments, connected by four repeaters, of which three segments are mixing segments.

A

Access Link: Any other port on the switch other than the uplink port.

Address Resolution Protocol (ARP): A system of rules that resolves IP addresses into the physical address or the Media Access Control (MAC) address.

Adware: A relatively harmless type of spyware that tracks the Internet sites that have been visited for the purpose of sending targeted advertisements to the user.

Anycast Traffic: The traffic sent from one host to one other host from a list of multiple hosts.

Application Layer: Layer 7 of the OSI Model, which interacts with the Presentation layer below it and the application running on the computer.

Application Layer Filtering: A firewall configuration where traffic is filtered based on an application or service.

Application Resilience: The capability of an application to maintain its own availability by detecting outdated, corrupted, or missing files and automatically correcting the problem.

Application-Specific Integrated Circuits (ASICs) Speed: The speed on the chassis where the modules plug in.

As-Built: A term that refers to the documentation of all changes made to the original plans during the network installation process.

Asynchronous Transfer Mode (ATM): A cell-based method of transferring data.

Authentication: A process where clients must provide credentials such as a username and password.

Authentication Header (AH) Protocol: The protocol IPsec uses to prove the identity of the sender.

Autoconfiguration Enabled: An indicator that APIPA is enabled and an APIPA address will be assigned if a DHCP server cannot be reached.

Autoconfiguration IPv4 Address: An address starting with 169.254 shows this is an APIPA address and that the DHCP client could not receive an address from a DHCP server.

Autonegotiation System: An optional Fast Ethernet specification that enables a dual-speed device to sense the capabilities of the network to which it is connected and then adjust its speed and duplex status accordingly.

B

Backplane Speed: The internal speed of a switch.

Back-to-Back Connection: A connection formed when cables are connected together with a DTE end joining to a DCE end.

Bandwidth: A measure of how much data a device such as a switch can process at a time.

Baseline: The starting point for comparisons of varying performance levels of a system.

Basic Rate Interface (BRI): A type of ISDN service that uses two 64 Kbps B channels and one 16 Kbps D channel to provide a 128 Kbps data link.

Basic Service Set (BSS): A wireless network composed of one WAP and one or more wireless devices.

Best Current Practice (BCP): A single-stage alternative to the previous stages of an RFC.

Bits: Bits are the data at the physical layer, or ones and zeros.

Botnet: An abbreviation for "robot network," implying an automated network.

Bridge: A network device that connects two or more network segments.

Broadband Cable: An Internet connection method that provides Internet access through the same cable that provides cable TV.

Broadcast Domain: A group of devices on a network that can receive broadcast traffic from each other.

Broadcast Storm: An occurrence when broadcasts generated by one switch arrive at other switches, all of which forward the broadcasts in turn.

Broadcast Traffic: The traffic transmitted by one computer that goes to all of the computers on the same subnet.

Broadcast Transmission: A transmission of data to every device on a network.

Buffer: An area of computer memory designed to hold incoming data as it is being processed.

Buffer Overflow: A condition in which a program sends too much data to a buffer, and it spills over into an area of memory intended for another purpose.

Building Automation System (BAS): A system that normally utilizes some form of network signaling to provide monitoring and control from a central location.

Built-in Redundancy and Fault Tolerance: The property that automatically replicates domain data with at least two domain controllers in a domain.

Bus Topology: A configuration in which each computer is connected to the next one in a straight line.

C

Cache: An area of memory used for short-term storage.

Carrier Sense Multiple Access with Collision Detection (CSMA/CD): The media access control (MAC) mechanism that early Ethernet systems used to regulate access to the network medium.

Challenge Handshake Authentication Protocol (CHAP): A method of authentication that provides encryption by using Message Digest 5 (MD5) to encrypt the password instead of passing the password in clear text.

Challenge Handshake Authentication Protocol version 2 (MS-CHAPv2): A method of authentication with more password security by providing mutual authentication; the server authenticates back to the client before the client passes the authentication data to the server.

Change Management: The practice of identifying and implementing necessary changes to a computer system.

Client-Server Network: A group of connected computers in which one or more, called servers, perform the function of transferring files to the others, called clients.

CLOSE_WAIT: A connection state that indicates that the system is waiting for a final packet from the remote system to close the connection.

Coaxial Cable: A central copper conductor (which carries the signals) surrounded by a layer of insulation.

Collision: An occurrence when two systems on the LAN transmit at the same time; also termed SQE.

Collision Domain: A group of devices on the same network segment that are subject to collisions.

Collision Detection Phase: The part of the CSMA/CD process in which systems detect when their packets collide, avoiding a situation where corrupted data reaches a packet's destination system and is treated as valid.

Command Prompt: A symbol on the screen that indicates the computer is ready to accept a command as input.

Connectionless Protocol: A set of rules that transmits messages to a destination without first establishing a connection to the receiving system.

Content Filtering: A firewall configuration where traffic is blocked based on the content.

Counters: The measurements of hundreds of different system and network performance characteristics.

Convergence: A process that stops an NLB cluster from sending clients to a server that is not functioning or offline.

Cracking: The process of penetrating an encryption protocol by discovering its cryptographic key.

Crossover Cable: A type of Ethernet cable that connects similar devices to each other.

Cyclical Redundancy Check (CRC): An error-checking process used by TCP to verify that the data is intact in each segment.

D

Data and Pad (46 to 1,500 Bytes): A field that contains the data received from the Network layer protocol on the transmitting system, which is sent to the same protocol on the destination system.

Data Collector Set: The name for the data logs captured by the Performance Monitor.

Data Communications Equipment (DCE): The equipment end that plugs into network components like a router or a modem and provides clocking for the interface, which is critical to the proper function of the interface.

Data Link Layer: Layer 2 of the OSI Model, which is concerned with data delivery on a local area network (LAN).

Data Terminal Equipment (DTE): The equipment end that plugs into the serial port on a terminal or server.

Default Gateway: The IP address of the router's interface on the local subnet.

Default Route: The path that IP traffic takes when another path is not identified.

Denial of Service (DoS) Attack: An attempt to overwhelm a server or an application with incoming traffic.

Destination: A piece of information that identifies the destination subnetwork.

Destination Address (6 Bytes): A field that contains the 6-byte hexadecimal MAC address of the network interface adapter on the local network to which the frame will be transmitted.

DHCP Enabled: An indicator that the system is a DHCP client.

Dial-up: A method of remote access where a client uses a modem and phone line to connect to a remote access server that also has a modem and a phone line.

Digital Certificate: A file that includes data used to encrypt the data for confidentiality.

Digital Subscriber Line (DSL): A method of Internet connection that uses telephone lines but sends the data digitally instead of using an analog signal.

Directly Connected Route: Any subnetwork that is directly connected to a router; in other words, the router has interfaces in those subnetworks.

Distributed Denial of Service (DDoS): A simultaneous attack on a single system or server by multiple attackers.

DNS Revolver Cache: Another name for the host cache, since many of the entries are created when DNS is queried to resolve a hostname.

DNS Root Servers: The DNS servers at the top of the hierarchy; there are only 13 in the world.

DNS Round Robin: A technique wherein multiple resource records are created using the same name, with a different server IP address in each record; when clients attempt to resolve the name, the DNS server supplies them with each of the IP addresses in turn; as a result, the clients are evenly distributed among the servers.

Domain: A set of network resources to which certain users are given access.

Domain Directory: A database of objects such as users, computers, and groups.

Domain Name System (DNS): The primary name resolution service that the Internet and Microsoft networks use.

Draft Standard (DS): The second official stage in the Standards Track category.

Dynamic Host Configuration Protocol (DHCP): A network protocol that provides IP addresses and other TCP/IP configuration information to all the devices on the switch's network.

Dynamic Routing: A technique that allows the routers to automatically learn about changes in the network and alter their forwarding decisions based on those changes.

E

E: The switch port label for the first 10 Mbps port, labeled as E0.

Eavesdropping: The practice of using tools such as protocol analyzers or packet sniffers to capture data transmitted in clear text and read it.

Electromagnetic Interference (EMI): A disruption to the functioning of an electrical device caused by other electrical equipment.

Encapsulating Security Protocol (ESP): The protocol IPsec uses to encrypt traffic.

Encapsulation: The overall process in which data from the higher layer protocols is encapsulated, or packaged and incorporated, by those at the lower layers.

Enterprise Mode: A compatibility mode that requires authentication with a back-end server known as an 802.1x server.

ESTABLISHED: A connection state that indicates that a TCP session is established.

Ethernet: The most commonly used LAN protocol at the Data Link layer.

Ethernet Frame: The packet format that Ethernet systems use to transmit data over the network.

Ethertype: A hexadecimal value that identifies the protocol that generated the data in the packet.

Ethertype/Length (2 Bytes): The field in the DIX Ethernet frame that contains a code identifying the Network layer protocol for which the data in the packet is intended.

Evil Twin: An unauthorized wireless access point deliberately configured to closely mimic an authorized one.

Extended Service Set (ESS): A wireless network with more than one WAP, with each WAP supporting one or more wireless devices.

Extensible Authentication Protocol (EAP): A method of authentication that supports additional methods, including Protected EAP (PEAP) and smart cards.

Extranet: An area between the Internet and an intranet that hosts resources for trusted entities.

F

F: The switch port label for the first 100 Mbps port, typically labeled F0 or F0/0, a Fast Ethernet port.

Failover Cluster: A collection of two or more servers that perform the same role or run the same application and appear on the network as a single entity.

Fault Tolerance: A failure can occur and a system can recover from it by itself.

Fiber-Optic Cable: A cable that uses transparent fibers to transmit light instead of electrical signals.

File Transfer Protocol (FTP): A protocol that transfers files to and from an FTP server.

Firewall: A part of a computer network that blocks unwanted traffic from the Internet, providing a layer of protection for internal clients.

Flowchart: A systematic diagram of the activities performed and the decisions made by a computer during a specific protocol task.

Form-Factor Switch: A switch type that has a set number of ports built into it.

Frame: A unit of data at layer 2, the Data Link layer.

Frame Check Sequence (4 Bytes): A single field that comes after the Network layer protocol data and contains a 4-byte checksum value for the entire frame.

Frame Relay: A WAN technology whereby systems convert data into variable-sized frames and transfer them over permanent virtual circuits.

Fresnel Zone: The area underneath and above the direct line of sight between two points.

Full-Duplex: A property where systems can send data and receive it at the same time.

Fully Qualified Domain Name (FQDN): The full computer name when it is a host and part of a domain.

G

Gateway: The IP address of the destination router's network interface.

Gi: The switch port label for the first 1,000 Mbps port, labeled a Gi0/0, a gigabit port.

Group Policy: A hierarchical structure used by administrators to configure, control, and manage users and computers.

Guest Network: An isolated portion of the internal network that can be used by guests or visitors.

H

Half-Duplex: A property where systems can send data both ways, but only one way at a time.

Hardware Redundancy: The addition of components to ensure that the failure of one component does not result in a complete failure.

Hardware Router: A dedicated hardware device that routes packets.

Hexadecimal Notation: A system that uses a base of 16 with the numbers 0 through 9 followed by a through f. Each hexadecimal number can be represented using four binary bits.

Homegroup: A special type of workgroup in new Windows operating systems that facilitates sharing files between computers in a home or small office network.

Host Cache: An area of memory on any computer that is dynamically updated with hostnames and their corresponding IP addresses.

Hostname: A user-friendly string of characters (or label) assigned to a computer or other network device.

Hosts File: A simple text file located in the c:\windows\system32\drivers\etc folder by default that maps the names of computers to IP addresses.

HTTP over Secure Sockets Layer (HTTPS): An Application layer protocol that handles encryption and decryption of secure data on the Internet.

Hub: A device that provides basic connectivity for other devices in a network.

Hypertext Markup Language (HTML): The format most web pages are created in.

Hypertext Transfer Protocol (HTTP): The primary protocol that clients use to transfer data to and from web servers on the Internet.

I

IEEE 802.11: The wireless LAN technology that is the most common alternative to Ethernet used today.

Implicit Deny Policy: A firewall configuration that specifies that all traffic that has not been explicitly allowed is blocked.

Integrated Services Digital Network (ISDN): A group of standards used for transmitting voice, data, and video.

Interface: A piece of information that identifies the network adapter that should be utilized to connect to the specified destination by name.

Interface Identifier: A piece of information that identifies an interface, such as a media type, a slot number, a port number, or a combination of these, depending on the router.

Internet Assigned Numbers Authority (IANA): An organization that assigns port numbers to protocols.

Internet Control Message Protocol (ICMP): A system of rules that carries error messages and diagnostic reporting messages between systems.

Internet-Facing Servers: Any servers accessible from the Internet.

Internet Group Management Protocol (IGMP): A specialized protocol for determining which systems are part of the multicast group that recognizes that address.

Internet Group Multicast Protocol (IGMP): A system of rules that is responsible for managing the groups that receive multicast traffic.

Internet Layer: The layer at which protocols on the Internet control the movement and routing of packets between networks.

Internet Message Access Protocol (IMAP): A system of rules that clients use to receive e-mail messages.

Internet Protocol (IP) Address: A unique identifier for each computer that directs data to its destination.

Internet Protocol v4 (IPv4): An addressing protocol that uses 32-bit addresses for the devices on the network.

Internet Protocol v6 (IPv6): An addressing protocol that uses 128-bit addresses.

Internet Service Provider (ISP): A company or organization that provides access to the Internet for households and businesses.

Intranet: A private LAN that uses TCP/IP protocols, the same protocols found on the Internet, to share resources within the network.

IPv6 Prefix: An indicator that identifies the type of IPv6 address.

K

Kerberos: The primary authentication protocol used within a Microsoft domain and managed as part of Active Directory.

L

Layer 2 Tunneling Protocol (L2TP): A protocol used with VPNs that often uses IPsec (as L2TP/IPsec) to encrypt the traffic.

Leased Lines: A number of lines for which an organization contracts with a communications provider, and the provider guarantees a specific level of service identified in a service level agreement (SLA).

Lightweight Directory Access Protocol (LDAP): A protocol that transmits queries and replies to and from a directory service, such as Microsoft's Active Directory Domain Services (AD DS).

Line Password: A password that requires administrators to authenticate when accessing the router through a console or network connection.

Link Layer: The layer that defines how data is transmitted on the media.

Link-Local Multicast Name Resolution (LLMNR): A method of resolving names that is similar to broadcast, but it can resolve both IPv4 and IPv6 addresses.

LISTENING: A connection state that indicates that the system is ready to accept a connection.

Local Area Network (LAN): A group of connected computers in the same geographical location, sharing the same level of connectivity.

Logical Design: A high-level description of how information moves through an entire network without reference to its physical components like switches, routers, and firewalls.

Logical Link Control (LLC) IEEE 802.2: The system of rules that interacts directly with the network layer.

Logical Ports: The numbers used to indicate how systems handle data when it reaches its destination; examples are TCP and UDP ports.

Loopback Address: An address that is used to test the TCP/IPv4 protocol stack (the software).

M

MAC Addresses: The media access control (MAC) address or physical address uniquely identifies the network adapter.

MAC Address Table: A list of the MAC address of each computer and maps to the port to which they're connected.

Malware: Any software that has a malicious intent, whether obvious or obscure.

Man in the Middle (MITM): An attack form in which the attacker interposes himself or herself between two individuals who think they are communicating with each other. The attacker receives the messages from each party in the transaction and relays them to the other party, but not without reading them (or even modifying them) first.

Managed Switch: A configurable switch.

Management Information Base (MIB): A database where managed objects are stored that agents gathered with specific information about the network devices.

Media Access Control (MAC) IEEE 802.3: A protocol that defines how systems place packets onto the physical media at the Physical layer.

Mesh Topology: A network configuration in which each computer is connected to two or more other computers.

Microsoft Point-to-Point Encryption (MPPE): A protocol that encrypts PPTP traffic.

Modular Switch: A switch type that typically starts with zero or a few ports and can expand to hundreds of ports.

Multicast Promiscuous Mode: A special mode that causes the network interface adapter to process all incoming packets that have the multicast bit (the last bit of the first byte of the destination hardware address) set to a value of 1.

Multicast Traffic: The traffic transmitted from one computer to many other computers.

Multiple-Input/Multiple-Output (MIMO): The antenna technology IEEE 802.11n uses.

Multiple Access Phase: The phase during which the station transmits its data packet when the network is free and all of the stations on the network are contending for access to the same network medium.

N

Names: An identifier a computer is assigned and can usually be reached by over a network.

Neighbor Discovery (ND): An IPv6 protocol that uses Internet Control Message Protocol version 6 (ICMPv6) messages to discover details about the network.

Neighbor Discovery Protocol (NDP): A new Data Link layer protocol that performs multiple functions, including local network system discovery, hardware address resolution, duplicate address detection, router discovery, DNS server discovery, address prefix discovery, and neighbor unreachability detection.

Network: A group of computers and other devices connected together.

Network Address Translation (NAT): The process where the public IP addresses used on the Internet are translated to private IP addresses on the internal network, and vice versa.

Network Basic Input/Output System (NetBIOS) Name: A 15 character-long unique identifier that NetBIOS services use to point to network resources.

Network Interface Adapter: A link between the Data Link layer and Physical layer.

Network Layer: Layer 3 of the OSI Model, responsible for determining the best route for data to travel to its destination.

Network Load Balancing: A type of clustering, useful when a web server or other application becomes overwhelmed by a large volume of users, in which multiple identical servers (also known as a server farm) are deployed and the user traffic is distributed evenly among them.

Network Mask: The subnet mask of the network ID.

NLB Cluster: A logical entity like a failover cluster with its own name and IP address; clients connect to the cluster, rather than the individual computers, and the cluster distributes the incoming requests evenly among its component servers.

Normal Link Pulse (NLP) Signals: Signals that standard Ethernet networks use to verify the integrity of a link between two devices.

O

Offline Files: A mechanism that individual users can employ to maintain access to their server files, even if the server becomes unavailable.

Open Shortest Path First (OSPF): A routing protocol that routers use to communicate with each other on internal networks.

Open Systems Interconnection (OSI) Model: A general framework or set of guidelines for data handling and network communication.

P

Packet: A unit of data at layer 3, the Network layer.

Packet-Filtering Firewall: A component that filters packets based on IP addresses, ports, and some protocols.

Password Authentication Protocol (PAP): An authentication method that passes the password across the wire in clear text.

Peer-to-Peer Network: A group of connected computers, each of which can act as a server to the others for transfer of the files stored on it.

Perimeter Network: An area between the Internet and an intranet that hosts servers accessible from the Internet.

Personal Mode: A mode that requires manual configuration, similar to WEP but without the upkeep of changing the key; also called Preshared Key (PSK) Mode.

Phishing: The practice of sending out an official looking e-mail or letter to users that points them to a website containing a form asking for personal information; the letter and website are bogus, and the confidential information the users supply goes right to the attacker.

Physical Layer: Layer 1 of the OSI Model that defines the physical specifications of the network, including physical media such as cables and connectors and basic devices such as repeaters and hubs.

Physical Layer Specifications: The details that define the various types of network media that can be used to build Ethernet networks, as well as the topologies and signaling types they support.

Physical Ports: The components of switches and routers into which cables are plugged.

Plain Old Telephone Service (POTS): The Internet connection method used with a dial-up method, which works by connecting a modem to a telephone line and then connecting to the remote network by dialing out through the phone line.

Point-to-Point Protocol (PPP): A system of rules used for dial-up networking.

Point-to-Point Topology: A simple configuration in which one computer is directly connected to another.

Point-to-Point Tunneling Protocol (PPTP): A VPN protocol that provides a secure connection over a public network such as the Internet.

Point-to-Point Tunneling Protocol v4 (PPTP): A system of rules commonly used with virtual private networks (VPNs).

Port: A physical connection between devices.

Port Address Translation (PAT): A popular way that NAT is implemented.

Port-to-Port Speed: The speed with which data moves between ports on the different blades of the same chassis.

Post Office Protocol (POP3): An e-mail protocol that clients use to retrieve e‚Äë-mail from POP3 servers.

Preamble (7 Bytes): The field that contains 7 bytes of alternating 0s and 1s, which the communicating systems use to synchronize their clock signals.

Presentation Layer: Layer 6 of the OSI Model, which interacts with the Session and Application layers by acting as a translator and determining how to format and present the data.

Preshared Key (PSK) Mode: A mode that requires manual configuration similar to WEP but without the upkeep of changing the key; also called Personal Mode.

Primary Rate Interface (PRI): A type of ISDN service that uses 23 64 Kbps B channels and one 16 Kbps D channel to provide a 1472 Kbps data link for a total of 1.536 Mbps.

Private (Home/Work): A network type with firewall configurations specified for a small, protected network where other devices on the network are known and trusted, so network discovery is enabled.

Privileged Mode Password: A password that prevents anyone not possessing the password from switching the router into privileged mode with the *enable* command; sometimes referred to as an enable password.

Promiscuous Mode: A setting where a network interface adapter reads and processes all the traffic that is transmitted over the network, not just the packets that are addressed to it.

Proposed Standard (PS): The first official stage of the Standards Track category, where an RFC starts.

Protocol Analyzer: A tool that captures a sample of the traffic passing over the network, decodes the packets into the language of the individual protocols they contain, and allows an administrator to examine them in minute detail; sometimes called a packet sniffer.

Protocol Data Unit (PDU): A unit of data packaged for transport on a network.

Proxy Server: A server type that acts on behalf of the client computers on the internal network to retrieve web content from the Internet.

Public: A network type with firewall configurations specified for a public location such as in a coffee shop or airport.

R

Remote Access: The ability for individuals working outside the company to access resources internal to the company.

Remote Desktop Protocol: The same protocol used by Windows for Remote Assistance, which enables a help-desk professional to take control of an end user's desktop (with permission) and provide assistance.

Remote Desktop Services (RDS): An additional role included in Microsoft Windows servers to host applications or entire desktops that are accessible to users on the network.

Requests For Comments (RFCs): The publication type that most of the documents published by the IETF are known by.

Reverse Lookup Zone: A tool that uses pointer (PTR) records to do reverse lookups.

Reverse Proxy Server: An additional server on the perimeter network that isolates the web servers from direct access by systems on the Internet, providing a layer of protection from Internet attackers.

Ring Topology: A configuration in which a signal transmitted by a computer in one direction circulates around ring, eventually ending up back at its source, like a bus with the two ends joined together.

Rogue Access Point: An unauthorized wireless access point connected to a network.

Router: A device that connects networks together.

Routing and Remote Access Service (RRAS): Windows Server software that allows a server to be configured as a router.

Routing Capabilities: The ability to move data from one place to another, illustrated by a built-in router routing data from the internal network to the Internet and from Internet data back to the internal network.

Routing Information Protocol (RIP): A basic routing protocol that routers use on internal networks.

Routing Information Protocol version 2 (RIPv2): The primary routing protocol used on Windows Server.

Routing Interface: The interface where packets are received and transmitted.

Routing Protocols: Rule systems that enable routers to communicate with each other and share routing information.

Routing Table: A table maintained within a router that identifies all known subnetworks and the paths to these subnetworks.

S

Second-Level Domain DNS Servers: These servers are authoritative in the second-level DNS namespace.

Secure LDAP (SLDAP): A protocol that uses SSL or TLS to prevent attackers from using sniffers to capture data.

Secure Shell (SSH): An encryption protocol that creates a secure encrypted session that other protocols can use.

Secure Sockets Layer (SSL): An encryption protocol used for a wide assortment of purposes.

Segment: A TCP unit of data at layer 4, the Transport layer.

Service Level Agreement (SLA): A contract that defines expectations for performance guaranteed by a service provider and often identifies penalties if the service fails to meet the expectations.

Session Layer: Layer 5 of the OSI Model, which is responsible for establishing, maintaining, and terminating sessions.

Server Message Block (SMB): A file transfer protocol that Microsoft networks use.

Service Set Identifier (SSID): A name that someone has assigned to the network.

Shadow Copies: A mechanism that automatically retains copies of files on a server volume in multiple versions from specific points in time. When users accidentally overwrite or delete files, they can access the shadow copies to restore earlier versions.

Shielded Twisted-Pair (STP): A type of cable that is twisted-pair cable that has a foil or mesh shield surrounding all four pairs.

Signal Quality Error (SQE): An occurrence when two systems on the LAN transmit at the same time; also called collision.

Simple Mail Transfer Protocol (SMTP): The primary protocol that transmits e-mail messages to and between mail servers.

Simple Network Management Protocol (SNMP): A protocol that systems use to manage network devices such as routers and switches.

Simple Network Monitoring Protocol: A TCP/IP application layer protocol and query language that specially equipped networking devices use to communicate with a central console.

Simplex: A characteristic when systems can send data only one way.

Simplified Management: A group of centralized tools administrators use to manage accounts in a domain.

Small Office and Home Office Network (SOHO): A network providing services to a small office on commercial or residential property.

Smurf Attack: A sneaky form of ICMP-based DoS attack that involves flooding a network with *ping* messages sent to the network's broadcast address.

SNMP Agent: A software module that is responsible for gathering information about a device and delivering it to a computer that has been designated as the network management console.

Social Engineering: the term used to describe a practice in which a friendly attacker contacts a user by telephone, mail, or e-mail and pretends to be an official of some sort; that attacker then gives some excuse for needing the user's password or other confidential information.

Software Router: A server that includes software used to route packets.

Source Address (6 Bytes): The field that contains the 6-byte hexadecimal MAC address of the network interface adapter in the system generating the frame.

Spanning Tree Protocol (STP): A system of rules that enables a switch to discover a subset of the network topology that does not contain loops, eliminating the endless propagation of broadcasts.

Spyware: A hidden program that gathers information about computer activities and sends it to someone on the Internet.

Standard (STD): The final stage of an RFC.

Star Topology: A configuration in which each computer or other device is connected by a separate cable run to a central cabling nexus (that is, a switch or a hub).

Start of Frame Delimiter (1 Byte): The field that contains 6 bits of alternating 0s and 1s, followed by two consecutive 1s, which is a signal to the receiver that the transmission of the actual frame is about to begin.

Static Routing: A technique that requires an administrator to manually add routes to different subnets.

Straight-Through Cable: A type of cable that connects computers to networking devices, as in the case of a connection from a computer to a hub, or a computer to a switch.

Stateful Filtering: A firewall configuration where traffic is filtered based on the state of the network connections.

Stateless: A configuration type performed based on router advertisements.

Stateless Address Autoconfiguration (SLAAC): A feature of IPv6 that starts with a self-assigned link-local address and then goes through a process to verify it and learn about the network by communicating with local routers.

Subnet: A group of computers separated from other computers by one or more routers.

Subnetting: A technique that divides a larger network into multiple smaller networks by taking bits from the host ID.

Supernetting: The opposite of subnetting, this combines multiple smaller networks into a single larger network by taking bits from the network ID.

Switch: A device that connects the computers in a network segment together.

T

T1: A line that combines 24 DS0 channels for a total of 1.544 Mbps.

T3: A line that combines 28 DS1 channels for a total of 44.736 Mbps.

TCP/IP Model: A four-layer communications model created in the 1970s by the U.S. Department of Defense (DoD).

TCP Sliding Window: The number of segments the computers can send at a time.

Telnet: A command-line interface that provides bidirectional communication with network devices and other systems on the network.

Teredo: A tunneling protocol that encapsulates IPv6 packets within IPv4 datagrams.

Thick Ethernet: The first Ethernet standard, which it described a network that used RG-8 coaxial cable in a bus topology up to 500 meters long with a transmission speed of 10 Mbps; also called 10Base5.

Thin Ethernet: A second version of the standard published in 1982, called DIX Ethernet II; this version added a second Physical layer specification, calling for RG-58 coaxial cable; also called 10Base2.

Third- and Lower-Level Domain DNS Servers: A server level that is possible but needed only when the FQDN includes these lower levels.

Token Ring: A technology where the computers pass a logical token between themselves; a computer can communicate on the network only when it has the token.

Top-Level Domain DNS Servers: These servers know the addresses of second-level domain DNS servers in their namespace.

Transmission Control Protocol (TCP): A set of rules that provides guaranteed delivery of data.

Transport Layer: The layer at which protocols control data transfer on the network by managing sessions between devices.

Transport Layer Security (TLS): A similar security protocol to SSL that can provide confidentiality, integrity, and authentication.

Transport Mode: A compatibility mode where only the data is encrypted instead of the entire packet.

Trap: A special message that an agent can generate in the event of a serious problem, which it transmits immediately to the console, causing it to alert the administrator of a potentially dangerous condition.

Trivial FTP (TFTP): A lightweight FTP protocol that transfers smaller files with less data overhead.

Trojan Horse: A non-replicating program that appears to perform an innocent function, but that in reality has another, more malicious, purpose.

Trunk Link: A connection type that carries traffic from multiple VLANs and, therefore, needs to be as high-capacity as possible.

Tunnel Mode: A compatibility mode where IPsec encrypts the entire IP packet (both data and headers).

Twisted-Pair Cable: A cluster of thin copper wires, with each wire having its own insulating sheath.

U

Unicast Traffic: The traffic sent from one computer to one other computer.

Uniform Resource Locator (URL): The address used to access Internet resources such as websites.

Unique Local Address: An IPv6 address used in an internal network.

Unmanaged Switch: A nonconfigurable switch.

Unshielded Twisted-Pair (UTP): A cable that contains four wire pairs within a jacket with no additional shielding.

Uplink Port: A special port on a switch used to connect the switch to another switch or to other devices.

User Datagram Protocol (UDP): A set of rules that provides a best-effort method of delivering data.

V

Virtual Private Network (VPN): A group of connected computers that provides access to an internal network using a public network such as the Internet.

Virus: a type of program that replicates by attaching itself to an executable file or a computer's boot sector and performs a specified action—usually some form of damage—at a prearranged time.

W

War Chalking: A practice associated with war driving in which the people discovering an unprotected network leave a mark on a wall or gatepost indicating its presence, so that future drivers can find it.

War Driving: The process of cruising around a neighborhood with a scanner, looking for unprotected wireless networks to which one can connect.

Wide Area Network (WAN): A group of connected computers where two or more LANs in separate geographical locations are connected.

Wi-Fi Protected Access (WPA): An improved security standard over WEP.

Windows Installer: The component in Windows Server that enables the system to install software packaged as files with an *.msi* extension.

Windows Internet Name Service (WINS): A service that can be added to a server to resolve NetBIOS names to IPv4 addresses.

Wired Equivalent Privacy (WEP): The first security model used on IEEE 802.11 wireless networks.

Wireless Access Point (WAP): A hardware device that provides connectivity for wireless clients, which the wireless device supports.

Witness Disk: A type of storage that holds the cluster configuration database.

Worldwide Interoperability for Microwave Access (WiMAX): A technology available in some cities that provides a wireless alternative for broadband cable and DSL and gets speeds up to 40 megabits per second (Mbps).

Worm: A program that replicates itself across a network by taking advantage of weaknesses in computer operating systems.

Workgroup: A collection of networked computers that share a common workgroup name.

WPA2 Enterprise Mode: A compatibility mode that uses 802.1x for authentication.

WPA2 Personal Mode: A compatibility mode that is for home users and small businesses that are not using an authentication server; also called WPA2-PSK for preshared key.

Z

Zero Compression: A technique that is used to identify a contiguous group of zeros.

Zombie: The term for an infected computer that becomes a member of a botnet.

INDEX

<00>, 212
5-4-3 rule, 78, 79f
6to4, 145
8-wire crossover cable, 36f
8P8C connector, 34f, 35
8P8C ports, 173
10 Gigabit Ethernet, 84–86
10 Mbps Ethernet, 77–78, 77f
10/100 Mbps, 184
10Base2, 72, 77f
10Base5, 72, 77f
10Base-FL, 77f, 78
10Base-T, 33, 77f
10Base-T network, 78
10Gbase-ER, 85f, 86
10Gbase-EW, 85f, 86
10Gbase-LR, 85f, 86
10Gbase-LW, 85f, 86
10Gbase-SR, 85f, 86
10Gbase-SW, 85f, 86
10Gbase-T, 85f
100 Mbps Fast Ethernet physical layer specifications, 80
100/1,000 Mbps, 184
100Base-FX, 80, 80f
100Base-TX, 80, 80f
100Base-X, 80
169.254, 266
192.168.1.1, 103
192.168.1.61 network ID, 126f
192.168.1.65 network ID, 127f
1000Base-CX, 82–83, 82f
1000Base-LX, 82, 82f
1000Base-LX10, 82
1000Base-SX, 82f
1000Base-T, 82f
1000Base-TX, 82
1000Base-X specifications, 82

A (host), 217f
AAAA (host), 217f
Access link, 184
Acknowledge (ACK), 154
Active Directory, 100
Active Directory Domain Services (AD DS), 9, 10, 166
Active Directory Users and Computers (ADUC), 10, 166, 166f
AD DS. *See* Active Directory Domain Services (AD DS)
Adapter configuration, 145f
Address bar, 161f

Address Resolution Protocol (ARP), 57, 114, 150, 150f
Administering the network. *See* Management and administration
Administrative privileges, 265
ADSL. *See* Asymmetric DSL (ADSL)
ADUC. *See* Active Directory Users and Computers (ADUC)
Advanced encryption standard (AES), 99
Advanced subnet masks, 127f
Adware, 285
AES. *See* Advanced encryption standard (AES)
AH. *See* Authentication Header (AH) protocol
"All People Seem To Need Data Processing," 52, 53f
Allowed programs, 236f
American Standard Code for Information Interchange (ASCII), 54, 55f
Amplifier, 66
AND, 120
ANSI-TIA-568-C standard, 45
ANSI-TIA-568-D standard, 45
ANSI-TIA-568-D.0, 45
ANSI-TIA-568-D.1, 45
ANSI-TIA-568-D.2, 45
ANSI-TIA-568-D.3, 45
ANSI-TIA-568-D.4, 45
Antivirus software, 231
Any cast transmission, 138–139, 154
APIPA. *See* Automatic private Internet protocol addressing (APIPA)
Application availability, 292–294
Application layer
    OSI model, 53–54
    TCP/IP model, 64
Application layer filtering, 236
Application layer protocols, 160–168
Application resilience, 293
Application-specific integrated circuits (ASICs) speed, 185
ARIN, 136
ARP. *See* Address Resolution Protocol (ARP)
arp -a, 151
ARP cache, 151
As-built condition of installation, 315
ASCII. *See* American Standard Code for Information Interchange (ASCII)
ASICs speed. *See* Application-specific integrated circuits (ASICs) speed

Asymmetric DSL (ADSL), 249
Asynchronous transfer mode (ATM), 60
ATM. *See* Asynchronous transfer mode (ATM)
Attenuation, 32
Augmented Category 6 (CAT6A), 33
Authentication
    ✓FACT sidebar, 167
    defined, 100
    enterprise, 100f
    IPSec, 164
    MITM attacks, 285
    RAS, 255, 256f
    RIPv2, 197
    SSL, 163
Authentication Header (AH) protocol, 164
Auto-MDIX, 184
Automatic private Internet protocol addressing (APIPA), 117, 130, 223
Automation network, 14–15
Autonegotiation, 84
Autosensing, 36f, 184
Availability
    application, 292–294
    data, 288–292. *See also* Data availability
    server, 294–301. *See also* Server availability

4B/5B signaling method, 80
B channels, 254
B-node, 225f
Back-to-back connection, 200
Backoff period, 76
Backplane speed, 185
Backup, 288
BACnet, 14
Bandwidth, 183
Bandwidth-throttling techniques, 250
BAS. *See* Building automation system (BAS)
Baseline, 317–318
Basic rate interface (BRI), 254
Basic service set (BSS), 104
BCP. *See* Best current practice (BCP)
Berners-Lee, Tim, 16
Best current practice (BCP), 16
Binaries and decimals, 118f, 310
Binary IPv4 addresses, 117–121
Binary-to-decimal conversion, 119, 310

Binary values by bit, 118f
Bits, 62
Bluetooth, 93
Boolean AND logic, 120
BootP broadcast, 24, 130
Botnet, 230–231
Branch office, 13, 13f
Branching tree topology, 39
BRI. *See* Basic rate interface (BRI)
Bridge, 28–29, 29f, 66
Broadcast cable, 248f, 250, 250f
Broadcast domain, 26
Broadcast name resolution, 215f, 223
Broadcast storm, 180
Broadcast transmission, 23–25, 138, 223
Browser-to-web server, 293
BSS. *See* Basic service set (BSS)
Buffer, 286
Buffer overflow, 286
Building automation system (BAS), 14–15
Bundling, 185
Bus topology, 39–40, 40f
Byte, 2

CA. *See* Certificate authority (CA)
Cable diagram, 314–315
Cabling standards, 44–46
Cache, 151
Caching, 233
Carrier Sense Multiple Access with Collison Avoidance (CSMA/CA), 92
Carrier Sense Multiple Access with Collison Detection (CSMA/CD)
 carrier sense, 75
 collision detection, 75–76
 CSMA/CA, contrasted, 92
 multiple access, 75
 switches, 175, 176
 usage of CSMA/CD in today's world, 76–77
Category 3 (CAT3), 33
Category 5 (CAT5), 33
Category 5e (CAT5e), 33
Category 6 (CAT6), 33
CAT6A. *See* Augmented Category 6 (CAT6A)
Category 7 (CAT7), 33
Category 7a (CAT7a), 34
Category 8 (CAT8), 33
Central office, 249
Certificate authority (CA), 164
Challenge Handshake Authentication Protocol (CHAP), 255
Challenge Handshake Authentication Protocol version 2 (MS-CHAPv2), 101, 255
Change management, 315–317
Channel service unit/data service unit (CSU/DSU), 200, 205, 254
CHAP. *See* Challenge Handshake Authentication Protocol (CHAP)

CIDR notation. *See* Classless inter-domain routing (CIDR) notation
Cipher text, 162
Cisco routers, 97, 190, 201, 202
Cladding, 32
Class A address, 115, 115f
Class A network, 125f
Class B address, 115, 115f
Class B network, 125f
Class C address, 115, 115f
Class C network, 121f, 124f
Class C subnet, 123f
Class D address, 116
Class E address, 116
Classful IP addresses, 115–117, 307
Classless inter-domain routing (CIDR) notation, 119–120, 120f, 308
Classless IP addresses, 121, 121f
Client/server application, 293
Client-server network, 4
Client-to-gateway VPN, 259
clock rate *XXXXX*, 205
Clone, 230
CLOSE_WAIT, 275
CNAME (alias), 217f
Coaxial cable, 36
Collision, 75, 76, 176f
Collision domain, 25–26, 175–177
Collision domain with hubs, 176, 177f
Collision domain with switches, 176–177, 177f
COM1, COM2, COM3, and COM4, 201
Command prompt, 264–265
Computer Name/Domain Changes dialog box, 214, 214f
Computer names, 211f, 214–215
Computer networks, 5–14
 automation network, 14–15
 client-server network, 4
 home computer network, 5–7
 large enterprise network, 12–14
 large office network, 11–12
 local area network (LAN), 2
 peer-to-peer network, 4
 small office and home office network (SOHO), 7–11
 wide area network (WAN), 2–3, 3f
Computer virus, 285
Confidentiality, 163
Connection-oriented protocol, 56
Connectionless protocol, 56, 151
Connectivity problems, 268–271
Console cable, 201
Console session, 201–202
Content checking, 233–234
Content filtering, 236
Control Panel, 237
Convergence, 298
Converting binary to decimal/decimal to binary, 119, 310
Copper cable, 32
copy running-config startup-config, 205

Corrective measures (disaster preparedness), 301
Cracking, 287
CRC. *See* Cyclical redundancy check (CRC)
Crossover cable, 36, 36f
CryptSvc, 277
CSMA/CA. *See* Carrier Sense Multiple Access with Collison Avoidance (CSMA/CA)
CSMA/CD. *See* Carrier Sense Multiple Access with Collison Detection (CSMA/CD)
CSU/DSU. *See* Channel service unit/data service unit (CSU/DSU)
Cyclical redundancy check (CRC), 156

D channels, 254
DAP. *See* Directory Access Protocol (DAP)
Data and Pad field, 73f, 74
Data availability, 288–292
 backup, 288
 disk mirroring, 291
 disk redundancy, 290–292
 Offline Files, 290
 RAID, 291–292
 Shadow Copies, 288–290
Data backup, 288
Data collector set, 318
Data communications equipment (DCE) end, 200
Data encapsulation, 152
Data Link layer, 58–60
Data-link layer protocols, 150–151
Data terminal equipment (DTE) end, 200
Data transmission types
 broadcast, 23–25
 multicast, 25, 25f
 unicast, 22, 23f
Datagram, 62, 63
DB-60, 200
DCE end. *See* Data communications equipment (DCE) end
DDoS. *See* Distributed denial of service (DDoS)
Default gateway, 31, 113–114, 114f, 191f, 192
Default routes, 191–192
Demilitarized zone (DMZ), 238. *See also* Perimeter network
Demodulation, 66
Denial of service (DoS), 284
Destination Address field, 73, 73f
Destination Host Unreachable, 270
Destination subnetwork, 194
Development environment, 191
DHCP. *See* Dynamic Host Configuration Protocol (DHCP)
DHCPv6. *See* Dynamic Host Configuration Protocol version 6 (DHCPv6)
Dial-up connection, 6, 14, 14f, 248, 248f

Index

Dial-up remote access, 256, 257f
Digital certificate, 163, 164
Digital signal lines, 253, 253f
Digital subscriber line (DSL), 248–249, 248f
Direct sequence spread spectrum (DSSS), 94, 94f
Directional antenna, 106
Directly connected routes, 192–193
Directory, 167
Directory Access Protocol (DAP), 166
Disaster preparation, 301
Discovery measures (disaster preparedness), 301
Dish type directional antenna, 106
Disk drive folders, 167
Disk duplexing, 291
Disk mirroring, 291
Disk redundancy, 290–292
Distance Vector Multicast Routing Protocol (DVMRP), 154
Distributed application, 293
Distributed denial of service (DDoS), 230, 284
DIX Ethernet, 72
DIX Ethernet II, 72
DMZ. *See* Demilitarized zone (DMZ)
DNS. *See* Domain Name System (DNS)
DNS console, 216f
DNS records, 217f
DNS root servers, 217f, 218
DNS Round Robin, 300–301
DNS Suffix and NetBIOS Computer Name dialog box, 214, 215f
do show int, 204
do show interface, 204
Documentation. *See* Network documentation
DoD model, 63. *See also* TCP/IP model
Domain, 9, 9f, 10
Domain directory, 166
Domain Name System (DNS), 11, 53, 158, 212
Domain Name System (DNS) server, 215f, 216–219
DORA process, 129, 129f
DoS. *See* Denial of service (DoS)
Dotted decimal format, 117, 118f
Double ring topology, 41–42, 42f
Downloading e-mail from POP3 server, 159
Draft standard (DS), 16
DS. *See* Draft standard (DS)
DS0 channel, 253, 253f
DS1 (T1) channel, 253f
DS3 (T3) channel, 253f
DSL. *See* Digital subscriber line (DSL)
DSL splitter, 248
DSSS. *See* Direct sequence spread spectrum (DSSS)
DTE end. *See* Data terminal equipment (DTE) end

Dual IP stack, 143–144
Duplicate NetBIOS names, 212, 214, 214f
Duplicate servers, 295
DVMRP. *See* Distance Vector Multicast Routing Protocol (DVMRP)
Dynamic Host Configuration Protocol (DHCP), 7, 54, 128–130, 204
Dynamic Host Configuration Protocol version 6 (DHCPv6), 139, 142, 144–146
Dynamic NAT, 233
Dynamic ports, 159f
Dynamic routing, 190, 195–197, 195f

E-carrier, 253, 253f
E-mail, 158–159, 158f
E0, 174
E0/0, 174
E1/E3 lines, 253f, 254
EAP. *See* Extensible Authentication Protocol (EAP)
Eavesdropping, 167
EBCDIC. *See* Extended Binary Coded Decimal Interchange Code (EBCDIC)
Electromagnetic interference (EMI), 32
EMI. *See* Electromagnetic interference (EMI)
Enable password, 203
enable password *XXXXX*, 203
enable secret *XXXXX*, 203
Encapsulated data, 62 62f
Encapsulating Security Protocol (ESP), 164
Encapsulation, 60
encapsulation frame-relay, 205
encapsulation hdlc, 205
encapsulation ppp, 205
Encryption, 164
Encryption protocols, 162
Enterprise authentication, 100f
Enterprise mode, 99
Enterprise network, 12–14
Errors. *See* Troubleshooting
ESP. *See* Encapsulating Security Protocol (ESP)
ESS. *See* Extended service set (ESS)
ESTABLISHED, 275
Ethernet, 70–87
  5-4-3 rule, 78, 79f
  10 Gigabit, 84–86
  10 Mbps, 77–78, 77f
  addressing, 74
  autonegotiation, 84
  CSMA/CD, 75–77
  DIX, 72
  Fast, 80–81, 81f
  frame, 73–75
  full-duplex mode, 81
  Gigabit, 80, 82–84
  half-duplex mode, 81

  historical overview, 70
  IEEE 802.3, 72–73
  normal link pulse (NLP) signals, 78–85
  physical layer specifications, 77–78
  standards, 72–73
  thick, 72
  thin, 72
  usage of term "Ethernet" in industry, 72
Ethernet adapters, 267
Ethernet addressing, 74
Ethernet connection, 200
Ethernet frame
  Data and Pad, 73f, 74
  Destination Address, 73, 73f
  Ethernet addressing, 74
  Ethertype/Length, 73–74, 73f
  Frame Check Sequence, 73f, 74
  overview, 73f
  Preamble, 73, 73f
  protocol identification, 74–75
  Source Address, 73, 73f
  Start of Frame Delimiter, 73, 73f
Ethernet packets, 75
Ethernet port, 174
Ethernet speeds, 183f
Ethernet standards, 72–73
Ethernet WAN, 253f, 254–255
Ethertype, 74
Ethertype/Length field, 73–74, 73f
EUI. *See* Extended unique identifier (EUI)
EUI-64 addresses, 140
Evil twin, 287
Extended Binary Coded Decimal Interchange Code (EBCDIC), 54
Extended service set (ESS), 104, 105f
Extended star topology, 39
Extended unique identifier (EUI), 140f
Extensible Authentication Protocol (EAP), 101, 256
Extranet, 242–243, 243f

F0, 174
F0/0, 174
F0/1, 174
Failover cluster, 295
Failover Cluster Management console, 295–297
Failover clustering, 295–297
Fast Ethernet, 80–81, 81f
Fast Ethernet port, 174
Fault tolerance, 186. *See also* Network fault tolerance
FCC. *See* Federal Communications Commission (FCC)
FDDI. *See* Fiber distributed data interface (FDDI)
fe80, 136
Federal Communications Commission (FCC), 95

FHSS. *See* Frequency-hopping spread spectrum (FHSS)
Fiber distributed data interface (FDDI), 41
Fiber-optic cable, 36–38
Fiber-optic connectors, 38
Fiber optic inter-repeater link (FOIRL), 77f, 78
Fibre Channel, 296
Field, 73
File Transfer Protocol (FTP), 53, 162
Filtering, 233, 234
Finish (FIN), 156
Firewall, 160, 160f, 235–238
  allowed programs, 236f
  packet-filtering, 235–236
  perimeter network, 238, 239f, 240–241
  WAP, 7
  Windows, 236–238
Flag, 154
Flowchart, 312
FOIRL. *See* Fiber optic inter-repeater link (FOIRL)
Form-factor switch, 174
Four-port bridge, 29f
Four-port hub, 26, 27f
Four-port switch, 28f
FQDN. *See* Fully qualified domain name (FQDN)
Frame, 23, 58, 62
Frame Check Sequence, 73f, 74
Frame relay, 60
Frequency-hopping spread spectrum (FHSS), 93
Fresnel zone, 106, 106f
FTP. *See* File Transfer Protocol (FTP)
FTP bounce attack, 285
Full-duplex mode, 81, 184
Full-duplex system, 55
Full mesh topology, 42, 43f
Fully qualified domain name (FQDN), 211

Gateway, 31, 194
Gateway-to-gateway VPN, 259, 259f
Gbps. *See* Gigabits per second (Gbps)
Geostationary orbit, 251
Gi0/0, 174
Gi1/0, 174
Gigabit backbones, 83
Gigabit Ethernet, 80, 82–84
Gigabits per second (Gbps), 2
Gigabyte (GB), 2
Global unicast address, 137f, 140–141, 141f
GlobalNames zone (GNZ), 219, 224
GNZ. *See* GlobalNames zone (GNZ)
GROUP <00>, 212
Group Policy, 10, 294
Guest network, 241–242
Guest network interface, 242f

H-node, 225f
Half-duplex mode, 81, 184
Half-duplex system, 55
Handshake, 156, 157
Hardware-based RAID, 292
Hardware configurations, 315
Hardware redundancy, 186, 295
Hardware router, 190, 191
HDLC. *See* High-level data link control (HDLC)
Heartbeats, 298
Hexadecimal characters, 57
Hexadecimal notation, 136–137, 137f
Hierarchical routing, 141f
Hierarchical star topology, 39, 39f
High availability. *See* Availability
High-level data link control (HDLC), 205
High-order bit identification, 116f
High-speed switch, 184
Home computer network, 5–7
Home wireless network, 101–103
Homegroup, 237
Hop, 270, 272
Host cache, 215f, 219–220
Host ID, 110–113
Host-to-host layer, 63f, 64
Hostname, 210–212, 210f
Hostname resolution, 223–224
Hosts file, 215f, 220–221, 221f
Hot-swappable drive array, 295
HTML. *See* Hypertext Markup Language (HTML)
HTTP. *See* Hypertext Transfer Protocol (HTTP)
HTTP over Secure Sockets Layer (HTTPS), 53, 160–162, 259
HTTPS, 53, 160–162, 259
Hub, 26, 27f, 66, 77, 175
Hybrid (node type), 225
Hybrid RAID technologies, 292
Hyper-V, 293
HyperTerminal, 201
Hypertext Markup Language (HTML), 16, 160
Hypertext Transfer Protocol (HTTP), 53, 56, 160–162

IANA. *See* Internet Assignment Numbers Authority (IANA)
IBSS. *See* Independent basic service set (IBSS)
ICMP. *See* Internet Control Message Protocol (ICMP)
ICMPv4, 153
ICMPv4 error messages, 153
ICMPv4 query messages, 153
ICMPv6, 153. *See* Internet Control Message Protocol version 6 (ICMPv6)
IEEE. *See* Institute of Electrical and Electronics Engineers (IEEE)
IEEE standards mapping, 46f

IEEE 802.2, 45, 46f, 58
IEEE 802.3, 45, 46f, 60, 72–73, 78, 183f
IEEE 802.3-2015, 72
IEEE 802.3an, 183f
IEEE 802.3bw—2015, 72
IEEE 802.3u, 183f
IEEE 802.3z, 84, 183f
IEEE 802.5, 60
IEEE 802.11, 46f, 59f, 94
IEEE 802.11a, 93, 95
IEEE 802.11ac, 93, 96
IEEE 802.11b, 93, 95–96
IEEE 802.11f, 93, 96
IEEE 802.11i, 8
IEEE 802.11n, 93, 96
IETF. *See* Internet Engineering Task Force (IETF)
IGMP. *See* Internet Group Multicast Protocol (IGMP)
IKEv2, 258f
IMAP. *See* Internet Message Access Protocol (IMAP)
IMAP4, 165
Implicit deny policy, 236
Incident response, 288
Increased availability, 282. *See also* Availability
Independent basic service set (IBSS), 104
Institute of Electrical and Electronics Engineers (IEEE), 17, 95
Integrated services digital network (ISDN), 253f, 254, 254f
Integrity, 163
interface fa0/0, 204
Interface identifier, 140, 140f, 204
Interface verification, 204
International Organization for Standardization (ISO), 52, 95
International Telecommunication Union (ITU), 17
Internet, 16
Internet Assignment Numbers Authority (IANA), 140, 159
Internet Control Message Protocol (ICMP), 57, 152–153, 270, 272
Internet Control Message Protocol version 6 (ICMPv6), 139
Internet Engineering Task Force (IETF), 15–16
Internet-facing servers, 238
Internet Group Multicast Protocol (IGMP), 25, 57, 153–154
Internet layer, 64
Internet Message Access Protocol (IMAP), 54, 165–166
Internet Options dialog box, 235, 235f
Internet Protocol (IP) address, 110. *See* IP address
Internet Protocol (IP) functions. *See* IP functions
Internet Protocol Security (IPsec), 60, 139, 164–165, 238, 259

Index

Internet Protocol v4. *See* IPv4
Internet Protocol v6. *See* IPv6
Internet service provider (ISP), 5, 6, 141
Internetwork layer, 64
Internetwork protocol layer, 64
Intranet, 231–235
IP address, 22, 110, 210, 230. *See also* IPv4; IPv6
IP address calculation (subtraction method), 311
IP address classes, 115–117, 307
IP address conflict, 111
ip address dhcp, 204
IP address masking, 120–121, 120f
IP functions, 151–152
   data encapsulation, 152
   IP routing, 152
   IPv4 addressing, 152
IP routing, 152
ipconfig, 110, 110f, 192f, 265–268, 268f
ipconfig /?, 264
ipconfig/all, 58, 59f, 130, 131f, 144, 225f, 265, 266f, 267f
ipconfig/displaydns, 219, 220f, 268, 268f
ipconfig/flushdns, 220, 268, 268f
ipconfig/registerdns, 268, 268f
ipconfig/release, 268f
ipconfig/release6, 268f
ipconfig/renew, 268f
ipconfig/renew6, 268f
IPsec. *See* Internet Protocol Security (IPsec)
IPv4, 57, 108–133, 152
   about 4 billion addresses, 136
   APIPA, 117, 130
   automatic assignment (DHCP), 128–130
   binary IPv4 addresses, 117–121
   CIDR notation, 119–120, 120f
   classful IP addresses, 115–117
   classless IP addresses, 121, 121f
   default gateway, 113–114, 114f
   DORA process, 129, 129f
   dotted decimal format, 117, 118f
   high-order bit identification, 116f
   host ID, 110–113
   ID errors, 112f
   IP address conflict, 111
   ipconfig/all, 130, 131f
   local and remote addresses, 114–115
   loopback address, 117
   manual assignment, 128, 129f
   masking an IP address, 120–121, 120f
   network ID, 110–113
   octet, 117
   reserved IP address ranges, 117
   same network ID, 113, 127
   subnet mask, 110–111, 111f, 120
   subnetting. *See* IPv4 subnetting
   transmission types, 138
   valid IP addresses, 110

IPv4-mapped IPv6 address, 143, 143f
IPv4 subnetting, 121–128. *See also* Subnet calculations
   advanced subnet masks, 127f
   class C subnet, 123f
   creating subnets, 122f
   local and remote addresses, 124–126
   number of hosts, 124, 125f
   number of subnet bits, 122–124
   troubleshooting, 126–128
   two high-order bits, 123f
IPv4 to IPv6 tunneling protocols, 143–144
IPv5, 16, 136
IPv6, 57, 134–147
   address identification, 143
   address interface, 136f
   automatic assignment (DHCPv6), 144–146
   components of IPv6 address, 139–143
   displaying IPv6 address, 138, 138f
   double colons, 138, 138f
   dropping leading zeros, 138f
   dual IP stack, 143–144
   easy host configuration, 139
   EUI-64 addresses, 140
   global unicast address, 140–141, 141f
   hexadecimal notation, 136–137, 137f
   improvements over IPv4, 139
   interface identifier, 140, 140f
   IPsec, 139
   IPv4-mapped IPv6 address, 143, 143f
   IPv4 to IPv6 tunneling protocols, 143–144
   link-local address, 141–142, 142f
   manual assignment, 144, 145f
   neighbor discovery (ND), 139
   network identifier, 140, 140f
   number of IP addresses, 136
   prefix, 136, 137f
   routing, 139
   stateful configuration, 142
   stateless configuration, 142
   Teredo, 144, 144f, 145f
   transmission types, 138–139
   unique local address, 142–143, 144
   zero compression, 138, 138f
IPv6 prefix, 136, 137f
iSCSI storage area network (SAN), 296
ISDN. *See* Integrated services digital network (ISDN)
ISDN modem, 254
ISO. *See* International Organization for Standardization (ISO)
ISP. *See* Internet service provider (ISP)
ITU. *See* International Telecommunication Union (ITU)

Jam pattern, 76

Kerberos, 167

"L" bit, 144
LACP, 185
LAN. *See* Local area network (LAN)
Large enterprise network, 12–14
Large office network, 11–12
Late collisions, 76
Layer 1 (Physical layer), 60–61
Layer 2 (Data Link layer), 58–60
Layer 2 switches, 179–180
Layer 2 Tunneling Protocol (L2TP), 60, 167, 258f, 259
Layer 3 (Network layer), 57–58
Layer 3 switches, 67, 179–180
Layer 4 (Transport layer), 55–57
Layer 5 (Session layer), 54–55
Layer 6 (Presentation layer), 54
Layer 7 (Application layer), 53–54
LDAP. *See* Lightweight Directory Access Protocol (LDAP)
Lease offer, 129
Leased lines, 252
Lightweight Directory Access Protocol (LDAP), 54, 166–167
line aux, 203
line console, 203
Line of sight, 104, 106f
Line password, 203
line vty, 203
Link layer, 64
Link-local address, 137f, 141–142, 142f
Link-local multicast name resolution (LLMNR), 215f, 223
LISTENING, 275
LLC. *See* Logical link control (LLC)
LLMNR. *See* Link-local multicast name resolution (LLMNR)
lmhosts file, 215f, 222–223
lmhosts.sam file, 223
Local and remote addresses
   IPv4, 114–115
   IPv4 subnetting, 124–126
Local area network (LAN), 2, 150
Logical link control (LLC), 58, 75
Logical network organization, 5
Logical port, 56, 174
Logical ring topology, 41, 41f
login, 204
Loopback address, 117, 137f
L2TP. *See* Layer 2 Tunneling Protocol (L2TP)
L2TP/IPsec, 60, 167

M-node, 225f
MAC address, 58–59, 59f, 150, 177–178, 210
MAC address table, 178, 178f
MAC filtering. *See* Media access control (MAC) address filtering
Malware, 230, 284–288. *See also* Network fault tolerance
   adware, 285
   antivirus software, 231

Malware (*continued*)
  botnets, 230–231
  buffer overflow, 286
  cracking, 287
  denial of service (DoS), 284
  evil twin, 287
  man in the middle (MITM) attach, 284–285
  mitigation techniques, 287–288
  phishing, 286
  routing access point, 287
  social engineering, 286
  spyware, 285–286
  trojan horse, 285
  virus, 285
  war chalking, 287
  war driving, 286
  wireless threats, 286–287
  worm, 285
Man in the middle (MITM) attach, 284–285
Managed switch, 179
Management and administration, 304–319
  documentation. *See* Network documentation
  network monitoring software, 311–312
  network process flowchart, 312, 313f
  SNMP communications, 306–307
  subnet calculations. *See* Subnet calculations
Management information base (MIB), 306
Manufacturer serial number, 59f
Masking an IP address, 120–121, 120f
MAU. *See* Multistation access unit (MAU)
Mbps. *See* Megabits per second (Mbps)
MD5. *See* Message Digest 5 (MD5)
MDI/MDI-X button, 184
Media access control (MAC). *See* MAC address
Media access control (MAC) address filtering, 101
Megabits per second (Mbps), 2
Memory
  NAT table, 232
  routing table, 198
Mesh topology, 42–44
Message Digest 5 (MD5), 255
MIB. *See* Management information base (MIB)
Micron (μm), 37
Microsoft Challenge Handshake Authentication Protocol version 2 (MS-CHAPv2), 101, 255
Microsoft domain, 9, 11
Microsoft enhanced B-node, 225f
Microsoft point-to-point encryption (MPPE), 167, 259
Microsoft RAS, 255, 256f
Mills, Dave, 269
Mirrored stripe sets, 292

Mirroring disks, 291
MITM attack. *See* Man in the middle (MITM) attach
Modem, 66, 249
Modular switch, 174
more command, 264
MOSPF. *See* Multicast Open Shortest Path First (MOSPF)
MPPE. *See* Microsoft point-to-point encryption (MPPE)
MS-CHAPv2. *See* Microsoft Challenge Handshake Authentication Protocol version 2 (MS-CHAPv2)
MSA. *See* Multisource agreement (MSA)
Multicast Open Shortest Path First (MOSPF), 154
Multicast promiscuous mode, 154
Multicast transmission, 25, 25f, 154
  IPv4, 138
  IPv6, 138
Multimode fiber-optic, 37–38, 37f
Multiple-computer connections, 172, 172f
Multiple network connections, 190–193
Multiple subnets, 308, 309f
Multisource agreement (MSA), 86
Multistation access unit (MAU), 41
Mutual authentication, 285. *See also* Authentication
Muuss, Mike, 269
MX records, 217f
MYBOOKWORLD, 277

NACK. *See* Negative acknowledge (NACK)
Name resolution methods, 215f
Names. *See* Resolving names to IP addresses
NAT. *See* Network address translation (NAT)
NAT Overload, 232
NAT table, 232, 232f, 233f
nbtstat -c, 222, 223f
nbtstat -n, 212, 213f
nbtstat -R, 222
ND. *See* Neighbor discovery (ND)
NDP. *See* Neighbor Discovery Protocol (NDP)
Negative acknowledge (NACK), 156
Neighbor discovery (ND), 139
Neighbor Discovery Protocol (NDP), 153
NetBIOS cache, 215f, 222
NetBIOS name, 210f, 212, 213f
NetBIOS name resolution, 224–225
NetBIOS over TCP/IP (NetBT), 224
NetBIOS sixteenth byte, 212, 213f
NetBT. *See* NetBIOS over TCP/IP (NetBT)
NetBT node types, 224–225, 225f
netstat, 275–278
netstat /?, 264
netstat ?, 276f
netstat 15, 276f
netstat -a, 276f

netstat -ano, 275, 277
netstat -b, 276, 276f
netstat -e, 276f
netstat -f, 276f
netstat -n, 276f
netstat -o, 276f
netstat -p protocol, 276f
netstat -p TCP, 276f
netstat -r, 276f
netstat -s, 276f
netstat *interval,* 276f
Network, 173. *See also* Computer networks
Network access layer, 63f, 64
Network adapter, 66
Network address port translation, 232
Network address subnetting, 308–309
Network address translation (NAT), 7, 231–233, 259
Network architectures, 3–5
Network diagram, 315, 316f
Network discovery, 237
Network documentation, 312–318
  baselines, 317–318
  cable diagram, 314–315
  change management, 315–317
  hardware configurations, 315
  network diagram, 315, 316f
Network fault tolerance, 282–303. *See also* Network security
  application availability, 292–294
  data availability, 288–292. *See also* Data availability
  disaster preparation, 301
  malware. *See* Malware
  server availability, 294–301. *See also* Server availability
Network hardware, 25–32
  bridge, 28–29, 29f
  hub, 26, 27
  router, 30–32
  switch, 27–28, 28f
Network ID, 110–113, 182
Network identifier, 140, 140f
Network interface adapter, 66, 177, 191
Network interface layer, 63f, 64
Network layer, 57–58
Network layer protocols, 151–155
Network line drawing, 173f
Network Load Balancing Manager console, 297
Network load balancing (NLB), 297–300
Network mask, 194
Network monitoring software, 311–312
Network policy and access services (NPAS), 100, 101f
Network process flowchart, 312, 313f
Network remote access, 13–14
Network security, 228–245. *See also* Network fault tolerance
  antivirus software, 231
  botnets, 230–231

Index

confidentiality, 163
extranet, 242–243, 243f
firewall, 235–238. See also Firewall
guest network, 241–242
integrity, 163
intranet, 231–235
malware, 230, 284–288. See also Malware
network address translation (NAT), 231–233
perimeter network, 238–242
proxy server, 233–235
reverse proxy server, 241, 241f
risks on the Internet, 230–231
summary/overview, 244
switches, 185–186
wireless networks, 96–101
zombie, 230
Network statistics, 275–278
Network topologies
bus topology, 39–40, 40f
mesh topology, 42–44
point-to-point topology, 44, 44f
ring topology, 40–42
star topology, 38–39
Network types, 5–14
Networking protocols, 148–169
application layer protocols, 160–168
FTP, 162
ICMP, 152–153
IGMP, 153–155
IMAP, 165–166
internally used ports and protocols, 160, 161f
IPsec, 164–165
Kerberos, 167
LDAP, 166–167
L2TP, 167
mapping protocols to OSI and TCP/IP models, 67f
network layer protocols, 151–155
physical and data-link layer protocols, 150–151
POP3, 165
ports, 158–161
PPTP, 167
RAS protocols, 255–256
RDS, 163
routing protocols, 190
SMTP, 165, 165f
SNMP, 167–168
SSH, 164
SSL, 163
TCP, 155–156
Telnet, 162–163
TFTP, 162
TLS, 163–164
transport layer protocols, 155–158
UDP, 157–158
VPN tunneling protocols, 258–259, 258f
New Inbound Rule Wizard, 238

NLB. See Network load balancing (NLB)
NLB cluster, 297
NLP signals. See Normal link pulse (NLP) signals
no shutdown, 204
Normal link pulse (NLP) signals, 78–85
NPAS. See Network policy and access services (NPAS)
NS records, 217f

Octet, 117
OFDM. See Orthogonal frequency division multiplexing (OFDM)
Offline Files, 290
Omnidirectional antenna, 106
Open Shortest Path First (OSPF), 57, 197
Open Systems Interconnection (OSI) model. See OSI model
Organizationally unique identifier (OUI), 59, 59f, 74, 140
Orthogonal frequency division multiplexing (OFDM), 94
OSI model
advantages, 52
Application layer, 53–54
Data Link layer, 58–60
encapsulated data, 62 62f
goals, 52
how it works?, 60–61, 61f
mapping devices to specific layers, 64–67
mapping protocols to specific layers, 67f
mnemonics, 52, 53f
Network layer, 57–58
origins, 63
Physical layer, 60–61
Presentation layer, 54
Session layer, 54–55
TCP/IP model, compared, 63f
Transport layer, 55–57
OSPF. See Open Shortest Path First (OSPF)
OUI. See Organizationally unique identifier (OUI)

P-node, 225f
P2P wireless. See Point-to-Point (P2P) wireless network
P2P wireless bridge, 253f
Packet, 23, 57, 62
Packet-filtering firewall, 235–236
Packet loss, 272
Packet sniffer, 287, 311
PAN. See Personal area network (PAN)
PAP. See Password Authentication Protocol (PAP)
Parabolic directional antenna, 106
Parity-based RAID, 291
Partial mesh topology, 42
Password, 202–204

Password Authentication Protocol (PAP), 255
password command, 203
password *XXXXX*, 203
PAT. See Port address translation (PAT)
Patch management, 287
pathping, 57, 272–275, 275f
pathping /?, 264
PC-to-router cabling, 201
PDU. See Protocol data unit (PDU)
PEAP. See Protected EAP (PEAP)
Peer-to-peer network, 4
perfmon, 278
Performance Monitor, 278, 317–318, 317f
Perimeter network, 238–242
Personal area network (PAN), 93
Personal mode, 99
Phishing, 286
Physical address, 150. See also MAC address
Physical layer, 60–61
Physical layer (PHY) module, 86
Physical layer protocols, 150–151
Physical media
coaxial cable, 36
fiber-optic cable, 36–38
network topologies, 38–44
twisted-pair cable, 32–36
Physical network infrastructure, 5
Physical port, 56, 174
PID. See Process ID (PID)
PIM. See Protocol independent multicast (PIM)
ping, 57, 152, 268–271, 284
ping /?, 264
ping localhost, 221
Pinout standards, 45
PIPv1, 196f
PKI. See Public key infrastructure (PKI)
Plain old telephone service (POTS), 248
"Please Do Not Throw Sausage Pizza Away," 52, 53f
Point-to-Point (P2P) wireless network, 104–106
Point-to-Point Protocol (PPP), 167
Point-to-point topology, 44, 44f
Point-to-Point Tunneling Protocol (PPTP), 167, 258–259, 258f
Point-to-Point Tunneling Protocol v4 (PPTPv4), 60
Pointer (PTR) records, 216, 217f
Policies and procedures, 287–288
POP3. See Post Office Protocol (POP3)
Port, 26, 56, 158–160
commonly used, 161f
dynamic, 159f
IANA, 159
internally used ports and protocols, 160, 161f
logical, 174
netstat, 275–277

Port (*continued*)
  physical, 174
  registered, 159f
  switches, 172–175, 185–186
  well-known, 159f
Port 0, 56
Port 80, 56
Port 135, 277
Port 1030, 277
Port 1060, 277
Port 2078, 277
Port 3389, 277
Port address translation (PAT), 232
Port mirroring, 179
Port numbers, 56
Port-to-port speed, 185
Post Office Protocol (POP3), 54, 165
POTS. *See* Plain old telephone service (POTS)
Power supply, 295
PPP. *See* Point-to-Point Protocol (PPP)
PPTP. *See* Point-to-Point Tunneling Protocol (PPTP)
PPTPv4. *See* Point-to-point tunneling protocol v4 (PPTPv4)
Preamble field, 73, 73f
Presentation layer, 54
Preshared key (PSK) mode, 99
Preventive measures (disaster preparedness), 301
PRI. *See* Primary rate interface (PRI)
Primary rate interface (PRI), 254
Print spooler service, 277
Private network location, 237
Privileged mode password, 203
Problems. *See* Troubleshooting
Process ID (PID), 277, 278, 278f
Promiscuous mode, 312
Proposed standard (PS), 16
Protected EAP (PEAP), 256
Protocol. *See* Networking protocols
Protocol analyzer, 311–312
Protocol data unit (PDU), 22, 23, 62
Protocol independent multicast (PIM), 152
Proxy server, 233–235
Proxy server filters, 233, 234
PS. *See* Proposed standard (PS)
PSK mode. *See* Preshared key (PSK) mode
PTR. *See* Pointer (PTR) records
Public key infrastructure (PKI), 164
Public network location, 237
PuTTY, 164

RADIUS. *See* Remote authentication dial-in user service (RADIUS)
RADIUS server, 260f
RAID. *See* Redundant array of independent disks (RAID)
RAID 5, 291–292
RAID 6, 292

RARP. *See* Reverse Address Resolution Protocol (RARP)
RAS. *See* Remote access service (RAS)
RAS authentication, 255, 256f
RDS. *See* Remote desktop services (RDS)
RDS server farm, 298–299, 299f
RDS servers load balancing, 298–300
Redundant array of independent disks (RAID), 290–292
Redundant disk, 290–292
Registered ports, 159f
Reliability and Performance Monitor, 317
Remote access, 13–14, 257
Remote access server, 14
Remote access service (RAS), 255–259
Remote authentication dial-in user service (RADIUS), 98, 259–260
Remote Desktop, 257
Remote Desktop Connection Broker, 298–300
Remote desktop services (RDS), 163, 293
Repeater, 66, 104, 105f
Replacement of failed router, 191
Request for comment (RFC), 15–16
"Request timed out," 270
Request to Send/Clear to Send (RTS/CTS), 92, 93f
Reserved IP address ranges, 117
Resolving names to IP addresses, 208–227
  broadcast name resolution, 215f, 223
  computer names, 211f, 214–215
  DNS server, 215f, 216–219
  duplicate NetBIOS names, 212, 214, 214f
  FQDN, 211
  host cache, 215f, 219–220
  hostname, 210–212, 210f
  hostname resolution, 223–224
  hosts file, 215f, 220–221, 221f
  LLMNR, 215f, 223
  lmhosts file, 215f, 222–223
  name resolution methods, 215f
  NetBIOS cache, 215f, 222
  NetBIOS name, 210f, 212, 213f
  NetBIOS name resolution, 224–225
  NetBIOS names derived from hostnames, 212–214, 214f
  NetBIOS sixteenth byte, 212, 213f
  URL, 211
  WINS, 215f, 221–222, 222f
Resource Monitor, 278, 279f
Resource usage, 279f
Reverse Address Resolution Protocol (RARP), 57
Reverse lookup zone, 216
Reverse proxy server, 241, 241f
RFC. *See* Request for comment (RFC)
RFC 791, 16
RFC 792, 153
RFC 1542 compatible, 130
RFC 1819, 16

RFC 1918, 117
RFC 2246, 163
RFC 2460, 16
RFC 4193, 144
RFC 4510, 166
RFC 5246, 163
Riflescope, 106
Ring topology, 40–42
RIP. *See* Routing Information Protocol (RIP)
RIPv2, 57. *See* Routing Information Protocol version 2 (RIPv2)
RJ45 connector, 34f, 35
Round-trip delay time, 75
Round-trip times, 272
route print, 198, 198f
Routed path verification, 272–275
Router, 12, 24, 30–32, 66–67. *See also* Routing networks
  home wireless network, 102, 102f
  IPv6, 139
  web interface, 103, 103f
  wireless, 91, 97, 97f
Router cabling, 199–201
Router configuration, 199–206
Router identification, 272
Router-to-CSU/DSU cabling, 200
Router-to-router cabling, 200
Router-to-switch cabling, 201
Router web interface, 103, 103f
Router#, 202
Router>, 202
Router(config)#, 202
Router(config-if)#, 204
Router(config-line)#, 203
Routing access point, 287
Routing and remote access service (RRAS), 193
Routing Information Protocol (RIP), 57
Routing Information Protocol version 2 (RIPv2), 196–197, 196f
Routing interface, 190, 204–206
Routing networks, 188–207. *See also* Router
  back-to-back connection, 200
  console session, 201–202
  default gateway, 191f, 192
  default routes, 191–192
  directly connected routes, 192–193
  dynamic routing, 195–197, 195f
  hardware router, 191
  hierarchical routing, 141f
  interface configuration, 204–206
  IP routing, 152
  multiple network connections, 190–193
  OSPF, 197
  password configuration, 202–204
  PC-to-router cabling, 201
  RIPv2, 196–197, 196f
  routed path verification, 272–275
  router cabling, 199–201

Index

router configuration, 199–206
router identification, 272
router routes, 193f, 195f
router-to-CSU/DSU cabling, 200
router-to-router cabling, 200
router-to-switch cabling, 201
routing table, 197–198
RRAS console, 197f
saving configuration changes, 206
software router, 191
static routing, 193–195, 195f
summary/overview, 206
terminology, 190
traffic routing, 193–198
transmission speeds, 198–199
trunk link, 201
Routing protocols, 190
Routing table, 190, 197–198
RRAS. *See* Routing and remote access service (RRAS)
RRAS console, 197f
RTS/CTS. *See* Request to Send/Clear to Send (RTS/CTS)
Run as Administrator, 265
Run dialog box, 264
Runt, 76

Same network ID, 102, 113
SAN. *See* Storage area network (SAN)
SATA. *See* Serial ATA (SATA)
Satellite Internet access, 248f, 250–252
SCOM. *See* Systems center operations manager (SCOM)
Screened twisted-pair (ScTP), 33
SCSI. *See* Small computer system interface (SCSI)
ScTP. *See* Screened twisted-pair (ScTP)
SDSL. *See* Symmetric DSL (SDSL)
Second-level domain DNS servers, 217f, 218
secret command, 203
Secret password, 203
Secure LDAP (SLDAP), 167
Secure Shell (SSH), 164, 279
Secure Socket Tunneling Protocol (SSTP), 258f, 259
Secure sockets layer (SSL), 259, 613
Secure wireless network, 8
Security. *See* Network security
Segment, 23, 56, 62
Sending SMTP e-mail, 158–159
Serial ATA (SATA), 291
Serial connections, 200
Server availability, 294–301
  DNS Round Robin, 300–301
  failover clustering, 295–297
  hardware redundancy, 295
  network load balancing (NLB), 297–300
  RDS server farm, 298–299, 299f
  RDS servers load balancing, 298–300
Server clustering, 294

Server farm, 295. *See also* RDS server farm
Server message block (SMB), 54
Server subnet, 31, 31f
Service level agreement (SLA), 252
Service set identifier (SSID), 91–92, 91f
Session, 55
Session layer, 54–55
SFP+, 86
Shadow Copies, 288–290
Shared storage, 296
Shielded twisted-pair (STP) cable, 33
show int, 204
show interface, 204
Signal quality error (SQE), 75
Simple Mail Transfer Protocol (SMTP), 54, 165, 165f
Simple Network Management Protocol (SNMP), 54, 167–168, 306–307
Simplex system, 55
Single sign-on (SSO), 9
Singlemode fiber-optic, 37f, 38
Sixteenth byte (NETBios), 212, 213f
SLA. *See* Service level agreement (SLA)
SLAAC. *See* Stateless address autoconfiguration (SLAAC)
SLDAP. *See* Secure LDAP (SLDAP)
Small computer system interface (SCSI), 291
Small office and home office network (SOHO), 7–11
SMB. *See* Server message block (SMB)
SMTP. *See* Simple Mail Transfer Protocol (SMTP)
Smurf attack, 284
Sniffer, 287, 311
SNMP. *See* Simple Network Management Protocol (SNMP)
SNMP agent, 306–307
SNMPv1, 307
SNMPv2, 307
SNMPv3, 307
SNMPv2c, 307
Social engineering, 286
Software router, 190, 191
SOHO. *See* Small office and home office network (SOHO)
SONET. *See* Synchronous optical network (SONET)
Source Address field, 73, 73f
Spanning Tree Protocol (STP), 180
spoolsv.exe, 277
Spyware, 285–286
SQE. *See* Signal quality error (SQE)
SRV records, 217f
SSH. *See* Secure Shell (SSH)
SSID. *See* Service set identifier (SSID)
SSL. *See* Secure sockets layer (SSL)
SSO. *See* Single sign-on (SSO)
SSTP. *See* Secure Socket Tunneling Protocol (SSTP)
Standalone application, 292–293
Standard (STD), 16

Standards organizations
  IEEE, 17
  IETF, 15–16
  ITU, 17
  W3C, 16
Standards Track category, 15–16
Star topology, 38–39
Start of Frame Delimiter field, 73, 73f
startup-config, 205
Stateful configuration, 142
Stateful filtering, 236
Stateless address autoconfiguration (SLAAC), 142
Stateless configuration, 142
Static NAT, 233
Static routing, 190, 193–195, 195f
STD. *See* Standard (STD)
Storage area network (SAN), 296
STP. *See* Spanning Tree Protocol (STP)
STP cable. *See* Shielded twisted-pair (STP) cable
Straight-through cable, 35, 35f
Streaming media, 157
Subnet, 11, 11f, 110, 173, 308. *See also* IPv4 subnetting; Subnet calculations
Subnet calculations, 307–311. *See also* IPv4 subnetting
  IP address calculation (subtraction method), 311
  multiple subnets, 308, 309f
  network address subnetting, 308–309
  subnet mask, 308
  subnetting between bytes, 309–310
Subnet mask, 110–111, 111f, 120, 308
Subnetwork, 173
Subtraction method (IP address calculation), 311
Switch, 12, 27–28, 28f, 66, 170–187
  backplane speed, 185
  bundling, 185
  collision domains, 175–177
  CSMA/CD, 175, 176
  form-factor, 174
  hardware redundancy, 186
  high-speed, 184
  layer 2 and layer 3, 179–180
  MAC address, 177–178
  managed, 179
  modular, 174
  multiple-computer connections, 172, 172f
  ports, 172–175, 185–186
  security, 185–186
  Spanning Tree Protocol (STP), 180
  speeds, 183–185
  transmission speeds, 184
  troubleshooting commands (command prompt), 264
  unmanaged, 179
  uplink port, 184
  virtual LAN (vLAN), 180–183
Switch port identification, 175

Switch ports, 172–175, 185–186
Switch speeds, 183–185, 198
Symmetric DSL (SDSL), 249
Synchronize (SYN), 154
Synchronous optical network (SONET), 253f
System backup, 288
System Center Configuration Manager, 294
System Monitor, 317
System problems. *See* Troubleshooting
Systems center operations manager (SCOM), 168

T1/T3 lines, 3, 253f, 254
T568A, 45
T568A straight-through Ethernet cable, 35f
T568B, 45
T568B straight-through Ethernet cable, 35f
TCP. *See* Transmission Control Protocol (TCP)
TCP/IP configuration/settings, 265–268
TCP/IP model
  alternate names for layers, 63f, 64
  Application layer, 64
  Internet layer, 64
  Link layer, 64
  mapping devices to specific layers, 64–67
  mapping protocols to specific layers, 67f
  origins, 63
  OSI model, compared, 63f
  Transport layer, 64
TCP/IP protocol, 230
TCP/IP statistics, 275–278
TCP/IPv4 stack, 116
TCP sliding window, 156, 157f
tcpdump, 312
Telnet, 162–163, 202, 278–279
telnet /?, 264, 279
telnet telnetservername, 279
Temporary Key Integrity Protocol (TKIP), 99
Teredo prefix, 144
Teredo tunneling protocol address, 137, 144, 144f, 145f
Terrestrial transmit, 250
TFTP. *See* Trivial File Transfer Protocol (TFTP)
Thick Ethernet, 72
Thin Ethernet, 72
Third- and lower-level domain DNS servers, 217f, 218
Threat mitigation, 287–288
Threats to the system. *See* Malware; Network fault tolerance; Network security
Three-way handshake, 156
Time to live (TTL), 270
TKIP. *See* Temporary Key Integrity Protocol (TKIP)

TLS. *See* Transport layer security (TLS)
Token ring topology, 41, 60
Top-level domain DNS servers, 217f, 218
tracert, 57, 272, 273f
tracert /?, 264
Training and awareness, 287
Transmission Control Protocol (TCP), 56, 155–156, 155f
Transmission speeds, 184, 198–199
Transmission types. *See* Data transmission types
Transport layer
  OSI model, 55–57
  TCP/IP model, 64
Transport layer protocols, 155–158
Transport layer security (TLS), 163, 256
Transport mode, 165
Trap, 306
Trivial File Transfer Protocol (TFTP), 53–54, 162
Trojan horse, 285
Troubleshooting, 262–281
  administrative privileges, 265
  case sensitivity, 265
  command prompt, 264–265
  connectivity problems, 268–271
  Help at command prompt, 264
  ipconfig, 265–268, 268f
  IPv4 subnetting, 126–128
  netstat, 275–278
  pathping, 272–275, 275f
  Performance Monitor, 278
  PID, 277, 278, 278f
  ping, 268–271
  Resource Monitor, 278, 279f
  resource usage, 279f
  routed path verification, 272–275
  router identification, 272
  switches, 264–265
  TCP/IP configuration/settings, 265–268
  TCP/IP statistics, 275–278
  Telnet, 278–279
  tracert, 272, 273f
Truncated binary exponential backoff, 76
Trunk link, 201
TTL. *See* Time to live (TTL)
TTL Expired in Transit, 270
Tunnel mode, 164
Tunneling protocols, 258–259, 258f
Twisted-pair cable, 32–36
Twisted-pair cable connector, 34f, 35
Twisted-pair wiring, 35–36
Two-firewall perimeter network, 238, 239f, 240

UDP. *See* User Datagram Protocol (UDP)
UDP ports 67 and 68, 130
Unicast address, 140–141, 141f
Unicast transmission, 22, 23f, 154
  IPv4, 138
  IPv6, 138

Uniform resource locator (URL), 211
UNIQUE <00>, 212
Unique local address, 137f, 142–143, 144
Unmanaged switch, 179
Unshielded twisted-pair (UTP) cable, 32–33
Uplink, 174
Uplink port, 184
URL. *See* Uniform resource locator (URL)
U.S. Postal Service (USPS) address, 111
User Datagram Protocol (UDP), 56, 157–158
UTP cable. *See* Unshielded twisted-pair (UTP) cable
UTP categories, 33–35

V.35 DEC connectors, 200
Valet Wireless Router (Cisco), 242
Valid IP addresses, 110
Validate a Configuration Wizard, 296
VDSL. *See* Very-high-bit-rate DSL (VDSL)
Very-high-bit-rate DSL (VDSL), 249
Virtual LAN (vLAN), 180–183
Virtual private network (VPN), 14, 15f, 255
Virtualization, 293
Virus, 285
vLAN. *See* Virtual LAN (vLAN)
VPN. *See* Virtual private network (VPN)
VPN server, 14
VPN tunneling, 256–259
VPN tunneling protocols, 258–259, 258f

W3C. *See* World Wide Web Consortium (W3C)
WAN. *See* Wide area network (WAN)
WAN connections, 150, 246–261. *See also* Wide area network (WAN)
WAN DSL, 253f
WAN interfaces, 205–206
WAN links, 252, 252f
WAP. *See* Wireless access point (WAP)
War chalking, 287
War driving, 8, 98, 286
Web browser, 161f
Web server-to-database server, 293
Well-known ports, 159f
WEP. *See* Wired equivalent privacy (WEP)
Wi-Fi, 47
Wi-Fi Alliance, 47, 95, 99
Wi-Fi protected access (WPA), 97f, 98–99
Wi-Fi protected access 2 (WPA2), 97f, 99–100
Wide area network (WAN), 2–3, 3f, 150, 246–261
  broadcast cable, 248f, 250, 250f
  dial-up connection, 248, 248f
  digital signal lines, 253, 253f
  DSL, 248–249, 248f

E1/E3 lines, 253f, 254
enterprise connectivity, 252–255
Ethernet WAN, 253f, 254–255
ISDN, 253f, 254, 254f
P2P wireless bridge, 253f
RADIUS, 259–260
remote access service (RAS), 255–259
rural areas, 248, 251
satellite, 248f, 250–252
SOHO connectivity, 248–252
SONET, 253f
T1/T3 lines, 253f, 254
VPN tunneling, 256–259
WAN DSL, 253f
WAN interfaces, 205–206
WAN links, 252, 252f
WiMAX. *See* Worldwide Interoperability for Microwave Access (WiMAX)
Windows Calculator, 119f
Windows command-prompt commands, 264–265
Windows Firewall, 236–238
Windows Firewall GUI, 238f
Windows Firewall with Advanced Security console, 237, 238, 239f
Windows Installer, 294
Windows Internet Name Service (WINS), 215f, 221–222, 222f
Windows Performance Monitor, 317–318, 317f
Windows Server 2012 R2, 128, 140, 191
Windows Server RAID 5, 291–292
WINS. *See* Windows Internet Name Service (WINS)
Wired equivalent privacy (WEP), 97f, 98
Wired LAN standards, 44–45

Wireless access point (WAP), 7, 90–91, 90f
Wireless bridge, 104
Wireless governing bodies, 95
Wireless LAN standards, 45–46
Wireless network, 88–107
  advantages, 101
  authentication, 100, 100f
  businesses, 103–104
  CSMA/CA, 91–92, 91f
  DSSS, 94, 94f
  FHSS, 93
  governing bodies, 95
  home network, 101–103
  IEEE 802.11, 94
  IEEE 802.11a, 93, 95
  IEEE 802.11ac, 93, 96
  IEEE 802.11b, 93, 95–96
  IEEE 802.11f, 93, 96
  IEEE 802.11n, 93, 96
  MAC filtering, 101
  OFDM, 94
  P2P networks, 104–106
  router, 91, 97, 97f
  security, 96–101
  service set identifier (SSID), 91–92, 91f
  speeds/distances, 95
  standards, 93–96
  theft, 98
  war driving, 8, 98
  Wi-Fi protected access (WPA), 97f, 98–99
  Wi-Fi protected access 2 (WPA2), 97f, 99–100
  wired equivalent privacy (WEP), 97f, 98

wireless access point (WAP), 90–91, 90f
wireless threats, 286–287
Wireless network theft, 98
Wireless router, 6–7, 91, 97, 97f
Wireless speeds/distances, 95
Wireless standards, 93–96
Wireshark, 312
Witness disk, 296
Workgroup, 8
World Wide Web Consortium (W3C), 16
World Wide Web (WWW), 16, 230
Worldwide Interoperability for Microwave Access (WiMAX), 248
Worm, 285
WPA. *See* Wi-Fi protected access (WPA)
WPA Enterprise, 104
WPA personal mode, 99
WPA2, 8. *See* Wi-Fi protected access 2 (WPA2)
WPA2 enterprise mode, 100
WPA2 personal mode, 99
WPA2-PSK, 99
WWW. *See* World Wide Web (WWW)
www.contoso.com, 300

X.500, 166
xDSL, 249
XENPAK, 86
XFP, 86

Yagi directional antenna, 106

Zero-based numbering, 174
Zero compression, 138, 138f
ZIP code, 111
Zombie, 230